AF771863

Energy, Climate and the Environment

Series Editor

David Elliott, Emeritus Professor of Technology Policy, Open University, UK

Titles include:

Giorel Curran
SUSTAINABILITY AND ENERGY POLITICS
Ecological Modernisation and Corporate Social Responsibility

Declan Kuch
THE RISE AND FALL OF CARBON EMISSIONS TRADING

Claire Dupont and Sebastian Oberthür (*editors*)
DECARBONISATION IN THE EU
Internal Policies and External Strategies

Espen Moe
RENEWABLE ENERGY TRANSFORMATION OR FOSSIL FUEL BACKLASH
Vested Interests in the Political Economy

Manuela Achilles and Dana Elzey (*editors*)
ENVIRONMENTAL SUSTAINABILITY IN TRANSATLANTIC PERSPECTIVE
A Multidisciplinary Approach

Robert Ackrill and Adrian Kay (*editors*)
THE GROWTH OF BIOFUELS IN THE 21ST CENTURY

Philip Andrews-Speed
THE GOVERNANCE OF ENERGY IN CHINA
Implications for Future Sustainability

Gawdat Bahgat
ALTERNATIVE ENERGY IN THE MIDDLE EAST

Ian Bailey and Hugh Compston (*editors*)
FEELING THE HEAT
The Politics of Climate Policy in Rapidly Industrializing Countries

Mehmet Efe Biresselioglu
EUROPEAN ENERGY SECURITY
Turkey's Future Role and Impact

Jonas Dreger
THE EUROPEAN COMMISSION'S ENERGY AND CLIMATE POLICY
A Climate for Expertise?

Beth Edmondson and Stuart Levy
CLIMATE CHANGE AND ORDER
The End of Prosperity and Democracy

David Elliott (*editor*)
NUCLEAR OR NOT?
Does Nuclear Power Have a Place in a Sustainable Future?

Neil E. Harrison and John Mikler (*editors*)
CLIMATE INNOVATION
Liberal Capitalism and Climate Change

Antonio Marquina (*editor*)
GLOBAL WARMING AND CLIMATE CHANGE
Prospects and Policies in Asia and Europe

Espen Moe and Paul Midford (*editors*)
THE POLITICAL ECONOMY OF RENEWABLE ENERGY AND ENERGY SECURITY
Common Challenges and National Responses in Japan, China and Northern Europe

Marlyne Sahakian
KEEPING COOL IN SOUTHEAST ASIA
Energy Consumption and Urban Air-Conditioning

Benjamin K. Sovacool
ENERGY & ETHICS
Justice and the Global Energy Challenge

Joseph Szarka, Richard Cowell, Geraint Ellis, Peter A. Strachan and Charles Warren (*editors*)
LEARNING FROM WIND POWER
Governance, Societal and Policy Perspectives on Sustainable Energy

Thijs Van de Graaf
THE POLITICS AND INSTITUTIONS OF GLOBAL ENERGY GOVERNANCE

Energy, Climate and the Environment
Series Standing Order ISBN 978–0–230–00800–7 (hb) 978–0–230–22150–5 (pb)

You can receive future titles in this series as they are published by placing a standing order. Please contact your bookseller or, in case of difficulty, write to us at the address below with your name and address, the title of the series and the ISBNs quoted above.

Customer Services Department, Macmillan Distribution Ltd, Houndmills, Basingstoke, Hampshire RG21 6XS, UK.

Renewable Gas

The Transition to Low Carbon Energy Fuels

Jo Abbess
Associate Research Fellow, Birkbeck College, University of London, UK

palgrave
macmillan

© Jo Abbess 2015

All rights reserved. No reproduction, copy or transmission of this publication may be made without written permission.

No portion of this publication may be reproduced, copied or transmitted save with written permission or in accordance with the provisions of the Copyright, Designs and Patents Act 1988, or under the terms of any licence permitting limited copying issued by the Copyright Licensing Agency, Saffron House, 6-10 Kirby Street, London EC1N 8TS.

Any person who does any unauthorized act in relation to this publication may be liable to criminal prosecution and civil claims for damages.

The author has asserted her rights to be identified as the author of this work in accordance with the Copyright, Designs and Patents Act 1988.

First published 2015 by
PALGRAVE MACMILLAN

Palgrave Macmillan in the UK is an imprint of Macmillan Publishers Limited, registered in England, company number 785998, of Houndmills, Basingstoke, Hampshire RG21 6XS.

Palgrave Macmillan in the US is a division of St Martin's Press LLC, 175 Fifth Avenue, New York, NY 10010.

Palgrave Macmillan is the global academic imprint of the above companies and has companies and representatives throughout the world.

Palgrave® and Macmillan® are registered trademarks in the United States, the United Kingdom, Europe and other countries.

ISBN: 978–1–137–44179–9

This book is printed on paper suitable for recycling and made from fully managed and sustained forest sources. Logging, pulping and manufacturing processes are expected to conform to the environmental regulations of the country of origin.

A catalogue record for this book is available from the British Library.

A catalog record for this book is available from the Library of Congress.

For the zero carbon world

Contents

List of Tables and Figures xiii

Series Editor's Preface xv

Acknowledgements xviii

1 An Introduction to Renewable Gas 1
1.1 What is Renewable Gas? 1
1.2 A rationale for Renewable Gas 5
1.2.1 Growth in the energy sector 5
1.2.2 The partnership and synergy between gas and power 6
1.2.3 Energy sector development rates and risks 7
1.2.4 The Limits to Growth: Peak Oil and Peak Natural Gas 7
1.2.5 Contrasting answers to a growing energy demand–supply gap 9
1.2.6 The decarbonisation of gas 9
1.3 The Natural Gas story 10
1.3.1 The "wonderfuel" development of Natural Gas 10
1.3.2 A trend towards dependency on Natural Gas 11
1.3.3 Risks of dependency on Natural Gas 12
1.3.4 Low carbon gas fuels to de-risk the scenario 16
1.3.5 A collection of low carbon gas fuels – from a range of technologies 16
1.3.6 A win-win-win situation 17
1.3.7 Challenging rates of progress 18
1.3.8 The natural end point for British and European energy policy 19

2 Energy Change and Investment Challenges 21
2.1 Navigating the nexus of the economic, climate and energy crises 21
2.1.1 Peak fossil fuel production 21
2.1.2 Global economic change 22
2.1.3 Managing the inevitable energy change – cross-cutting issues 23
2.2 Energy Change: the energy paradigm shift 23
2.3 Challenge to invest: the engineering and resource life cycle 24

2.3.1 Losing investment from energy engineering 26
2.3.2 Why oil and gas supply investment might not be forthcoming 28
2.3.3 Peak Fossil Fuels: fossil fuel supply depletion severity needs to become acknowledged 31
2.3.4 Technological blind alleys and the economics of new technologies 39
2.3.5 Can carbon pricing or taxation be effective in stimulating investment in the low carbon economy? 41

3 Energy Transitions and Renewable Gas 45
3.1 What can work is what is already working for energy change 45
3.1.1 What is already working? Strong continued investment in renewable electricity 45
3.1.2 What is already working? Natural Gas-fired power generation 46
3.2 Why gas? 48
3.2.1 Flexibility in use, multiplicity in source and flexibility in composition 48
3.2.2 Nature's own energy solution 49
3.2.3 Industrial gas solutions 50
3.2.4 Energy storage – because renewable electricity is sustainable, but variable 51
3.3 Why Renewable Gas? 52
3.3.1 Gas is our flexible friend 52
3.3.2 Renewable Gas can decarbonise the gas supply 53
3.4 Why methanate? 54
3.4.1 Methanation permits a gradual transition in gas energy systems 55
3.4.2 Methanation can upgrade a mix of gases from a range of sources 56
3.4.3 Methanation can help to resolve problems arising from changing Natural Gas quality 57
3.4.4 Methanation can enable the recycling of fossil carbon 57
3.4.5 Methanation could avoid exploration for Natural Gas in sensitive regions of the world 58
3.4.6 Methanation and methane storage could regulate gas systems 58

3.5 Practical measures in an orderly gas transition 58
3.5.1 Efficient management of Natural Gas resources 59
3.5.2 Accurate assessments of Renewable Gas production potential 59
3.5.3 Management of gas market competition 59
3.6 Pathways to Renewable Gas 61
3.6.1 Space heating and cooling of buildings 61
3.6.2 The Prime Mover of transportation: the importance of fixing transport options 62
3.6.3 Electricity generation 68
3.6.4 Industrial chemistry 69
3.7 Can we stop digging yet? Technological advance in Renewable Gas 69
3.7.1 Renewable Gas is "cuspish" – on the verge of major industrialisation efforts 70
3.7.2 Vanguards of research 70
3.7.3 The forefronts of knowledge 70
3.7.4 Industrial gas systems design 72
3.8 Advances and the way ahead 73
3.8.1 Market development – volumising 73
3.8.2 A summary of the future state of gas 74

4 A Brief History of Gas 77
4.1 Eternal flames 77
4.2 Coal Gas and Town Gas 77
4.3 Manufactured gas and the conversion to Natural Gas 79
4.4 Research into Natural Gas substitutes 81
4.5 The parallel lives of syngas during the 20th Century 81
4.6 SNG and synthetic liquid fuels from coal and biomass 82
4.7 Carbon Capture and Storage (CCS) and Enhanced Oil Recovery (EOR) 83
4.8 SNG made from heavy oils and petrorefinery waste 84
4.9 Other gas manufactured in petrorefineries 84
4.10 Gas for power generation 85
4.11 The chemistry of manufactured gas 85
4.12 Biomass gasification and carbon recycling 86
4.13 A brief future of gas 87

5 Renewable Gas Systems 91
5.1 The evolution of gas energy systems 91
5.1.1 The existing energy system 92
5.1.2 Carbon Capture and Storage (CCS) 93

5.1.3 Carbon Capture and Utilisation (CCU) 93
5.1.4 Increased bioenergy 94
5.1.5 Bioenergy with Carbon Capture and Storage (CCS) 95
5.1.6 Bioenergy with Carbon Capture and Utilisation (CCU) 96
5.1.7 Carbon recycling 96
5.1.8 Renewable Gas 97
5.1.9 Hydrogen Economy 99
5.2 Key process engineering 99
5.2.1 The importance of gasification 100
5.2.2 Syngas production 101
5.2.3 Methane production 102
5.2.4 Hydrogen production 103
5.3 Generic Renewable Gas power plant system design 104
5.3.1 Mode A: the production of Renewable Hydrogen 104
5.3.2 Mode B: the gasification of (hydro-) carbonaceous material 106
5.3.3 Mode C: Methanation: Type 1 108
5.3.4 Mode D: power generation 111
5.3.5 Mode E: Methanation: Type 2 114
5.3.6 Mode F: Methanation: Type 3 114
5.3.7 Mode G: the production of synthetic fuels 115
5.3.8 Mode H: gas grid injection 116
5.4 Alternative routes to syngas besides gasification 117
5.5 Thermal and gas balance 117
5.6 Carbon Capture and Storage (CCS) 118
5.7 Activation energies in thermochemical reactions 118
5.8 Renewable Gas for Iran 118
5.9 The carbon problem: an ocean solution? 123
5.10 Transitions and choices: major branching choices: alternative narratives 124
5.10.1 Option A: conventional or unconventional Natural Gas? 126
5.10.2 Option B: utilise or vent carbon dioxide? 129
5.10.3 Option C: separation or in situ methanation? 130
5.10.4 Option D: fossil gas or Renewable Gas? 132

6 The Technology of Renewable Gas 133
6.1 Key areas of research and development 133
6.1.1 Mode A: the production of Renewable Hydrogen 133
6.1.2 Mode B: gasification 135

6.1.3 Mode C: Methanation 138
6.1.4 Mode D: power generation 139
6.2 Transition pathways: technology adoption 142
6.2.1 Mode A: the production of Renewable Hydrogen 143
6.2.2 Mode B: gasification 143
6.2.3 Mode C: Methanation Type 1 144
6.2.4 Mode D: power generation 144
6.2.5 Mode E: Methanation Type 2 and Mode F: Methanation Type 3 144
6.2.6 Mode G: the production of synthetic fuels 145
6.2.7 Mode H: gas grid injection 145
6.3 Coal routes to low carbon gas (CtG) 146
6.3.1 Carbonisation, gasification and carbon capture 147
6.3.2 Co-firing 147
6.3.3 Fuel cell conversion 147
6.3.4 Hydrogen, steam, catalysts and sorption 148
6.4 Waste routes to low carbon gas energy (WtE, EfW) 148
6.5 Biological routes to low carbon gas (BtG) 149
6.6 Hydrobiological (hydrogen-biological) routes to low carbon gas (H+BtG) 150
6.7 System integration 151
6.7.1 Candidates for process configuration for carbon capture 151
6.7.2 Doubling down on doubling up: peaker plant design 154
6.7.3 Doubling down on doubling up: Sabatier reaction versus syngas methanation 156
6.7.4 Doubling down on doubling up: heat management and heat integration 156
6.7.5 Further heat integration 158
6.8 Pinch Points 158
6.8.1 Pinch Point: carbon slip 158
6.8.2 Pinch Points: the production of Renewable Hydrogen 159
6.8.3 Pinch Point: methanation efficiency 165
6.8.4 Pinch Point: economics 165

7 The Energy Policy Context for Renewable Gas 167
7.1 The policy waymarker 167
7.1.1 Why we cannot be "technology-neutral" or "technology agnostic" in transforming and re-investing in energy systems 168
7.1.2 Why we might need to develop gas as public infrastructure assets 178

7.1.3 Why energy demand management is critical 181
7.1.4 Carbon balance 185
7.1.5 Smooth transition 193
7.1.6 The economics of Renewable Gas 194

8 Reflections and Conclusions 203
8.1 The hydrogen-carbon question 203
8.2 The Natural Gas and ammonia question 204
8.3 Sustainable business 204
8.4 Renewable Gas is imminent 205
8.5 Renewable Gas can already displace some fossil fuels 205
8.6 Renewable Gas and uranium 206
8.7 Development pathways 206
8.8 The hydrogen numbers game 206
8.9 Good "black swans" 208
8.9.1 Hydrogen catalysis 208
8.9.2 Electromagnetic radiation-assisted hydrogen and methane production 208
8.9.3 A breakthrough in biomass gasification 209
8.9.4 A breakthrough in the methanation of sour/acid Natural Gas 209
8.10 The policy ambit 209
8.10.1 Retaining gas-to-power capacity 209
8.10.2 Enabling the transition to Renewable Gas 210
8.11 Recommendations 210

References 213

Index 257

List of Tables and Figures

Tables

1.1 Estimate of risks posed by increased and continuing dependency on Natural Gas in industrialised nation energy systems 13

1.2 Brief analysis of the BP Statistical Review of Energy 2014 Workbook, Revised version (BP, 2014c) 18

2.1 A selection of organisations modelling energy in the economy 33

3.1 Degrees of recycling of carbon dioxide 54

3.2 Simple medium-term guesstimates for carbon dioxide recycling applications 54

4.1 Typical chemistry of various manufactured, natural and biologically sourced gases (% percentage) (see also: Biogas Renewable Energy, 2009; Larminie and Dicks, 2003, Tables 8.4, 8.5) 78

5.1 Sour and acid gas and nitrogen composition of a selection of Iranian gas fields 120

7.1 Estimates of hydrogen production and consumption 188

7.2 Analysis of the Levelised Cost of Energy (LCOE) for Renewable Gas and Natural Gas 195

8.1 Estimates of annual global hydrogen production (all markets – captive and merchant) 207

Figures

5.1 Existing energy system 92

5.2 Carbon Capture and Storage (CCS) 93

5.3 Carbon Capture and Utilisation (CCU) 94

5.4 Increased bioenergy 95

5.5 Bioenergy with Carbon Capture and Storage (BECCS) 95

5.6 Bioenergy with Carbon Capture and Utilisation 96

5.7 Carbon Recycling 97

5.8 Renewable Gas 98

5.9 Hydrogen Economy 99

5.10 Mode A: The Production of Renewable Hydrogen 105

5.11 Mode B: The Gasification of (Hydro-) Carbonaceous Material 107

5.12 Mode C: Methanation: Type 1 109

5.13 Mode D: Power Generation 112

5.14 Mode E: Methanation: Type 2 114

5.15 Mode F: Methanation: Type 3 115

5.16 Mode G: The Production of Synthetic Fuels 115

5.17 Mode H: Gas Grid Injection 116

5.18 Renewable Gas: The Transition to Low Emissions Energy Gas and Major Choices 125

Series Editor's Preface

David Elliott

Concerns about the potential environmental, social and economic impacts of climate change have led to a major international debate over what could and should be done to reduce emissions of greenhouse gases. There is still a scientific debate over the likely *scale* of the severity of climate change, and the complex interactions between human activities and climate systems, but, global average temperatures have risen and the cause is almost certainly the observed build up of atmospheric greenhouse gases.

Whatever we now do, there will have to be a lot of social and economic adaptation to climate change – preparing for increased flooding and other climate related problems. However, the more fundamental response is to try to reduce or avoid the human activities that are causing climate change. That means, primarily, trying to reduce or eliminate emission of greenhouse gases from the combustion of fossil fuels. Given that around 80% of the energy used in the world at present comes from these sources, this will be a major technological, economic and political undertaking. It will involve reducing demand for energy (via lifestyle choice changes – and policies enabling such choices to be made), producing and using whatever energy we still need more efficiently (getting more from less), and supplying the reduced amount of energy from non-fossil sources (basically switching over to renewables and/or nuclear power).

Each of these options opens up a range of social, economic and environmental issues. Industrial society and modern consumer cultures have been based on the ever-expanding use of fossil fuels, so the changes required will inevitably be challenging. Perhaps equally inevitable are disagreements and conflicts over the merits and demerits of the various options and in relation to strategies and policies for pursuing them. These conflicts and associated debates sometimes concern technical issues, but there are usually also underlying political and ideological commitments and agendas which shape, or at least colour, the ostensibly technical debates. In particular, at times, technical assertions can be used to buttress specific policy frameworks in ways which subsequently prove to be flawed.

The aim of this series is to provide texts which lay out the technical, environmental and political issues relating to the various proposed policies for responding to climate change. The focus is not primarily on the science of climate change, or on the technological detail, although there will be accounts of the state of the art, to aid assessment of the viability of the various options. However, the main focus is the policy conflicts over which strategy to pursue. The series adopts a critical approach and attempts to identify flaws in emerging policies, propositions and assertions. In particular, it seeks to illuminate counter-intuitive assessments, conclusions and new perspectives.

The present text is no exception in exploring new approaches to the use of gas as a major energy source. Fossil gas use has boomed in recent years, in part since it is less carbon intense than coal, but of course burning it still produces carbon dioxide, and reserves, even with shale gas included, are not infinite. That is a worry, since gas has attractions. Unlike electricity, it is easily stored and can be transmitted over long distances with low energy losses. Gas-fired power plants can be quite flexible, so some see them as an important complement to variable renewables. Peak fossil gas may be some way off, and it may be possible for carbon dioxide emissions from gas-fired generation to be captured and stored. However, in terms of low carbon options, there is also another approach: using biomass stocks of various kinds to make "renewable gases". If done right, with the right feedstock and sustainably managed sources, producing and then burning biogas can be almost carbon neutral over time. And with carbon capture and storage, perhaps even carbon negative.

Switching to renewable gas does not just involve a simple low-carbon fuel change, it also opens up some interesting new possibilities, including an alternative to decarbonisation by electrification. One currently quite dominant view in the UK is that domestic heating should no longer be done using fossil gas, but by heat pumps using electricity from renewable sources and/or nuclear. Renewable gas offers an alternative, which might make more sense, given that the UK gas grid currently carries 3–4 times more energy than the electricity grid. Switching from gas to electric heating would put a huge strain on the power transmission and distribution system and entail constructing high levels of new electricity generation capacity that would only be used for a few months of the year.

There are land-use and biodiversity limits to how much biogas can be produced from biomass, but, in addition to using farm and food wastes, there may be a novel solution – the production of synthetic green gases, using renewable electricity and carbon dioxide captured from the air or from power plant exhausts. The simplest approach is just to make hydrogen gas by electrolysis of water, but this can also be converted into methane using captured carbon dioxide. The conversion processes will of course add to the costs, but that can be offset if use is made of surplus electricity from wind or photovoltaic (PV) solar power, which would otherwise just be dumped. This so-called "power to gas" system, with energy being stored as gas for use when wind or PV were less available, would provide a way to balance variable renewables. Like biogas, it can also replace or augment fossil gas in the gas mains and be used as a green vehicle fuel.

In addition to the fully renewable biogas and power to gas options, there are also many other intriguing "green gas" options, including the use of a range of industrial gases and fossil gas sources, which can be decarbonised, with, in some cases, the carbon being recycled and used, with hydrogen, to make syngases. Although not strictly renewable, these options offer substantial amounts of low-carbon energy for as long as fossil reserves last. They could be an important interim or transitional gas option, for use while fully renewable sources are developed.

The wide-scale adoption and use of these various "green gas" options would have significant implications for how energy systems were developed and for energy policy generally. Although some of these options may still be relatively new, the policy issues that are raised need to be considered urgently. The technical and strategic analysis in this text provides a good starting point.

Acknowledgements

The author acknowledges with gratitude the encouragement and support of Dr Rebecca Briant and Dr Paul Elsner of the GEDS group at Birkbeck, University of London, and the commitment to enquiry and information-sharing amongst the Claverton Energy Research Group, especially the research support of Dr Frederick Starr FIMMM.

1
An Introduction to Renewable Gas

1.1 What is Renewable Gas?

Comfort-loving citizens in the Northern Hemisphere have good reason to take an interest in the future of gas, particularly since the demand for energy from Natural Gas in winter can be several times higher than the consumption of electrical energy (DECC, 2014a), and the rates of insulation renovations for old, draughty buildings can be slow. Unlike electricity, it is possible to store gas from season to season, making it a practical energy vector; extended storage means that gas production can be averaged out throughout the year. However, the use of Natural Gas in the long term is in some doubt as it is a fossil fuel and its combustion disturbs the deep geological carbon cycle, thus contributing to global warming. It is therefore appropriate to consider whether there might be viable low carbon alternatives to Natural Gas. Biogas, naturally occurring from the microbiological decomposition of biomass, has much to offer, but its advancement may well be hampered by changing patterns of land use, including constraints imposed by climate change. Enhanced and advanced biogas processing techniques could compensate, giving higher yields of gas from biomass (e.g. Luo and Angelidaki, 2012), although gas produced with any biological processing steps could remain slow. Consequently, industrially manufactured low carbon gas holds the most promise in terms of production volumes, although its development depends on adaptations and integration across several sectors of the economy. The manufacture of some types of low carbon gas does not even require biomass as an input feedstock, or may only need a recyclable cache of carbon from biomass as an enabler, or catalyst, to produce hydrogen from water.

For a gaseous phase energy fuel to be truly renewable, it should be produced directly from biomass (such as biogas), or indirectly from other renewable resources (such as hydrogen produced using renewable electricity). However, in what follows, a broader definition of what constitutes sustainable gas technology is adopted, covering a range of low carbon gas options (Abbess, 2014, Table 10). Some of these low carbon gases are transitional, wholly or partly manufactured from fossil fuels, in the interim. The technology platform put in place will permit the transition to fully Renewable Gas energy systems in the future, by altering the input feedstocks or feed materials. The options for manufacturing low carbon gas are, for the most part, not new technologies but a repurposing of those already in use, with the aim of making strategic, long-lasting investments, and thence a decades-long transition to a fully sustainable gas energy supply. What follows is not the promotion of an individual technology but a recommendation for an evolution in energy systems, with gas fuels taking centre stage.

The three factors that most concern policymakers in the energy sector are: sustaining energy supplies, decarbonising energy systems – thus preventing mainly carbon dioxide and methane emissions – and ensuring that energy bills don't spiral upwards. In the last fifty years, the world's engineers have been working to perfect devices that can capture wind, sunshine and moving water energy for zero carbon renewable electricity production, and its mass deployment has contributed to a major reduction in the cost whilst contributing to energy security. Now, it seems like the right time to look at Renewable Gas – a range of low net carbon emissions gas energy fuels. Ironically, around fifty years ago, at about the same time they took up renewable electricity pursuits, North America and Europe – although not China (Yang et al., 2014) or Hong Kong – gave up manufacturing gas and switched to Natural Gas; but now, some of that chemical processing knowledge must be resurrected, in order to prevent greenhouse gas emissions from gas energy.

For the next few decades, fossil gas (Abbess, 2014, Table 10) is going to remain vital in industrial economies which are anchored to Natural Gas; even with a firm commitment to Renewable Gas, it will take some time to displace a significant quantity of Natural Gas with low carbon gas delivered via advanced energy networks. Furthermore, replacing the use of coal with Natural Gas for power generation gives an immediate reduction in carbon emissions (Venkatesh et al., 2011); thus, Natural Gas can have enormous value in the work to mitigate near-term climate change. However, for this to be meaningful progress, gas grid and fugitive emissions, particularly from hydraulic fracturing, will need to be eliminated (Biello, 2014; Dlugokencky et al., 2011; Marston, 2013; Miller et al., 2013; Nisbet et al., 2014), gas flaring and venting to the atmosphere must be prevented (EPA, 2014; GGFR, 2013; UNFCCC, n.d.), and gas well integrity improved (Cherry et al., 2014; IEA, 2012b), particularly subsurface trespass. This need to improve methane containment will become more urgent not only because of the increased demand for gas, which will increase distribution flows, but also because of ageing gas pipeline networks, and perhaps also because of the nature and accessibility of the gas resources themselves.

An added spur to making progress on Renewable Gas could be that the resources of Natural Gas brought to production in twenty years' time are likely to be of a lesser quality and more complicated to drill than those we are using today (Patterson, 2012). This will be partly because something like a fifth of "conventional" Natural Gas reserves are associated with crude oil fields (GEA, 2012, Section 7.3.3), some of which are depleting (Aleklett et al., 2010; Hook, 2014; McGlade, 2014; Sorrell et al., 2010). Good-quality conventional Natural Gas fields (GEA, 2012, Section 7.3.2), whether or not associated with crude oil, will begin to deplete due to continuing high production, a phenomenon seen in both North America in the past and the North Sea today. They will show a peak in supply volumes in a pattern that will in all likelihood be similar to the decline in crude petroleum oil production (Laherrere, 2013a, 2013b; Rutledge, 2011, 2013). We will turn to alternative fossil gas resources, even though they might not be exploited rapidly enough to prevent a peak in total Natural Gas production. The alternatives will include deeper gas, and more acid gas, gas with more vaporised oil in it, non-free-flowing gas and gas from more complex deposits (IEA, 2011, Page 50, Box 2.1; IIASA, 2012, Tables 7.10–7.14; McGlade et al., 2012a, 2013a, 2013b; Mohr and Evans, 2011; Skorobogatov et al., 2000; Soderbergh et al., 2010). These "unconventional" Natural Gases, and the manner in which they are mined, could present a range of problems, such as higher greenhouse gas emissions (Cathles et al., 2012; Glancy, 2013; Howarth et al., 2011; O'Sullivan and Paltsev, 2012; Wigley, 2011), unless they are deliberately curtailed (MacKay and Stone, 2013).

Ultimately, however, even with methane management measures, owing to ever-tightening carbon budgets set by climate protection policies, the carbon dioxide emissions from the burning of Natural Gas will begin to be the main concern. We will not be able to continue to burn Natural Gas "unabated" (UCS, 2013) – the fossil gas that we will use, within twenty to thirty years from now, will need to have the carbon taken out, and either permanently sequestered (IEA, 2011, Page 121; IPCC, 2014a, Sections 7.5.1, 7.6.3) or recycled, for instance, into fresh gas fuels. Permanent sequestration of carbon dioxide resulting from fossil fuel combustion is likely to remain costly, and so it can be expected that there will be a gradual shift towards low carbon gas supply technologies to prevent emissions at source. The first progress could possibly be seen in a variety of methods to reduce the carbon impact of gas carried by the grid networks: some carbon-rich waste, flue and exhaust gas will begin to be recycled into manufactured gas through chemical engineering. However, this will only be a proof-of-concept beginning, as there is much that can be done additionally. Key developments are likely to arise in crude petroleum oil refineries and Natural Gas processing plants – initially to answer processing problems with the worsening quality of fossil fuels, and to reduce the environmental impact of refined fuels – but these will eventually lead to manufacturing low carbon gas. Petroleum refineries could transition to being low carbon gas suppliers, and the refiners of biofuels.

The engineering and research communities are already occupied with the technologies of Renewable Gas, but as with every major transition, governments, and engineering and energy companies have yet to make strategic decisions about forks in the road ahead. What follows will venture to lay out the map for the journey to low carbon gas, to cover the technological and process engineering ground, and create a snapshot of the choices that will present themselves, and the options for scaling the obstacles in this dynamic field. The remainder of this chapter looks at trends, opportunities and risks in the use of gas energy, and why low carbon gas is important. Chapter 2 explores the need for energy sector investment and discusses the limitations of various technology choices and policies. Chapter 3 addresses the potential for gas energy and a transition to Renewable Gas, and how Renewable Methane, in particular, can be of significant value. Chapter 4 reviews the history of manufactured gas and Natural Gas to illustrate previous transitions in gas energy. Chapter 5 describes the evolution of energy systems, to show the natural transition possible from fossil fuels to the hydrogen economy. It ends with an analysis of choices for the next few decades. It also introduces a generic design for a Renewable Gas system, and where each component's technology is already used. Chapter 6 analyses the technology of Renewable Gas, weaknesses, pinch points and opportunities for system integration. Chapter 7 covers the policy framework for the introduction of Renewable Gas in the context of other changes. Chapter 8 is a collection of reflections and conclusions.

1.2 A rationale for Renewable Gas

In order to comprehend why a transition to Renewable Gas warrants consideration, it is necessary to unpack trends in gas energy, and make some reasonable conjectures about future changes.

1.2.1 Growth in the energy sector

Energy is widely regarded as a pivotal sector of the global economy, and likely to grow, as markets, under governance, answer energy security concerns with new resources, refreshed and expanded infrastructure, energy efficiency services and improved access (BP, 2013a, 2014a; IEA, 2003a, 2006, 2008, 2010a, 2012a, 2014a; IRENA, 2014; SE4ALL, 2014; Shell, 2008, 2011, 2013; Statoil, 2013; WBCSD, 2012; WEC, 2007a, 2007b; World Bank, 2013). Industrialised regions have accelerated the deployment of renewable energy, mostly in the form of wind power and solar power. The lead times for adding renewable electricity capacity are short, and renewable electricity can be rapidly integrated into gridded power networks. Accordingly, there are some projections that renewables can eventually supply very significant proportions of power demand (AEMO, 2013; CAT, 2013; CCCC, 2009, 2010; CSC, 2013; Denmark, 2011, 2012; Greenpeace, 2010; Jacobson and Delucchi, 2009; Moser et al., 2014; Peter and Lehmann, 2008; PwC, 2010; WWF, 2011). Home-grown, sustainable, low operational cost (Czisch, 2011a, 2011b), climate friendly and pollution free, renewable power is clearly to be viewed as a genuine asset, providing sound returns on investment; but experience is showing that there are two residual and connected issues with its adoption. The provision of renewable electricity is variable, at all timescales; and in addition, it is unlikely to become possible to use batteries, capacitors or other solid state technology to store power at low cost and at large scale for long periods. As the uptake of renewable electricity continues, the geographical spread of the power generation equipment can help to balance supply (Archer and Jacobson, 2013). For example, in the European Union, it is calculated that variations in weather-dependent and season-dependent wind power in the north of the region can be largely balanced by diurnal solar power from the south (Brouwer et al., 2014; Creutzig et al., 2014; Czisch and Schmid, 2006; EWEA, 2005; Heide et al., 2011; Rodriguez et al., 2014; Santos-Alamillos et al., 2014; Trieb and Muller-Steinhagen, 2009). In Germany, winter wind is complemented by summer sun (Fraunhofer, 2013, 2014). Yet, no matter how extended the grid, there will always be times when this instant's power demand cannot be met by the current moment's renewable electricity supply. Conversely, there will always be times when not all renewable electricity can be used by the power grids. Therefore, looking beyond electricity into the wider energy economy is essential in addressing these concerns.

1.2.2 The partnership and synergy between gas and power

Renewable electricity generation from zero-cost wind, sun and marine energy and other sustainable resources will become much more highly sought after – but it will take some time to make its contribution truly significant. Since renewable electricity is variable in output, flexible, "dispatchable" gas-fired power generation will be suitable to back up, or "load balance" renewable electricity generation. With increasing levels of variable renewable power generation, the concept of "baseload", or "always on" generation plant becomes outdated. Those power sources that cannot flex will wither – for example, nuclear power only has a certain amount of flexibility in output, and this comes at a cost (NEA, 2011). Plans to capture carbon dioxide from coal-fired power stations and permanently store it in caverns under the sea (Carbon Capture and Storage, or CCS) are not advancing rapidly, and it may remain costly, even with a strong carbon price, market or tax. Therefore, emissions-"abated" coal-fired power with CCS is unlikely to provide sufficient flexible generation capacity in the near future. Without other low carbon energy coming into play in the near future, this means that we need to continue building new gas-fired power stations, and strengthening gas supply lines, which amounts to a commitment to gas fuels.

Many developed countries are in a decades-long process of transitioning to a combined gas and power system to supply the energy needs of their built environment, and some of their transportation needs too (e.g. the European "Energy Union"). Natural Gas and electricity are highly complementary energy vectors. Gas is straightforward to store and is used for distributed space heating as well as centralised power generation. Besides the growth in battery electric vehicles (BEVs), in future, a range of gas-fuelled vehicles will become more prevalent. The demand profile for electricity can be very similar year-round, and that would remain the case even with more electric drive vehicles. However, the total amount of energy that must be supplied in colder months as heat, mostly provided by gas, is usually several times greater than the combined energy consumption of power and gas in the warmer months. It would not make any sense to supply the colder months' extra seasonal energy demand in the form of electricity, primarily as this would require major investment. Secondly, if these major additions in generation capacity were from nuclear fission or renewable electricity, this would all need emergency cover to back it up, and so could require the building of more gas-fired or coal-fired power plants. Thirdly, the timetable for the construction of all the new power generation plant required to convert all space heating and cooling to electricity could be unfeasible to meet; moreover, this super fleet of power generators would then lie idle for most of the year. The increasing use of Natural Gas in heating applications is therefore helping to moderate the complexity and extent of the energy supply systems, and minimising the total scale of committed capital. Additionally, it would make sense to retain the invested value of the gas infrastructure into the future, rather than abandoning it completely to go all-electric.

1.2.3 Energy sector development rates and risks

Energy security questions are being answered at the present time by a narrative that proposes the exploitation of unconventional and previously inaccessible fossil fuels, alongside a renaissance in the use of nuclear fission energy. Large centralised nuclear power plants are slow to construct, and given the projections for increased global power demand, atomic electricity might never be able to get beyond supplying only a few per cent of total energy demand, even with a significant new build programme – much of it only replacement for existing capacity, as older units are decommissioned. Proposals have surfaced for a network of small modular reactors (SMRs), which are almost entirely unproven, and would in any case face possibly insurmountable problems with the safe transport of fuels and waste. Problems with the supply market for Natural Gas may well develop during the construction phase of new nuclear power plants, and so it is unclear if new atomic generation could be ready in time to help prevent disruption to the linked gas and power networks. Although the development of unconventional and remote fossil fuels raises the prospect of greater threats to the environment, and even poses risks for meeting climate change policy targets, the principal concern is whether these resources can be exploited at sufficient rates to compensate for the depletion of conventional reserves, whilst at the same time meeting increased demand from expanding markets.

1.2.4 The Limits to Growth: Peak Oil and Peak Natural Gas

In considering the future of energy in general, Peak Oil, or a peak in capacity to supply oil and oil-based products, remains a spectre at the feast. North America may have experienced "peak demand" for petroleum-derived hydrocarbon vehicle fuel – unless this was simply a recession-driven dip (Lewis, 2014a) – but other parts of the world are still building their motor fleets. This strategy could be damaged by uncertainties in the supply of refined oil products: although some oil refineries have closed (OPEC, 2012, Chapter 10; 2013, Chapter 6; 2014, Chapter 6), refining capacity has increased (OPEC, 2012, Figure 6.4; 2013, Figure 6.10; 2014, Figure 6.14), and yet the projections of additional refined product output are contracting when compared year on year (OPEC, 2012, Figure 6.9; 2013, Figure 6.12; 2014, Table 6.3), even as demand increases. Investments will need to be made not only in petrorefinery but also in the supply of fossil fuel commodities. Reporting of crude oil production, besides conventional crude oil, now includes Natural Gas Liquids and other condensate, plus shale oil (light tight oil), heavy oil and bituminous oil (oil sands, tar sands); yet not all of this addition to "total liquids" can be processed into vehicle fuel. If the economics of light tight oil (shale oil), heavy oils and other resources such as oil shale (kerogen oil) extraction are demonstrated as questionable, especially where they are prone to rapid depletion, or are at a low density in supporting sediments, it might not be possible to incentivise oil supply to keep up with demand.

If a peak in raw conventional crude oil production is acknowledged, and unconventional oil demonstrates it cannot grow fast enough to compensate for the decline, and the supply of oil-based refined products is admitted to be in permanent decline, about fifteen to twenty years later there could also be a peak in the global capacity to supply Natural Gas. The timing of this would be mostly a geological phenomenon, a natural consequence of an overall peak in oil production; since much Natural Gas is associated either with petroleum oil deposits or in rock formations that could have matured crude oil had they been subject to different temperatures and pressures. That a peak in conventional Natural Gas could arrive before the widespread development of alternative gas resources would pose a significant risk to global energy security. For example, shale gas has been found to suffer higher field depletion rates than the usual Natural Gas resources (EIA, 2012), so without continued drilling, and much wider uptake, which depend on economic recoverability (Sovacool, 2014), it cannot prevent Peak Natural Gas, merely delay it, although maybe not significantly (de Castro et al., 2009; Weijermars, 2014). Other unconventional fossil gas resources may not be subject to the same depletion scenarios, but they do not yet have high production volumes. In some cases, the energy return (the energy value returned on the energy expended to produce the energy) may keep some unconventional gas cost-inefficient to produce, leading to questions of sustainability of supply. With rapidly exhausting shale gas, and slow growth in other unconventionals, Peak Natural Gas may be inevitable within a few decades.

1.2.5 Contrasting answers to a growing energy demand–supply gap

Despite the lack of evidence that fossil fuels can be produced in high volumes sustainably into the future, there is still strong demand globally for Natural Gas, or substitutes for Natural Gas, not only due to expanding energy demand, but also due to a "dash for gas" – the transition out of other fossil fuels to Natural Gas. This is happening due to a combination of environmental, economic and practical reasons. The multifaceted nature of gas – in both how it is sourced and how it is used, as well as its store-ability – makes it an obvious choice for building into energy futures scenarios: gas can easily be applied to any energy use, and so it is versatile.

The publicly expressed narrative from the oil and gas industry is that we should not concern ourselves about the risks of Peak Natural Gas, as unconventionals such as shale gas, tight gas, Arctic gas, gas from Alaska, coal seam gas (coalbed methane), deepwater gas, deep water-dissolved gas, gas from condensate, gas from heavy or shale oil, gas from oil sands refinery, and, possibly, methane hydrates can fill any widening supply–demand gap. This storyline is arguably sensitive to many factors out of the control of the energy industry. On the other hand, the alternative of Renewable Gas production could be within management scope, whilst ticking all the policy boxes on future sustainability of the current energy companies, carbon dioxide emissions reduction, methane emissions prevention, and energy supply security. The development of Renewable Gas is therefore perhaps inevitable.

1.2.6 The decarbonisation of gas

Not only can Natural Gas help to decarbonise an economy, Natural Gas can itself be decarbonised, an example of which is the injection of biomethane, upgraded from biogas, into the UK gas supply grid network. It appears unlikely that gas derived solely from the microbiological decomposition of biomass will be able to replace all gas supplied; however, projects of this kind demonstrate that it is possible to displace Natural Gas with low carbon alternatives. Questions about the overall shape of the energy system, and the pathways for transition, may not yet have been answered, but bringing new low carbon gas to market in the short term is important for bringing about the long-term vision of a fully decarbonised economy, proving concepts, capabilities and capacities. Bringing on low carbon gas can also offer practical solutions for very pressing problems in the short term.

Renewable Gas could be considered merely "additional" in its infancy, making up for shortfalls in Natural Gas production resulting from high global demand or regional production difficulties. However, it could also help to address worsening chemistry from conventional Natural Gas fields. It is estimated that 40% of remaining conventional Natural Gas is sour – with high levels of hydrogen sulphide, and acid, with high levels of carbon dioxide (ARI, 2013; Burgers et al., 2011; Foster Wheeler, 2008, Slide 5; Lallemand et al., 2012; Lokhorst et al., 1997, Section 4.5). The sulphur needs to be removed for a variety of reasons, and there is a financial benefit in the by-products. However, there is currently no significant market for carbon dioxide, yet it need not go to waste if production facilities for Renewable Gas are constructed in the short to medium term, and are used to make a gas that is partway decarbonised. The carbon dioxide from acid/sour Natural Gas can be made into methane, the primary constituent of Natural Gas. This can be done by reaction with Renewable Hydrogen, and Renewable Hydrogen can be made from water by the use of renewable electricity, for example. Choosing Carbon Capture and (Re)Utilisation (CCU) over Carbon Capture and Storage (CCS) (where carbon is permanently sequestered underground) means that valuable carbon is maintained in the energy system. Building up processing capacity for Renewable Gas will permit the major oil and gas companies to avoid the costs of permanent carbon burial by CCS, necessitated by the use of acid/sour gas, and encourage the continued use of secure gas resources (even if acid/sour), whilst the energy sector transitions out of fossil fuels.

1.3 The Natural Gas story

Natural Gas is perhaps a perfect fuel, despite its humble origins. It is non-corrosive, there are no solid or liquid wastes to deal with, and few noxious, toxic or polluting by-products, apart from carbon dioxide. It is not surprising that economies are increasingly relying on it; however, this dependency poses considerable risks, which need to be mitigated.

1.3.1 The "wonderfuel" development of Natural Gas

The development of Natural Gas as a fuel for energy has been an important evolution in energy systems for the industrialised nations, having useful and appropriate applications in space heating, electricity generation and, recently, transport, whilst being low in carbon dioxide and particulate emissions upon combustion, and increasingly being made safe to store, distribute and utilise by end consumers. It has progressed in favour from being a virtually unwanted vented or flared by-product of petroleum oil production to being the energy fuel of choice of most developed economies, and it is set to remain a key component in government and commercial plans for energy futures. Mostly composed of the simplest hydrocarbon, methane, adulterated with a few more complex constituents, Natural Gas is usually "thermogenic" (Abbess, 2014, Table 10) – the final stage of the decomposition of millions-of-years-old animal and plant remains, through high subterranean temperature and pressure. Natural Gas, or rather, Natural Methane, is also produced in quite young organic sediments, close to the surface of the Earth, via biological activity, for instance, in swamps and marshes, or river deltas, and this is known as "biogenic". However, some Natural Gas does not derive from the decomposition of once-living organisms: as with Natural Hydrogen seeps (Abbess, 2014, Table 10), some Natural Methane results purely from subterranean inorganic chemistry – perhaps to be termed "chemogenic". In addition, as during the aeons of the origins of life, some surviving ancient organisms (archaea) may feed on inorganic crustal gas seeps, for example, at deep-sea hydrothermal vents, and both consume and produce hydrogen and methane – perhaps to be termed "archaeogenic". Using Natural Gas is the ultimate in recycling, returning life waste into a valuable commodity, and in many aspects, it is an ideal fuel. The only potentially superior choice for energy storage and conversion is hydrogen gas, and future energy systems can be anticipated to make use of both hydrogen and methane in a variety of configurations. Deep-time research suggests that these two simple gases were companions in the development of early life on Earth, and microbiological communities of the present era indicate these energy carriers have remained preferential, as exhibited by the co-production of hydrogen and methane in the mechanised decomposition of biomass under a range of different conditions. The ubiquity of methane, not only in fossil Natural Gas, but also in surface-produced, biologically produced or manufactured gas, and the lack of complexity in the chemistry of combustion of methane, means that gas can provide a clean and efficient energy delivery system.

1.3.2 A trend towards dependency on Natural Gas

Technological improvements have underpinned distribution of Natural Gas through vast piped networks, and its wide use by power plants and industrial chemical infrastructure. This has led to an increase in the volumes of Natural Gas employed, embedding it in advanced nation economies and regions by making it ever more accessible. Along with the development of gas grids has been an expansion of supply routes, including Liquefied Natural Gas (LNG) shipments, large interconnectors and international pipelines, which have promoted economic development through export and import. Regional diplomatic and commercial arrangements could be said to have secured a safer, more equitable world on the basis of trade in Natural Gas. Many energy-intensive countries are now committed to an increased proportion of Natural Gas in their energy portfolio. For example, in 2012 in the United Kingdom, roughly 68% of all residential energy, and roughly 80% of residential space heating (the heating of the insides of private homes), was provided by Natural Gas (ECUK, 2014, Table 1.08). By contrast, in the services sector (offices, commercial properties, etc.), gas supplied roughly 46% of the energy used, and 66% of the space heating (ECUK, 2014, Table 1.08). In 2013 in the UK, Natural Gas formed around 35% of final energy consumption (ECUK, 2014, Table 1.02), and was used for roughly 24% of electricity generation (DUKES, 2014, Table 1.1). In 2012, this was roughly 25% (DUKES, 2014, Table 1.2), and in 2011, roughly 34% (DUKES, 2014, Table 1.3). With the expected closure of unabated coal-fired plants in the European Union (UKGOV, 2013b), and similar global action on coal, for example, in China, Natural Gas could become the main primary energy fuel for power generation in most developed and rapidly developing countries. Yet the very popularity of Natural Gas is leading to a greater dependency on this fossil fuel and its supply chain, and this trend holds the potential for significant risks (see Table 1.1).

1.3.3 Risks of dependency on Natural Gas

1.3.3.1 Economic Turmoil: economic risks from Natural Gas dependency

The economic risks (see Table 1.1) from the increasing exploitation of Natural Gas resources derive mostly from questions about investment, at both ends of the pipeline. To produce new Natural Gas flows, even from already-worked fields, requires capital injection, especially to harness Natural Gas that would otherwise be vented or flared; and from the consumer end of trade, new gas-fired power plants must be financed, to supplant coal combustion, as essential air quality and global warming legislation is enacted. Economic volatility could threaten these investments, and conversely, inefficient energy systems investment could contribute to a loss in economic vibrancy.

Table 1.1 Estimate of risks posed by increased and continuing dependency on Natural Gas in industrialised nation energy systems

Risk factor	Details	Time frame	Risk level	Probability
Economic Turmoil	Continued lack of economic confidence leading to lack of energy investment	0 years from now	High	Medium
Geopolitical Instability	Chaos in diplomacy, social unrest, military conflict disrupting supply chains	0 years from now	Medium	Medium
Market Competition	Strong competition between countries for the same fuel supplies and supply chains	5–10 years from now	High	High
Resource Depletion	Conventional fossil fuels experience depletion of large fields, whilst growth in unconventional fossil fuel production is slow	15–25 years from now	High	Medium
Carbon Emissions	Embedded and "locked in" carbon emissions, contributing to dangerous climate change	20–30 years from now	Medium	High

1.3.3.2 Geopolitical Instability: geopolitical risks arising from Natural Gas dependency

Geopolitical risks (see Table 1.1) to the global supply of Natural Gas are not meagre, and high dependency on Natural Gas could impact global security. The United Nations has had economic sanctions in place against several oil- and gas-producing countries in the last twenty years, and violent conflict is ongoing in some places. In addition, quasi-unilateral decisions made by members of the United Nations Security Council have led to military intervention in the "energy corridor" of the Middle East and North Africa (MENA). Together, these have resulted in some volatility in the supplies of so-called conventional fossil fuels. Besides the effects of trade embargo and socio-economic instability, destruction of valuable energy infrastructure and production capability has become a routine aspect of insurrections, civil wars and inter-nation conflicts, causing another perturbation to fossil fuel supplies. New gas pipelines feeding towards the European Union, from both MENA and the Russian federal region, could be seen as easy targets for malevolent parties, just as the security of oil and LNG tanker passage through the Strait of Hormuz has been seen as sensitive to disruption (EIA, 2014b). Gas pipeline flows could also be restricted by diplomatic stalemates over a range of issues.

1.3.3.3 *Market Competition: compounding risks*

As a result of increasing dependence upon and increased consumption of Natural Gas, there could be mounting competition experienced between countries for the same resource (see Table 1.1). If geopolitical factors were to present a restriction in some sections of the Natural Gas supply chain, this could potentially have the effect of creating scarcity in such a market. Geological resource depletion, whether in indigenous production or in source nations, could exacerbate the stress on the supply of the fuel from within-market competition, particularly for countries at the geographical fringes of Eurasia, such as the UK, and amplify any geopolitical tourniquet.

1.3.3.4 *Resource Depletion: the potential dead end of unconventional fossil fuel exploitation*

The development of unconventional fossil fuels has recently become a hot policy topic, particularly following high levels of drilling for shale gas in the US and concomitant high production volumes – although it might not be possible to replicate this success in other regions of the world – which would mean it does not become the "game changer" as publicised. The drive for shale gas, and other gas locked in sediments, could be interpreted as tacit acceptance that the days of conventional Natural Gas supply growth are numbered. Although mining for indigenous unconventional fossil fuel resources at home may increase energy supply security to some extent, and answer the risks from geopolitical upheaval abroad, environmental and climate change considerations may restrict the development of shale gas and other complex fossil fuel resources, such as coalbed methane (coal seam gas), coal gasified underground, methane hydrates, Arctic oil and gas, heavy oils and tar sands. It is also not clear to what extent unconventional fossil fuels can replace the conventional ones, where production begins to decline because of geological depletion (see Table 1.1), as dispersed and "tight" resources can be difficult to access and energy-costly to produce.

1.3.3.5 *Carbon Emissions: climate change risks*

Fuel switching from coal to gas in the power generation sector is one of the key recommendations of the Intergovernmental Panel on Climate

Change (IPCC) (IPCC, 2013, Section 7.5.1) to address climate change (see Table 1.1). However, in economies with very high levels of renewable electricity penetration, where combined cycle gas turbine (CCGT) generation could be seen as the least carbon-intensive "last resort" backup for variable wind and solar power, the return on investment from gas-fired power plants could plummet, owing to infrequent and intermittent hours of fire. Even though renewable electricity generation has not reached the high levels of this scenario, several countries have experienced this very setback in the last few years, partly due to low coal prices and partly due to the rise in renewable generation, both of which have had the effect of crowding out gas-fired generation. Special incentives have therefore become necessary, centred on a rationale of energy security. Various parties have developed "capacity" instruments to subsidise generation plant kept on standby (Caldecott and McDaniels, 2014). These could perhaps be justified on the basis that the marginalisation of fossil fuel combustion by zero carbon renewable power helps to meet carbon targets, so capacity payments to underused gas-fired power plants would be in recognition of the value they bring in reducing carbon dioxide emissions. If they can no longer be operated commercially, gas-fired power plants may eventually become seen as infrastructure assets for load following and supply balancing power grids, and thus could be state-financed and publicly owned.

Over and above market conditions and physical resource realities, it is the increasing embedded dependence on Natural Gas, resulting in "locked in" carbon emissions from power plants built to last several decades, that could be the highest risk to enacting British, European and other climate change policy in the thirty-year time frame. As climate change begins to bite, carbon control can be expected to rise in urgency. For example, without strong energy efficiency legislation leading to enforced, statutory change, such as mandated building insulation for existing housing stock as well as new build, the UK may exceed its Carbon Budgets and pay a range of financial and other costs. For example, it could be that coastal and low-lying inland communities that begin to be ravaged by too-frequent inundation from storm, flood and sea level rise are permitted to pursue legal challenges to the energy sector for greenhouse gas emissions, in order to compensate them for damages and loss of insurance and equity. It is to be regretted that policies for building insulation have in some cases been so lamentably unsuccessful in their aims.

With increasing acceptance of the genuine risks and damages arising from global warming-induced climate change, shareholders, investment funds and institutional investors may divest themselves of coal, oil and gas portfolios, and it may even become politically and economically untenable to continue to mine and combust fossil fuels. Many scenarios map out that the use of fossil fuels must decrease; whether this comes about as a result of strong climate change sanctions, or from depletion in high-quality fossil fuels, or from excessive fossil fuel production costs in an unstable economic regime, it makes no difference to the outcome. This cessation in the mining of fossil fuels would mean that Natural Gas is not a destination in the journey to a low carbon future, but because we can make Renewable Gas to take its place, we can continue to use our gas infrastructure assets nonetheless. Natural Gas then forms part of the bridge to genuinely sustainable energy systems. We can continue to make good use of all of the gas fuel grid infrastructure and plant whilst working towards replacing Natural Gas with sustainable and maintainable zero or low carbon gas.

1.3.4 Low carbon gas fuels to de-risk the scenario

All the risks from a dependency on Natural Gas (see Table 1.1), apart from the energy sector threats from continued economic instability and possible long-term economic contraction, can be circumvented by increasing the quantities of low carbon "green" gas fuels to displace the use of fossil fuel gases, both in power generation stations and in gas grids. It will take time to reach the stage where meaningful volumes of renewable, sustainable clean-burning gas fuels can be produced from biological and chemical resources. For less-developed countries, where energy consumption is low, biogas produced from the anaerobic digestion of food and animal waste could offset expensive imported fossil fuels within a short time frame and support economic development. However, regions of the world with much higher energy consumption intensities may struggle to curb their fossil gas dependency, as Renewable Gas may initially seem expensive, as much of it will need to be chemically manufactured. An industrial approach to gas manufacture will be necessary, since the supply of biologically derived gas by itself will not be comparable to Natural Gas consumption in much of the world.

1.3.5 A collection of low carbon gas fuels – from a range of technologies

Renewable Gas is a range of processes and techniques – it will not simply come from biomass origins, and it will not come solely from microbiological processing, and some routes look likely to be an order of magnitude more successful than biogas. Researchers, engineers and policymakers are turning their attention to the next logical step in the renewable energy story – the displacement of fossil fuel gas in piped grids, power plants and vehicles with renewably sourced gaseous phase fuels. In this, northern European countries are leading the way – particularly Germany which has a comprehensive strategy for Energy Transition (*Energiewende*, Energy Change, Energy Delta) – and various government agencies are crafting policy with three mutually stabilising elements: energy efficiency, renewable electricity and Renewable Gas. The simplicity of the suite of Renewable Gas products comes with a multifunctionality – it can address not only Carbon Emissions risks but also Resource Depletion and even Economic Turmoil (see Table 1.1), by providing a growth area in economies through the development of the energy sector, and also, very importantly, by reusing gas infrastructure and maintaining advanced engineering prowess.

1.3.6 A win-win-win situation

Renewable Gas offers a way for the major oil, gas, coal and perhaps even nuclear power enterprises to build a future and retain their position as pillars of the economy, maintaining the confidence of their investment community and individual shareholders; and this is significant because energy is a major investment portfolio. The development of Renewable Gas is essentially an adaptation strategy demanded by the changing landscape in fossil fuels. Natural Gas is not monolithic – every reservoir has different characteristics. Although there is plenty of gas out there (see Table 1.2), there are many challenges with the chemistry of recent prospects. More chemistry means less energy return on processing Natural Gas to make it suitable for pipeline and gas turbine. It may be normal in future to accept raw Natural Gas with high levels of nitrogen, or carbon dioxide and hydrogen sulphide. Making the best use of the increasing carbon dioxide in raw Natural Gas will increase the energy return by adding Renewable Gas to gas supplies – "gas processing gain". Low carbon manufactured gas fuels developed in different regions of the world would be to everyone's advantage – not only the oil and gas corporations, but also the Organization of the Petroleum Exporting Countries (OPEC) and the Gas Exporting Countries Forum (GECF), the environmental organisations, climate change negotiators, economically depressed states, energy consumers, energy suppliers and policy leaders. Efforts to increase renewable resources of gas fuels point the way to a smooth transition away from fossil fuels to sustainable, maintainable (and cleaner) supplies of energy for the future. However, we all need time to adapt if we want to join the great gas transition – building

Table 1.2 Brief analysis of the BP Statistical Review of Energy 2014 Workbook, Revised version (BP, 2014c)

Gas reserves – trillion cubic metres (tcm)		
8 tcm or greater	**4–8 tcm**	**2–4 tcm**
Iran	UAE	Australia
Russian Federation	Venezuela	Iraq
Qatar (recent write-down)	Nigeria	China
Turkmenistan (recent write-down)	Algeria	Indonesia
US		Norway
Saudi Arabia		Canada

Gas production – billion cubic metres (bcm), annual			
50 bcm or greater	**15–50 bcm**	**10–15 bcm**	**5–10 bcm**
US	Iran	China	Algeria
Russian Federation	Qatar	Norway	Indonesia
	Canada	Saudi Arabia	Malaysia
			Netherlands
			Turkmenistan
			Mexico
			Egypt
			UAE
			Uzbekistan

a new mindset will be difficult, even though many of the Renewable Gas technologies were devised a century or more ago. The realisation that with the stretch into unconventional resources, Natural Gas production is in danger of becoming fragmented, dispersed and insecure could be the tipping point towards Renewable Gas – a durable solution to gas demand.

1.3.7 Challenging rates of progress

An economist with standard training would ask the question – if Renewable Gas is so good, why is it not being done already? If they saw the fabled $100 bill abandoned on the sidewalk, they would think they were dreaming – somebody else should have already picked it up by now. Progress in energy can be slow and awkward. This is partly due to the incumbency of the oil, gas, coal and nuclear "dynasties". It is also partly to do with the fact that new or immature energy technologies and systems require upfront investment, and a certain amount of speculation. Anything that is "First Of A Kind" needs investment and planning concord – it needs facilitated decision-making amongst a large group of very disparate actors. In fact, much of the basic technology of manufactured gas is not new and there is ongoing work in the area of Renewable Gas. As an example, since the 1980s, there have been a variety of high temperature manufactured gas ("gasification") projects launched – some very successfully – but recycling of research into commercial enterprise can be difficult. This is not generally due to relative costs per unit of energy delivered to market, but a "faith threshold" that needs to be stepped over, as in the case of wind power. The centres of activity in the promotion of Renewable Gas are widespread, and for the wider picture to emerge, one has to connect the dots. India and China are pursuing the pragmatic benefits of biogas, and in the US, there is strong Department of Energy support for the "Hydrogen Economy". In the European Union, policymakers are still struggling with issues over biofuels and biomass, and the debates can sometimes reach a deadlock. Germany has a structured plan to introduce both Renewable Hydrogen and Renewable Methane into its energy provision. In many ways, a transition to Renewable Gas is already being implemented, and there is ongoing research and development, and deployment. However, there is much that could be done to increase take-up of the technologies, and promote the visibility of the transition.

1.3.8 The natural end point for British and European energy policy

It is almost universally recognised that Natural Gas is the perfect partner to electricity in grid networks, and this is central to energy policy in many industrialised countries. The aim for engineers and researchers is to show that just as electricity generation can be stripped of emissions, it is also possible to decarbonise gas – to develop gas fuel resources that have lower net carbon dioxide emissions to air than Natural Gas. This will be especially important for the medium-term future, because as conventional Natural Gas sources deplete, and more unconventional and sour gas supplement the supply, with their changing chemistry, the use of fossil gas could cause higher carbon dioxide and methane emissions. A parallel problem is likely to occur in the refinery of crude petroleum oils – as resources degrade in quality, the danger of rising carbon dioxide emissions from refinery can only be answered by decarbonising the processing – and here, too, low carbon gas will be essential (Abbess, 2014).

There is a distinct prospect that mining for fossil gas might taper away, as renewable, sustainable gas fuels can be made above ground from plants, microbes and other sources, that have a climate-neutral impact, and can even be recycled. Yet this is not just wishful thinking or a hypothetical conjecture. Energy engineers have been working with the agricultural sector and industrial chemical enterprise for decades to put otherwise waste "surface" gas by-products to use, sometimes by recycling the gas; and vanguard developments are now underway across a range of forefronts to increase the production volumes of all kinds of Renewable Gas up to grid- and economy-wide scale. Mining carbon from the Earth and burning it into the sky is a failed narrative, and although we may not be able to stop digging for energy yet, that day can surely come.

2
Energy Change and Investment Challenges

2.1 Navigating the nexus of the economic, climate and energy crises

Climate change is an energy problem – we cannot solve climate change without solving energy. Energy change is not just a matter of convincing the world's energy producers to alter the resources they use; or manufacturers to focus on energy efficiency; or consumers to alter their energy use behaviours. To meet greenhouse gas and particulate emissions targets, and manage water consumption, can never be merely a political, regulatory question, and arguably this would anyway require more organised, committed global governance than currently in evidence. The privately owned energy sector cannot be assumed to be in sufficient good health to respond to regulatory or market change, despite excellent stock market performance. It could be said that the strong value of shares in energy companies and a recently terminated decade of mounting energy commodity prices have created a distraction, a diversion from fundamental fault lines. Studies into energy increasingly alight upon concerns about the projections for future global energy production and consumption, and questions about the life cycle renewal of energy supply and infrastructure in an atmosphere of low economic growth.

2.1.1 Peak fossil fuel production

Regardless of any statutory demands on greenhouse gas emissions, there are strong indications that fossil fuel resources of all kinds are tending towards lower quality, smaller fields and seams, and greater inaccessibility. Net energy return on energy invested (EROI or EROEI) may sink: exploiting a resource makes decreasing sense when it takes the energy equivalent of a barrel of oil just to raise a barrel of oil out of the ground (e.g. Hall et al., 2014). The costs of production of complex, remote and deep fossil fuels may rise – owing to third-party equipment supply chain factors. Furthermore, even though global demand for energy will rise, the rates of production of fossil fuels may reduce, owing to segmentation of production from smaller and more dispersed reserves, strong market competition and technological complexities. All of these factors could tend to produce a peak in fossil fuel production. This would then point to a logical outcome – an ambition to shift away from coal, oil and Natural Gas – not only the depleting high-quality resources, but also the difficult, complex, so-called unconventional fossil fuels. These, in addition to being poorer quality, pose risks of higher levels of resulting pollutants, both in their production and from their consumption. However, solving the ultimatum of a change in energy is not just about substituting one set of energy resources for another, for most renewable energy is diffuse, low density and often variable, and despite the technological achievements in some renewable sources, such as hydropower, and rapid progress in wind and solar power, renewable energy has yet to be made to harness a relatively significant amount of investment and government support compared to fossil fuels.

2.1.2 Global economic change

The global economy is at a crossroads on energy. The International Energy Agency (IEA) and others are warning that significant new investment in energy infrastructure needs to be made. With economic turbulence, it is entirely possible that expensive options, such as nuclear power, Carbon Capture and Storage (CCS) and the exploitation of low-quality or hard-to-reach gas and oil, will not be pursued at very large scale worldwide. Private energy corporations are likely to resist large new investment yielding only a slow rate of return, without the promise of public finance support – yet many developed countries talk of austerity measures to cap or lighten national deficits. The most realistic and pragmatic way forward is to make the most of what we have already got – power and gas – and the grids, pipes and plant we have already invested in. We also need to think of energy solutions that require the shortest possible time from design to production. For example, renewable electricity from solar farms and wind farms is very rapidly adding electrical generation capacity to the power grids. Gas-fired power plants can be built in several years. Although we need to abandon coal burning to meet climate stabilisation demands, we can continue to burn fossil Natural Gas for a couple of decades.

2.1.3 Managing the inevitable energy change – cross-cutting issues

An added complexity for investment decisions is that to make the energy change, we need to solve a number of related things at once: how to harness bioenergy resources without compromising food security, for example, where bioethanol derived from staple crops is liable to compete with the human food supply chain, and genetically modified energy crops may compete for land with food farming, or even threaten it by affecting agricultural biosecurity. We also need to address problems related to technological failure – to review progress in "advanced biofuels" indicates that optimism is not always justified, in areas such as algae farming for biodiesel, and so-called second generation cellulosic bioethanol. And beyond the technological barriers, we need to work out how to deliver industrial scale volumes of alternative energy resources in a sustainable manner. For example, it is not clear if liquid biofuels can replace significant quantities of refined fossil fuels for transport. Another example is that soil fertility and water supply in energy crop production is a major concern, and will almost certainly place a natural upper limit on production volumes of biomass that can be made available.

2.2 Energy Change: the energy paradigm shift

Beyond technological, production and economic questions, there are the strategic issues – how to bridge between the investments we currently have and the assets we need to develop for the future. Much of the world's capital is locked into oil, gas and coal (and nuclear power), and the electricity generated from them, and is chasing more of the same, because it may be seen as too risky to jump to the renewables track. The inevitability of a renewable energy world, arising from dwindling pools of fossil fuel resources, the limits of uranium mining, the risks of fragmentation in the global economy and the likelihood of stringent climate change targets, has not yet fully percolated the consciousness of large investment funds, major corporations and many parts of national and international governance. Economists do not believe geologists – however, parts of the global economy, and some lawmakers, recognise the mandate for change.

It could be that big oil and gas companies, such as BP and Royal Dutch Shell, and big utility companies (Vasagar, 2014a, 2014b, 2014c) need to be reformed, from the inside out, in order to survive. The good fortunes of companies in the energy sector could be assured if they were to build their business projections on the adoption of bridging energies, such as low carbon gas, that can be used in the interim on the road to a fully zero carbon energy future. Bridging positions that are not dependent on public money (which is probably the case for CCS) and are not just public relations "greenwash" – whitewashing coal and unconventional fossil fuels – could procure global accord, and allow for multi-scale, multi-sectoral and, importantly, multiple ownership solutions to be developed. For example, Renewable Gas can create employment in the energy production supply chain, and also at the consumer end. Renewable Gas technologies can be small or large scale, and distributed to the most appropriate locations. And Renewable Gas can have different forms of ownership, and therefore more shared responsibility for its production, spreading the workload in maintaining supplies. Decentralised Renewable Gas can have a multiplicity of sources, and a multiplicity of uses, and depending on its purpose, a flexibility in its exact content; it is also potentially a bringer of efficiency to energy systems.

Large Renewable Gas production facilities and biochemical refineries or "biorefineries" may well replace petrochemical facilities, and because of the similar chemistry, this leaves plenty of space for traditional oil and gas companies to operate as large players in the new chemistry. However, besides new resources, a renewable energy future must have a significant tranche of energy conservation embedded into it, because efficiency is essential with low density, dispersed renewable resources, and that can only be achieved with co-operation from all sectors, including those whose business model until now has been based on the sole prime directive of selling increasing amounts of energy. This will be an entirely new paradigm, and the energy sector will need to operate as the sellers of energy services rather than simply energy itself.

2.3 Challenge to invest: the engineering and resource life cycle

The engineered components of energy systems, such as power plants, petroleum refinery processing units and grid networks, are not only replaced when something better comes along, such as new technology, or more efficient models. Each part of an engineered system has a designed lifetime, after which it must be significantly renewed or decommissioned. It may be possible to keep some energy system units in operation past their allotted hour, but there are usually costs and risks attached to this decision, and could make final decommissioning or replacement more expensive. Maintenance and replacement costs are not restricted to engineered components of energy systems – the raw resources that are the inputs, coal, Natural Gas and crude petroleum oil (and uranium), also need to be refreshed. As old seams, wells and fields deplete or depressurise, new reserves must be targeted for exploitation, or new mining and drilling techniques applied, and both options come at a cost. Energy systems come with a reinvestment commitment, not just an original price tag. The IEA (2014a) in its 2014 World Energy Investment Outlook projected that globally, investment in energy efficiency, energy system components and energy resources must rise from \$1.65 trillion in 2013 to \$2.5 trillion per year in the period 2031–2035 (IEA, 2014a, Table 1.2). Of the \$48 trillion cumulatively that the IEA project needs to be invested in the period to 2035, \$8 trillion will be for energy efficiency. Of the \$40 trillion remainder, less than half will be needed to meet increased energy demand. The rest will be required simply to keep the current energy systems operational, in both engineering and resources – running to stand still. Making the decision to accentuate investment in renewable energy to keep global warming below two degrees Celsius (2°C) would take the total global expenditure on energy up to \$53 trillion (IEA, 2014a) up until 2035, but this would create savings. The IEA publication *Energy Technology Perspectives 2014* (IEA, 2014b) estimated that \$44 trillion would need to be invested by 2050 to decarbonise the entire energy system and meet climate change targets, but that this would produce input energy cost savings of \$115 trillion. Reducing fuel costs to the global energy system by increasing the use of renewable resources would automatically affect the viability of oil, gas and coal supply companies unless they were to change their business strategies.

Regardless of whether the world decides to head for zero carbon or not, reinvestment in energy is recognised as essential, however there are developments taking place that may significantly constrict energy system spending.

Perhaps the most important and primary consideration is the financeability of engineered energy system components. This is felt most keenly in the market economies, where the hunt for profit, the rate of return on investment, will not appear as healthy for some energy system investments compared to others, or other sectors of the economy. For example, reinvestment in electricity grid networks in industrialised countries, where some cables, wires and breakers can be far older than power plants (e.g. Barrett et al., 2013), does not offer the promise of high rates of return, although it will maintain the integrity of the energy system as a whole. The utility of a supply of power is extremely high, but few want to sink capital into its provision.

Secondly, although the fundamental issues are different, the question of financeability also applies to developing primary energy resources. For example, the depletion of high-quality, high-volume fossil fuel resources and continued high demand are causing higher levels of spending on exploration and discovery by privately owned oil and gas companies, but with some evidence of a lower success rate (Ahmed, 2014; Carbon Tracker, 2014a; Evans-Pritchard, 2014; OPEC, 2014, Figure 1.4). By contrast, although lower-quality fossil fuels are likely to have expanding markets and so appear to be bankable, this all depends on the EROEI in their production, and the shifting landscape in commodity market prices. In light of trends, it may be that private enterprise oil and gas companies need to reassess their core strategies in order to maintain their profit base. One factor will be which resources they choose to invest in.

Thirdly, it is important to recognise the genuine geophysical limits of oil and gas (and coal and uranium).

Fourthly, it is likely that global capital will rein in its enthusiasm for novel and expensive technologies, particularly in the arena of carbon abatement. Whilst the Intergovernmental Panel on Climate Change (IPCC), IEA and other bodies are keen on a nuclear power renaissance and the deployment of CCS, the world's banks and other potential investors may well take a different view.

Fifthly, it is important to consider the potential effectiveness of setting a monetary value on carbon dioxide emissions, which is assumed to be a key method of stimulating investment in low carbon energy, but might not create the right incentives for major change.

2.3.1 Losing investment from energy engineering

The scope for investment in engineered energy system components is somewhat different under the paradigm of a market economy compared to a state-led economy. Regardless of the prevailing economic model, however, investment is never assured. In the privatised energy economy, investment seeks a return for the investors, and faces problems if loan recovery rates are not sufficient. For example, privatised electricity generation companies in the European Union may not be sufficiently capitalised to achieve a transition to low carbon power plants, and are likely to baulk at taking on the high levels of debt in order to finance it, an aversion to investment that is historical (Kuzemko, 2013). By contrast, in a planned economy, there could be a natural resistance to energy investment, even if there are gross inefficiencies in the current installations, or the installations are ageing and need replacement, simply because central policies have other priorities.

The global market economy is searching for growth, for profits to return from investment, in order to stimulate greater productivity, and yet it is showing signs of debt, equity and capital lockdown – which could restrict the flow of finance to all but the most promising sectors. To create genuine growth, it is necessary to draw new resources into the economy, something that can be done with renewable energy, for example, which is essentially cost-free solar power in various forms, but somebody needs to pay to harness it.

Unless it can show evidence of genuine returns, investment may fail to be directed towards the energy sector, and in addition, potential for an investment gap could become progressively worse as engineered energy system components age. For some system components, renewal with the promise of returns on investment is technically impossible, such as fission reactors at nuclear power plants. Even if state support enables a replacement programme for nuclear reactors, it may be hampered by the burden of decommissioning the older installations and disposing of the spent radioactive fuel and radioactive waste, depending on the apportionment of liabilities.

This may partly explain why – as of the end of 2013 – the International Atomic Energy Agency (IAEA) counted thirty-five operational nuclear reactors younger than ten years, but 312 nuclear reactors aged between twenty and forty years in operation (IAEA, 2014) – and there are only somewhere between sixty-three and seventy-two new reactors currently under construction (IEA, 2015; Kee, 2015). The pace of the much-lauded nuclear fission power renaissance is still lethargic, and there is a danger of competition for remaining good-quality uranium resources if many countries head in this direction simultaneously (Garcier, 2009).

Renewal in electricity grid networks has obvious merits for functionality, and although there would appear to be no return on investment in like-for-like replacement, the case for investment could be made on the basis of additional functionality – for example, widening power grids to allow more transmission access to new renewable electricity, and adding in smart demand and supply controllers in power networks. Governments can however decide to simply finance renewed and new power grid investments themselves, or coerce all electricity generators to co-fund upgrades, considering the grids as quintessential infrastructure that ensures energy security.

Investment in new electricity generation plant, outside of renewables, is under tension in the Organisation for Economic Co-operation and Development (OECD) (IEA, 2014a, Executive Summary). By contrast, in planned economies, for example, China, funding decisions are made differently, and a very large fleet of new coal-fired power plants has been installed (Finkenrath et al., 2012). Resistance to reinvestment is likely to be experienced in this case – for example, in retrofitting CCS to coal-fired power plants.

As with the electricity generation sector, ageing infrastructure and plant in the gas sector demand new investment. A critical indicator is seen in the monitoring of fugitive methane emissions from gas grid infrastructure, frequently spiking in urban consumption areas and rural/marine production locations. In places with underdeveloped gas grids, new uses for Natural Gas could compound gas supply stressors. Investment programmes for maintenance and replacement in distribution infrastructure, and the improvement of trade mechanisms and markets, are therefore essential. Again, as with power grids, responsibility for this could be delegated to governments by default.

In addition, owing to continuing uncertainties in the global economy, the investment appetite of major financial institutions for "green" energy projects that increase efficiency may well be lacking; because of low confidence levels in entirely new technologies, or doubts about the financial benefits of replacing inefficient old with efficient new; and in particular, renewables being cast as unreliable, instead of being recognised as genuine assets.

State intervention may well become essential if new asset investment is hesitant, as it is politically imperative to keep the lights on and the hot water flowing. Whether this is entirely privately or publicly led, the payback for participation in this energy revival will be that companies in the sector, and their supply chain, used in resourcing new energy infrastructure and plant, will experience growth and create employment. New energy sector projects may even reinvigorate banking itself, regardless of financing method.

2.3.2 Why oil and gas supply investment might not be forthcoming

"If there is demand, there is supply" is a fairly common sense assumption in market economics. However, the energy sector, and in particular the power, oil and gas part of it, is not a free market. Government intervention in the form of end product price control (e.g. fossil fuel subsidies) and revenue support is commonplace, and something like 80% of proven-plus-probable (2P) crude oil reserves and 60% of Natural Gas reserves are held by National Oil Companies (NOCs) and their host governments (IEA, 2014a, Chapter 2), whose policies may not be conducive to high levels of production. Since around the year 2000, global capital expenditure (capex) has roughly doubled in securing production of oil, gas and coal (IEA, 2014a, Chapter 2); partly because of third-party service company costs (IEA, 2014a, Chapter 2; Mitchell, 2012), partly because of the disappointing size of new discoveries (Mitchell et al., 2012, Chapter 5) and partly because of producing less economic resources, such as in onshore North America. Regardless of whether they are "conventional" or "unconventional", fresh hydrocarbon resources may well be more complex and more expensive to produce; so although proven reserves of oil and Natural Gas may be equivalent to 53.3 and 55.1 years of today's consumption (BP, 2014c), affordable production may become too slow to meet this rate of demand, despite megaprojects (Ernst & Young, 2014).

Aside from oil and gas commodities production capacity, the most important metric is return on investment, vital for private sector International Oil Companies (IOCs). There appears to be a residual assumption that energy demand will continue to increase, and that therefore energy sales revenues will continue to increase, and so always be high enough to enable internal or external production project finance on favourable terms. Refinery capacity utilisation is running at high levels, so demand for refined products is healthy – whether they are sold as fuel or go to storage as stocks. However, with a weak global economy, there is no reason to expect stronger volumes of refined hydrocarbon sales, and therefore no reason to expect high prices for refined products, and no need for high oil and gas commodity prices. Future oil and gas demand is uncertain (Mitchell et al., 2012), especially in the context of strong carbon dioxide emissions policies (IEA, 2014a, "Trends in the 450 Scenario"), and so recovery on capex through sales could become slower, which could cause oil and gas companies to pull in their investment horns. Some oil and gas companies could fail in these circumstances.

Crude petroleum oil commodity price volatility may cause oil and gas production companies to refocus on a narrower range of resources, where there are "easy wins" (Marcel, 2012), or switch away from complex oil to gas (Gismatullin, 2014; Reed, 2013, 2014; Shell, 2014; Vidal, 2014), which could bring about a decline in the supply of gasoline, diesel, kerosene and jet fuel products. This could damage the prospects for global expansion in transport and transportation vehicle fleets unless these are hydrogen-, Natural Gas- or electricity-fuelled. On the other hand, applying new fuel efficiency (e.g. the US CAFE Standards) and emissions standards (e.g. the European Euro 5 and 6 standards) through renewing/replacing existing vehicle fleets, would become easier to promote with less oil flowing. To avoid the risk of crude oil "stranded assets" (Carbon Tracker, 2014b), there might be contraction in oil and gas companies, or a fragmentation of their businesses, building pressure for a restructuring of the oil and gas sector. This could be problematic, as the private oil and gas companies (IOCs) might then lack the capacity to take on partnerships with NOCs in developing as yet untapped resources (Marcel, 2012). These arrangements will be crucial as the centres of future oil and gas production gravitate towards geographical locations where the market economy is less prevalent, and technical expertise or political management of energy is as yet less well developed. Regardless of the general health of the global economy, or the trends in oil and gas production volumes and revenues, hydrocarbons will continue to support economic development, and so new formulations of partnership will be needed between organisations with oil and gas expertise and resource-rich nations. More profit-sharing between the resource owners and the resource producers may be necessary, which may limit future investment capability.

In order to invest, oil and gas commodity producing companies with lower revenue prospects who were previously self-reliant for capex (funds for capital expenditure) might need to reach out progressively for external finance and government assistance – both of which could continue to be suffering the tourniquet of the "credit crunch". Banks, even development banks, are for a combination of reasons becoming more risk-averse, and could pull away from investment in oil and gas (and coal) in regions seen as having poor loan recovery rate conditions (IEA, 2014a, Chapter 2, "Implications for financing"). The "upstream" fossil fuel supply sector will have to bring in new sources of investment capital, for example, from individual and institutional investors, perhaps developing new financial products such as energy bonds or energy supply credit notes – promissory notes denominated in energy supplied rather than the usual currency. Financial instruments such as energy bonds – and energy efficiency bonds – could help acknowledgement of emerging bilateral and multilateral international arrangements. New treaties between energy supply-export countries and energy demand-import countries could be articulated in the flow of energy in one direction, and the flow of health, education and technological development in the other. Energy note markets, once established, could also be the basis of building in the "technology transfer" recognised in the Kyoto Protocol Article 10 as vital to achieving climate change targets through clean energy development (UNFCCC, 1997).

Although it is perhaps hard to imagine a significant charge for carbon dioxide emissions – whether through carbon taxes on trade frontiers, a widescale emissions market, imposed emissions quotas on either energy supply or demand, or some other artificial market perturbation intended to implement climate change policy – it is easier to see the possibility of increased regulatory control on energy use and the types of energy production encouraged. For example, without significantly impacting the price of power, the European Union has greatly promoted the deployment of renewable electricity, largely through subsidy arrangements, although this has had a side effect impact on the rates of recovery on investment in flexible and efficient gas-fired power plants, and has warranted deep consideration of a market in generation capacity (DECC, 2013c; DG ENER, 2013; Honore, 2014; IEA, 2014a, Chapter 3, "Market Reforms"; Keay-Bright, 2013; Vassilopoulos, 2013), a top-down regulatory mechanism. Despite the fact that bank lending interest rates are currently extremely low, lending volumes can also be low, and there may be a case for a range of interventions to lower the cost of capital for new low carbon power plants if investment cannot be justified solely on the basis of future power pricing and sales volume trends – just to keep the lights on. Were it to become accepted that climate change policy will be implemented by regulatory framework, regardless of energy market developments, there could potentially be a large "carbon bubble" that emerges, whereby oil, gas and coal energy suppliers could be seen as owning "unburnable carbon" resource assets, therefore devaluing their companies, which could affect their investment capability. Consumer energy prices could be massaged downwards in many regions of the world in order to protect economic stability, so it may not be possible to entirely eliminate subsidies on fossil fuels, which still amount to around 80% of all energy consumed globally. It may never become politically acceptable to add carbon emissions charges to energy sales. By contrast, there could be many forces striving to keep the price of fossil fuels low. However, artificially low prices for fossil fuel energy need not affect the decarbonisation of the energy supply, which could be stimulated by new forms of direct investment, particularly those outside the conventional banking sector. Regardless of price controls on energy, the cheapest form of energy overall (even where it has slow initial recovery on investment) is energy efficiency, and this may well become an international policy focus, which would also detract from investment in new energy supply, particularly oil and gas and coal.

2.3.3 Peak Fossil Fuels: fossil fuel supply depletion severity needs to become acknowledged

In both oil and gas (and coal in the slightly longer term), there are risks of "peak" supply, not because the Earth has run out of fossil fuel hydrocarbons, but because the rates of production become slower, and the economic advantage of production becomes smaller. The potential severity of Peak Oil is perhaps still not fully recognised; as yet it is not clear if exploiting unconventional resources could compensate for decline in conventional wells and fields into the medium term.

2.3.3.1 Conventional oil and gas depletion

In the petroleum oil and Natural Gas supply sector, irreversible regional depletion of the best-quality fossil fuels, such as in the North Sea province, is gathering recognition, but deserves more transparency, especially its impact on production rates. It is important to improve projections of crude oil and Natural Gas depletion and production in order to have a better picture of the future of energy. The IEA, the US's Energy Information Administration (EIA) and the oil and gas companies BP, Royal Dutch Shell and ExxonMobil, amongst others, regularly publish scenarios and reports on energy futures, usually projecting three to five decades of continued oil and gas dominance (see Table 2.1). These publications perhaps have an underlying intent of conveying that fossil fuel resource depletion is not a serious concern, and that even if it were, alternative resources can straightforwardly be brought to production with similar magnitudes and volumes as today. For example, BP added considerable amounts to the heavy oil fields of Venezuela and the oil sands (tar sands) of Canada in the "proved reserves" listings for oil in their annual Statistical Review of Energy in June 2012 (BP, 2011b, 2012b), after official announcements, which has influenced the trendlines for global oil "reserves growth". Calculated from BP's figures, oil reserves growth was 767 million barrels per year in 2013 (BP, 2013c), or 653 million barrels per year in 2014 (BP, 2014c), compared to only 187 million barrels per year in 2011 (BP, 2011c). Global reserves growth in gas from BP's figures appears to be largely down to recent North American unconventional additions (BP, 2011c, 2013c, 2014c). These increases in oil and gas reserves may appear healthy, but the reserves growth measure does not give consideration to the added refinery processing capacity that might be required, nor the energy penalty or infrastructure needs in production of the more complicated or remote resources, nor the rates of production, which might be significantly lower, especially with heavy oil.

There are other factors of concern. For example, the troubled short-term situation in several oil- and gas-rich countries in the Middle East has arguably been dampening crude oil production, and long-term governance policies in part of the region prevent the higher levels of production

Table 2.1 A selection of organisations modelling energy in the economy

BP	"Statistical Review of World Energy"
	"Energy Outlook"
Deloitte	"Vision 2040: Global Scenarios for the oil and gas industry"
EIA	"International Energy Outlook"
Eni	"World Oil and Gas Report" (2014)
	"World Oil and Gas Report" (2013)
EC	"Energy 2020"
	"Energy Roadmap 2050"
EREC	"Renewable Energy Scenario to 2040"
ExxonMobil	"The Outlook for Energy: A View to 2040"
Greenpeace	"Energy Revolution 2012"
IEA	"World Energy Outlook" (Executive Summary)
	"Energy Technology Perspectives" (2014)
	"Tracking Clean Energy Progress" (2014)
IIASA	"Global Energy Assessment"
IPCC	"Climate Change 2014" (Working Group 3, Chapters 6 and 7)
IRENA	"REmap 2030: A Renewable Energy Roadmap"
JODI	"JODI Gas"
	"JODI Oil"
MIT	"Energy and Climate Outlook"
OPEC	"Monthly Oil Market Report" (MOMR)
Practical Action	"Poor People's Energy Outlook"
PwC	"Energy Transformation: The Impact on the Power Sector Business Model"
Shell	"Shell Energy Scenarios to 2050"
	"New Lens Scenarios"
Statoil	"Energy Perspectives"
WBCSD	"The Energy Mix: Low-Carbon Pathways to 2050"
WEC	"Deciding the Future: Energy Policy Scenarios to 2050"
	"World Energy Scenarios: Composing Energy Futures to 2050"
	"World Energy Scenarios 2007: Energy Policy Scenarios to 2050"
WEF	"Energy Vision 2013: Energy Transitions: Past and Future"
World Bank	"Toward a Sustainable Energy Future for All"
UKERC	"Review and Analysis of UK and International Low-Carbon Energy Scenarios"

thought to be sought by transnational energy companies. In the case of Russian gas, a lot of gas resources are in remote parts of the country and could be described as "stranded", with no access to production, let alone markets. A further issue could be where an oil and gas company works to maintain a Reserves to Production Ratio (R/P) in the region of a decade. Although this "reserves replacement" might offer investors at least ten years of confidence, it does not guarantee that the situation from Year 11 onwards will be so comfortable, particularly if production volumes slip, or reserves are discovered to have been overinflated or simply in error (Leggett, 2013; McGlade, 2013; Meng and Bentley, 2008; Rogers, 2013). Individual companies may have inflated reserves figures, but so might countries. For example, BP's June 2013 Statistical Review of Energy significantly wrote down the 2011 end-of-year proven gas reserves for the Russian Federation (32.9 trillion cubic metres (BP, 2013b)), compared to the Review of 2012 (44.6 trillion cubic metres (BP, 2012b)).

Unless entirely new resources are included, reserves growth is only within the limits already set by 2P resources (McGlade, 2013, Section 3.4; Meng and Bentley, 2008, Section 2.4), and do not imply that production rates can be maintained or increased. Rates of production of fossil fuels are perhaps a more relevant metric than reserves or resources, as, historically, supply has usually been encouraged to meet demand, unless there has been some major crisis. There is evidence of diminishing returns on exploration investment (Fantazzini et al., 2011; Hall et al., 2014), and a falling EROEI indicates the risk of a gap opening up between demand and supply because of economic tension – that there might be no monetary or energy gain in exploiting some of these resources. This possibility makes measures of "economically" recoverable resources (ERR) contested, and could put some "technically" recoverable resources (TRR) out of reach, so projections for ultimate cumulative production continue to be made on the more reliable basis of proven reserves, which is a much slimmer figure. Depletion of resources by production may have been matched – at much cost – by new additions to proven reserves (Mitchell et al., 2012, Chapter 5); however, a much-repeated sanity check is that to keep up the current levels of fossil fuel production, the world would need to find the production equivalent of a Saudi Arabia every few years (e.g. Ahmed 2013). Recent large oil and gas discoveries are sub-Saudi volume (e.g. Miller and Sorrell, 2014, Figures 7 and 8), and sometimes sub-Saudi quality, and in some cases may only contribute a few per cent to annual production requirements – as finding significant reservoirs of heavy oil and complex gas does not mean that they can be produced as easily as light sweet crude, or high-pressure Natural Gas wells. Undeveloped "supergiant" fields of conventional resources are concentrated, for example, gas in Iran, and it is unlikely that previously unknown or unscoped sizeable conventional fossil fuel resources are to be found elsewhere. If there is anything significant to find in the untrodden Arctic, much of it will be under Russia, which could have a struggle in bringing this remote oil to market (USGS, 2008). "Peak oil supply" is a realistic emergence, well before "peak geological oil" is reached. It would be helpful if this were reflected in all energy sector reports; especially since when "peak oil supply" is experienced, a decade or so later "peak Natural Gas supply" is not out of the realms of possibility.

2.3.3.2 Projections of unconventional fossil gas reserves production

In order to be sure of future energy production levels, given indicators of depletion in conventional reserves, oil and gas companies have chosen to exploit previously discounted unconventional fossil fuels, a strategy which poses several risks. The current debating point about the hydraulic fracturing of sedimentary rocks to liberate shale gas, shale oil and other "tight" gas or condensate (Abbess, 2014), tends to focus on the potential to cause earthquakes, but there are other serious environmental issues at stake. Although shale gas may be chemically similar to Natural Gas, its distribution in sedimentary rock can be very different. Crucially, geology dictates that some areas of shale strata can be scant in gas, meaning that there could be high levels of empty well abandonment. Conversely, richer resources could be clustered in "sweet spots", meaning that high numbers of wells could need drilling close together, which would present its own problems. Also, as shale layers can be already naturally fractured, this poses drilling risks, and could lead to high levels of contamination in the produced gas, and damage to the Earth's surface, such as subsurface methane dispersal, disruption and pollution of water tables, heightened requirements for gas and drilling water filtration, local atmospheric pollution, and landslip and unstable land depressions. If a borehole passes through an interleaved coal seam or coal lens, there could be a localised zone of higher gas pressure, which could lead to methane gas seepage or leakage, or even blowout.

Shale gas is just one end of a spectrum of unconventional fossil fuels. In the US, the focus on indigenous oil and gas production has caused a glut of lighter hydrocarbons to be available, whereas many of the petrorefineries are designed for heavier oils (Fitzgibbon and Rogers, 2014). If the legal situation permits, much exported American condensate (Gardner and Leff, 2014) will undoubtedly be used to dilute heavy oil processed from tar sands (oil sands, bituminous sands) in Canada to import back into the US in diluted or synthetic forms known as syncrude (Gray, n.d.), dilbit or syndilbit. A relatively small level of investment would be needed to make new tar sands available; however, to continue to pursue this course could have devastating impacts on local environments, not only on Canadian lands, but also anywhere in the US where this oil is refined, spilled or refining waste is disposed of.

The wide range of environmental and technical issues should not obscure the principal concern – that low grade fossil fuels cannot resolve problems of pollution, only potentially exacerbate them.

Environmental risks could be considered matched by economic risks. It is not clear how depletion scenarios of what are known as conventional fossil fuels could be compensated for by growth in alternative, or unconventional, fossil fuels, despite improvements in technology, so investments in these alternatives may lead to a dead end. Unconventional fossil fuel growth projections could be considered contentious, and difficult to validate, as these resources can be quite dispersed, or randomly concentrated, more complex to produce, more inaccessible, more rapid to deplete and more expensive to refine; all of which could harm the profit baseline. Refreshed geological surveys of unconventional fossil fuels in place (Andrews, 2013; BGS, 2010; EIA, 2013; USGS, 2013) have led to confidence that there are large resources, particularly of shale seam gas, in various regions of the world, but it is not always certain where and when these can be added to tables of ERR, as uncertainties about rates of return on investment could risk their development.

In the US, there has been policy and regulatory focus on "energy independence" which has fostered growth in the exploitation of indigenous resources of shale gas and light tight oil (shale oil) reserves, and permitted mining companies to commandeer large tracts of public land (Bureau of Land Management (BLM)) for shale/tight gas and light tight oil (shale oil) exploitation; in addition, waiving various forms of environmental legislation to encourage gas and oil production from the commercial leasing of private mineral rights. As a result, there has been strong growth in onshore oil and gas drilling, and strong growth in production in some areas (Neff and Coleman, 2014). In contrast, it is not yet known if this type of federally supported development in the US can be replicated elsewhere. For example, European shale gas resources (gas in the ground) are considered an order of magnitude smaller than North and South America combined (ARI, 2013; McGlade et al., 2012a, Table 2-3), and downgrades in production volumes assessment have been made in Poland and China, both considered promising new regions (Koper, 2014; SXCoal, 2014). China has a relatively undeveloped interior where there are good shale gas prospects (Kang, 2014), although local water resources may not be suitable for production (Stevens, 2012). Technically recoverable (TRR) shale gas in Europe might be comparable to that for the remaining resources in the US (McGlade et al., 2012b, Table 4), or more than two times smaller (McGlade et al., 2012a, Table 2-4); however, the higher population density in Europe compared to the US, combined with a lack of historical low grade energy engineering (Howell, 2014), is likely to impair progress in shale gas growth (Deloitte, 2013, Figure 5; Ofgem, 2012, Section 2.25, Figure 2.6), despite overt attempts to create economic incentives (UKGOV, 2013a, Sections 1.176–1.182). Added to these factors, it is not certain if the decades of technical learning in shale gas production in the US could be easily translated and adopted into other geological and socio-economic contexts.

Gas condensates, where gas is "wet" with high levels of condensable vapour, are also being developed, but this probably warrants some revision of reserves analysis, as technical recovery rates from these strata and reservoirs might be quite different from conventional fields. This analysis would need to be based on very modern data in most cases, and the results may show that new metrics need to be devised. It could be necessary to have several sets of recovery estimates for gas condensate wells, for example, as more or less gases and liquids may be produced under different operating conditions (e.g. Khodapanah and Tabatabaei-Nejad, 2014).

Meanwhile, the American "shale gale" itself could be facing a slowdown. It is not certain if the current production pattern can be sustained, as without enduring tariff, subsidy or policy support, shale gas and oil production could in future become an unpalatable enterprise because of the poor rates of return – depletion in shale fields can be rapid, and so production from a particular field might not be sustained if new well drilling becomes unaffordable. The IEA suggests that unconventional oil production in the US could plateau in the 2020s (IEA, 2014a, Chapter 2), but recent oil price volatility, were it to continue, could bring an earlier peak. It could be argued that recent success has largely been due to the drilling contract structure, and that contracts could falter at the point of renegotiation.

The US Department of Energy projects an annual increase of at most a few per cent in gas production of all kinds – apart from fresh development spurts from coalbed methane and gas from Alaska – but this is dependent on their projection of an annual increase in the number of wells drilled (EIA, 2014a, Table A14), growth which is perhaps unfeasible. Combating productivity declines necessitates high replacement rates (Hughes, 2013, Table 2), and shale gas production may have more or less plateaued (Hughes, 2013, Figure 59). Despite technical advances, the economics of unconventional gas may not have become more favourable since the inception of shale exploration (Hammond, 2013; Wang and Krupnick, 2013, Page 3), and the contribution of shale gas to the American energy economy may be short-lived. If the shale gas bubble bursts, it may not be the scale of remaining resources that determines the date, or even their technical recoverability, but production contract length (Hughes, 2013, "Shale Gas", "Key Takeaways", Page 50).

It seems it will be essential to have more accurate projections of the levels and rates of economic production of various unconventional fossil fuel reserves – to have a better grasp of whether the production of gas from unconventional fossil fuel reserves and gas condensate can effectively offset depletion rates of conventional Natural Gas in the medium to long term, and so shift the date of global production decline significantly farther into the future. If it cannot, then this will add to the pressure to reduce demand for Natural Gas, and add to the rationale for increasing the production of Renewable Gas.

2.3.3.3 The risk of Peak Natural Gas and the greenhouse gas implications of alternatives

At the same time as demand for Natural Gas is projected to rise, maximum annual volumes of Natural Gas production may be experienced in some regions, as multiple fields in some provinces could begin to show signs of depletion. The economically recoverable amount of Natural Gas from a field, as with petroleum oil, has limits, as depletion in large reservoirs and the consequential failing pressure is not certain to be adequately compensated for in a reliable fashion by improvements in technology. Depletion of an individual well is inevitable, and when no new well drilling is done, and the balance of wells in a field are showing signs of depletion, a consequent decline in production from the entire field is probably irreversible, despite some success with re-pressurising gas wells ("Enhanced Gas Recovery", Hussen et al., 2012; Vandeweijer et al., 2011), which could theoretically reduce the risks of seismicity from heavily depleted fields, such as Groningen in The Netherlands (Van Eijs et al., 2006). An overall depletion effect is already being encountered in the North Sea area (Sorrell et al., 2012); the United Kingdom is experiencing a decline in offshore Natural Gas production (POST, 2004); and within around twenty years, despite their more prudent policy to exploit small onshore fields first, The Netherlands too will become a net importer of Natural Gas (NL, 2011). Whilst the impacts of these changes can be offset to a certain extent by international trade in Natural Gas, anticipation of a plateau in total global production within a few decades – Peak Natural Gas – has already caused a shift in ambition towards exploiting previously neglected unconventional resources of gas. Perhaps controversially, this potentially conflicts with climate change demands placed on the energy sector. The more deep and inaccessible or chemically complex a gas, especially if it has a composition high in Natural Carbon Dioxide (Abbess, 2014, Table 10), the higher the potential greenhouse gas emissions from its exploitation. These would be fugitive emissions of both methane and carbon dioxide from the complex mining, and carbon dioxide emissions from chemically treating the low value gas for market. What could also be witnessed within a few decades is "peak high value Natural Gas", which would have economic side effects, as less energy could be gained from pumping the same volume of gas, because of the degradation in its energy value due to non-methane compounds. It is important that the EROEI of Natural Gas also becomes an indicator in energy reporting.

2.3.4 Technological blind alleys and the economics of new technologies

There is the risk that energy policy frameworks could be designed to support technologies with the least potential to sustain energy supply, and the least capability to drive forward low carbon ambitions in an economy with low investment appetite. If these technologies are pursued, public funds, and the private funds it leverages, could be overcommitted to suboptimal choices.

The attraction of technology and the shine of the new has been a factor in the uptake of renewable electricity; wind turbines and solar panels have been installed in a decentralised, piecemeal fashion in many places. This rapid but dispersed deployment is sustainable, even in a low investment economy, as individual wind and solar power generators are not as difficult to finance as a large new thermal power plant, burning coal or Natural Gas, or splitting atoms. Hopes are still pinned on new forms of nuclear fission, but it seems possible that only significant state financing will carry them forward, and they could end up like the nuclear fusion project, a government-supported scientific endeavour with a goal that still appears as remote as it ever was. Nuclear power only supplies a few per cent of total global energy needs, and yet requires considerable investment, both upfront and in decommissioning and radioactive waste and radioactive spent fuel disposal at end of life. It may turn out not to be a pragmatic choice for potential investors.

CCS is called for in many technical and policy reports from governments and the energy sector, and would permit the continued burning of coal and Natural Gas way into the future if it were universally applied in power generation plant. Troublingly, however, there does not appear to be a way to make it attractive for investment, not least because retrofitting ordinary thermal power plants with CCS would cause an increase in fuel consumption for capturing and sequestering the carbon dioxide. Even if state coffers were dredged for the purpose of supporting CCS, there will almost certainly remain other more cost-efficient ways to reduce carbon dioxide emissions to air. Global projects for the permanent sequestration of carbon dioxide underground miss out on the opportunities that carbon recycling hold out. Most projections for low carbon futures do not include a major role for CCS in the next twenty or so years, and most assume that CCS can only be made affordable through the relativising effect of a high carbon dioxide emissions price or charge. The main problem with this is that there is no evidence that the global economy can sustain a high carbon emissions cost, which is, after all, a tax on unwanted vapour; and in fact, one would expect emissions pricing to remain as low as possible in order to protect trade and industry. Measures to set charges for carbon emissions would expect to encounter political challenge, and markets in carbon would be designed to deliver a business-optimal charge on emissions, probably enough to create upwards pressure on energy prices, and drive energy efficiency measures, but not necessarily high enough to influence major investment decisions.

Strategies to use biomass to displace coal in thermal power plants may prove misguided, as the combustion of biomass is perhaps the most inefficient way to convert it into useful energy; plus, some biomass supply chains might cause more carbon dioxide emissions than coal. These drawbacks may deter investment. Biomass could almost certainly be better and more efficiently exploited in gasification or some form of Chemical Looping Combustion (CLC), or via Direct Carbon Fuel Cells (DCFCs), to produce gas fuels for flexible use, as these can be stored, leading to system efficiencies. It would make no sense to promote biomass-fired power generation without also adopting a strategy to maximise heat recovery – Combined Heat and Power (CHP) coupled with District Heating (DH) – as the chemical energy value of biomass is inferior to most fossil fuels. In addition, substituting biomass for coal in large centralised power plants in the Global North will in some regions entail significant international trade, and owing to the long distances for transportation of biomass from the Global South, and the agricultural and forestry impacts it has where it is grown, may degrade its carbon neutrality. This would be parallel to the situation where the theoretical emissions benefits from the life cycle of liquid biofuels, in particular, biodiesel, have been cast into doubt and shown to have serious disadvantages in practice.

Other energy conversion routes for biomass also demonstrate problems. Gas of biological origin, such as biogas and biohydrogen, produced through biological means from biomass, is likely to remain slow to produce, despite progress in the treatment of algae, hydrogen-focused fermentation and similar research advances. Biogas production from anaerobic digestion (AD, or advanced AD (AAD)) might have natural limits for deployment, because microbial and algal productivity is just not very rapid. By contrast, thermochemical and catalytic chemical processes can be scaled up to produce industrial volumes of gas at speed. Additionally, some digestible or compostable biomass resource streams, such as food waste, especially those extracted from municipal waste (municipal solid waste (MSW)), are likely to be restricted in future, owing to a greater efficiency in materials flows. A similar contraction is likely in the stream of non-biodegradable biomass and non-biomass waste that could be used for thermochemical or electrochemical gas production, owing to restrictions placed on product packaging at source. So even if petrochemically sourced plastics were replaced by bioplastics, there would still be less waste overall that could be used as a source of gas energy. Biomass might be withheld from gas production as it could be argued that the best use of all for organic waste is in the treatment of agricultural soils; although this could be answered by creating a supply chain of digestate and biochar fertiliser, arising from biomass digestion and biomass gasification facilities, respectively.

Advances in the production of and strategy for hydrogen gas are typically in the area of transportation, such as in work on fuel cell vehicles, which is anticipated to take a good decade to roll out; yet Renewable Hydrogen could be so much more useful in the immediate time frame, for decarbonising electricity generation through load balancing variable windfarm output, by its combustion in gas turbines. Additionally, Renewable Hydrogen could make carbon capture profitable by reacting the hydrogen with carbon-rich waste gases to produce methane: a route for carbon recycling.

2.3.5 Can carbon pricing or taxation be effective in stimulating investment in the low carbon economy?

There are many socio-economic forces likely to be in opposition to carbon pricing, a traded carbon market or carbon tax, and evidence to date on the effectiveness of charging for carbon emissions is mixed. In economic models, an environmental pollutant can be taxed or priced out of the marketplace, as long as the offending chemicals are a niche product in the market, and are easily substitutable – yet carbon dioxide emissions meet neither requirement. Carbon dioxide emissions are produced largely in step with the consumption of energy because the vast majority of energy provision remains based around fossil fuels – in effect, the price of energy is the price of carbon – so why propose an additional charge? Although the growth in solar power and wind power could be considered exponential, it is not yet possible for every actor in the economy to drop the use of fossil fuel energy in exchange for renewables – they are simply not widespread enough. Only a small proportion of energy consumers have the option to elect to "buy green", and so carbon emissions are not easy to displace by changes in end use choices. Since everyone who consumes energy is responsible for carbon dioxide emissions (since fossil fuels still make up the overwhelming majority of energy supplied), a carbon price or carbon tax amounts to an added cost for everyone. It is hard to see how carbon price charges or tax revenue can be effectively recycled into cleaner energy; and energy markets can probably only support minor "cost-efficient", or marginal, emissions pricing. Polluters will continue to pollute, and pay, but pass the costs and responsibility for emissions on to their customers, thus leaving the original problem chemistry in place. An appropriate question could be: would charging for carbon dioxide prevent emissions in the same way that fining oil and gas companies for oil spills has prevented oil spills, or fining for illegal debt and loan practices has prevented bank misbehaviour? Surely, a more regulatory approach could effectively curb carbon, in much the same way it could have prevented the credit crunch?

Furthermore, carbon costs are simply an added burden within the overall profile of energy prices, which could climb or sink owing to other factors. It is entirely reasonable to suggest that rising or falling energy prices will obscure any signal for behaviour change and low carbon energy investments intended by the application of emissions pricing, markets or taxation. In other words, charging for carbon could only have an impact with relatively stable energy prices. With low economic growth, there would be several reasons why energy prices could fall, making an emissions charge a higher percentage in the cost of energy – in which case, carbon pricing could be rejected in order to maximise the opportunities for economic growth. If energy prices are massaged downwards by social policy or negative growth – economic contraction – a carbon price could only be justified if it is an insignificant charge on the use of polluting energy. In an alternative scenario of steep energy price rises, a possible outcome of a range of factors, governments would be under significant pressure to remove carbon pricing and carbon taxation, and adopt energy price controls. Neither of these pathways would lead to lower carbon dioxide emissions. Even if energy prices adopt a fulcrum position, there would be no political capital in implementing significant emissions charges, because everybody would be paying them. It does seem as if direct subsidies for, and directed investment into, non-emitting renewable energy deployment is more certain of achieving the aim of reducing greenhouse gas emissions from energy systems, rather than charging for carbon.

Carbon budget setters should have no reason to be wary of low commodity market prices for sweet, light crude oils, which should be passed through as cheaper vehicle fuel prices. Although cheaper vehicle fuels carry the risk of some greater vehicle fuel use, low fossil fuel commodity prices of themselves could keep carbon dioxide emissions under control to some extent. They will be beneficial in preventing the use of heavier crude oils in petrorefinery, as these have a higher processing cost (in the expenditure of energy) – and a higher sourcing cost (energy expenditure) if they need treatment at the point of production before their distribution to refinery. If there are supply constraints on lighter crude oils, and there is less profitability from the exploitation of heavier crudes, this could stall energy sector growth, which would also help with emissions control. However, regardless of commodity market fluctuations, growth in petrorefinery throughput continues to be low, and linear (BP, 2014c), despite fluctuations in crude oil prices, and this is the real determinant of flows of hydrocarbon fuels into the economy. Demand affects supply without the pressure of price, it would seem.

Carbon pricing is anyway meaningless in a political environment where the oil and gas sector, and other carbon-intensive industries, can safely make a case for special fiscal treatment. More importantly, in the time frame of two decades that the world has to make inroads into carbon dioxide emissions, it is unlikely to be possible to assert or implement an effective price on carbon to control carbon dioxide emissions.

3
Energy Transitions and Renewable Gas

3.1 What can work is what is already working for energy change

To minimise the risk of low investment levels in energy systems, it could be optimal to chart a course based on what is currently working, rather than overreaching. In a low growth global economy, big shifts in investment targets and large hikes in capital expenditure are painful, and so it is most pragmatic to work with what we have already got and expect only low threshold changes.

3.1.1 What is already working? Strong continued investment in renewable electricity

There is the potential for low velocity in the trajectories for investment in new thermal and nuclear power plants, the expensive adoption of Carbon Capture and Storage (CCS), and scaling up of unconventional fossil fuel production outside North America. By contrast, it can be anticipated that investment in renewable electricity could remain strong, as wind and solar power in particular are relatively fast to deploy. The rates of installations of electricity generation equipment using the power of the wind and the sun have accelerated in several regions of the world. Despite the financial crisis of the late 2000s, partly caused by the collapse of inverted pyramids of unsustainable debt secured on property (which has arguably led to a permanent step-wise contraction of the global economy), investment in new renewable power and the energy efficiency of buildings and manufactured goods have continued relatively unabated.

This is not only because of the speed at which windfarms and solar parks can be constructed, and insulation installed; it is also due to region-dependent government subsidies and other kinds of regulatory control and promotion, which mean that returns on outlay in this asset class are reasonable in size, and only take a couple of years to mature. As the amount of investment in renewable energy increases, equipment manufacture costs are falling, and subsidies can be safely degressed, or stepped down, as they will not be necessary to ensure a return on investment within a reasonable time frame. In particular, payback lead times for wind power, whether measured in energy, carbon or financial return, are short (Guezuraga et al., 2012; NREL, 2013). Even though solar photovoltaic power payback periods have been longer, this is shortening, as solar panels are becoming much cheaper, and are expected to remain fully functional for decades with little maintenance required to systems. Energy return on energy invested (EROEI or EROI) metrics and Levelised Cost of Electricity/Energy (LCOE) figures for solar power are contested, but appear already comparable with fossil fuels (Arvizu et al., 2011; Fthenakis, 2012; Fthenakis et al., 2011; Hsu et al., 2012; Inman, 2013; NREL, 2004; Palmer, 2013; Peng et al., 2013; Raugei et al., 2012). In addition, there are other forms of payback, including increased energy supply security, in an era when much privatised energy plant and infrastructure needs costly renewal; and countries need protection from energy import costs by utilising native renewable resources.

In some regions, government subsidies and tax breaks have stimulated uptake of the new technologies, but in others, utility companies have been the key drivers of investment, with their focus on long-term asset value. A shale gas field might be depleted within twenty years, but a wind turbine could produce power for forty, even though some parts of the equipment might need "repowering" in that time. Future strong carbon emissions policy could be imagined to have a highly enabling impact on the financeability and deployment of renewables. Renewable energy has the potential to be a growth area in all economies, even if no other sector thrives, and even in regions with absent or poorly functioning power networks. The reasons for investing in and deploying renewable electricity are accumulating, such as preventing the stranding of carbon dioxide emissions-based assets, rapid returns on investment, short lead times to generation, and the diversification of energy portfolios by national and regional power networks. Because of the exponential growth of renewable electricity, in some regions and some technologies another energy world is now not only possible, it is practical.

3.1.2 What is already working? Natural Gas-fired power generation

One of the outcomes of increasing levels of renewable electricity capacity is that at times there will be excess generation – in excess of demand.

At others, there may well be less generation than required. Owing to the increasing penetration of renewable electricity, grid operators will have more responsibilities in load balancing, as renewable electricity will introduce variability into supply. So, despite the success of renewable power, there remains the problem of the balancing of supplies, "load balancing", given the variable nature of the resources. Until and unless the world is powered, heated, moved and cooled by one hundred per cent renewable electricity, there will still be a need to fill in the gaps in the variable and intermittent production of resources such as wind and solar power.

Displacing coal from developed and developing world national power supplies, and replacing it with high levels of renewable capacity, can be done if the stopgap comes from Natural Gas-fired generation. Natural Gas will therefore likely play a major role in systems to match supply to demand, as the percentage of renewable electricity generation increases. This is because Natural Gas-fired generation is flexible, far less in carbon dioxide emissions than the combustion of coal, and likely to remain economically competitive, as global production remains inexpensive and copious enough to keep the fuel relatively inexpensive to burn for now.

Affordability, flexibility and efficiency will reinforce a growing dependency upon Natural Gas in electricity generation, which will be heightened as generators are already looking to Natural Gas to go some way towards addressing climate change concerns. However, despite the already successful development of high efficiency gas-fired power generation, long-term reliance on the fossil fuel Natural Gas and its potentially more polluting unconventional alternatives, such as coal gasified in situ underground (Underground Coal Gasification (UCG)), is not a climate change solution. It is not even an energy solution – as some unconventional resources, such as shale gas, could peak early. Emissions must continue to be eroded – so Natural Gas backup for renewable electricity is not the end point in decarbonisation of the power and gas grids.

Emissions mitigation is essential if industrialised countries want to continue to rely on gas and power for a large proportion of their energy; and yet, the only major proposal so far, CCS, does not appear to be financeable, at least for now. Regardless of the progress or otherwise of CCS, there is much that can be done otherwise to rein in emissions. With a strong renewables tranche in the supply of electricity, power from gas will not be needed all the time, reducing carbon emissions even further than simply switching away from coal; in addition, this would drastically reduce local environmental pollutants and therefore improve public health.

Supply Side and Demand Side Management (SSM and DSM or Demand Side Response, DSR) and smart grid networks will also help to avoid unnecessary carbon emissions. For example, in the UK, the National Grid is developing a suite of solutions called STOR – Short Term Operating Reserve – as part of their Balancing or Reserve Services, to meet shortfalls in power production ranging from a few hours to a few days. Reserve Services have to be enough to cover outages from the largest centralised generation equipment, such as nuclear power plants. Reserve Services can also be used for optimisation in matching supply with demand. Special contracts that enable switching out load during peak demand, and pulling in standby generation, can in future give higher efficiency to the use of variable renewable electricity. If nuclear reactors are not permitted to have lifetime extensions, the UK will lose the liability of needing backup for these huge generators by the mid-2020s, at which point, much of the STOR capacity can be put to use in helping to make the most of variable renewable generation. Some of the STOR plants are gas-fired, and offering this facility attracts a premium payment. It can be imagined that early uses of decarbonised gas supplies – Renewable Gas – would be in STOR plant, where the rewards could compensate for the initial expense of building plant equipment to manufacture the gas.

3.2 Why gas?

3.2.1 Flexibility in use, multiplicity in source and flexibility in composition

Gas fuels are a pragmatic choice. Gas fuels have been popular over the last hundred years or so, because of their simplicity – they are easy to transport, clean to burn and can be kept safe in storage. Gas fuels are likely to remain popular now, and into the future, and a higher proportion of energy supplies is likely to be gas-based, increasing the incentive not to flare or vent Natural Gas from oil wells (Brimont et al., 2012), and to prevent fugitive emissions from the gas fuels industry, gas grids, landfill and water treatment plant. It would be prudent to switch attention from liquid hydrocarbons to gas from now on – as petroleum oil production operations drill deeper into the Earth's crust for new fields, they find higher proportions of Natural Gas. The deeper they go, the higher the temperature and pressure in hydrocarbon reservoirs, and the greater probability that hydrocarbons have "cracked" to gas (e.g. Dyman and Cook, 1998). Importantly, gas fuels as a whole are flexibly produced and utilised – historically, mixes of gas of various origin were used industrially and in utility and domestic supply. Early gas fuels used in urban settings were based on coal gasification – but this "Town Gas" could be augmented by other resources such as wood gasification, if necessary. Clearly, different gas fuel compositions demand that gas fuel burning equipment can be adjusted or be flexible, or that the gas fuels are cleaned, upgraded and standardised before being supplied to their users. Despite these drawbacks, the ability to produce and use variable sources and mixes of gas fuel is a vital advantage in developing sustainable Renewable Gas resources.

3.2.2 Nature's own energy solution

Decomposition to gas and dirt is Nature's way of recycling once-living beings back to the soil of the Earth, replenishing the nutrients required for the next generations. Gas that comes from rotting organisms is either recycled through the inspiration of growing plants, taken up by water, or by rocks and soils through weathering, or decomposes or dissipates in the atmosphere. Nothing is wasted. Two of the important gas components of decomposition (biogas) are methane and hydrogen – lighter than air molecular compounds with some of the simplest gas chemistry – and two very important energy fuels if they can be harnessed. Natural Gas is very largely composed of methane, for example. Of the two, hydrogen and methane, each gas is important in its own right, and burned together they could be even more valuable, and "Hythane" would be a suitable goal for profitable Renewable Gas production. The proportion of hydrogen currently produced by the usual forms of microbiological decomposition is not significant; so biomass and organic waste might not be anticipated to deliver a "Hydrogen Economy", although microbiology will certainly provide meaningful methane gas. There are, however, important projects to use "spare" wind and solar power capacity, and other means, to produce Renewable Hydrogen by various processes, including the electrolysis of water. There is ongoing significant research into improving the output and optimising the costs of water electrolysis, and other ways of splitting water to produce hydrogen gas. The engineers of the semiconductor industry who brought us solar power are at work to bring us Renewable Gas, too. The potential for increasing the energy derived from cleaning up biogas to biomethane, or uprating it via the simple addition of Renewable Hydrogen, or from chemically reacting hydrogen with carbon-rich biogas to make Renewable Methane begins to look highly desirable.

3.2.3 Industrial gas solutions

The biological decomposition of expired plants and animals is not the only route from biomass to useful gas feedstocks. Gasification of biomass, organic waste and even some inorganic waste, has interesting potential. In addition, all thermal treatment of gas and liquid fuels results in varying quantities of carbon dioxide and carbon monoxide, and although carbon dioxide is generally considered to be a "waste" gas, it can be useful. As significant resources of Renewable Hydrogen are developed, any gas stream with a high carbon content will be valuable in producing methane and other hydrocarbons, by reaction with the hydrogen. The input fuels for Renewable Gas will include carbon dioxide from industrial chemical transformations, and the carbon dioxide emissions from power stations burning all forms of fossil fuel, including Natural Gas, and Renewable Gas (IPCC, SRCCS, 2005, Chapter 2).

Strictly speaking, carbon oxides from the combustion or gasification of fossil fuels and industrial chemistry used to make new fuel will not be truly Renewable Gas – however, this recycling of carbon gases from electricity generation and the chemicals industry, by using Renewable Hydrogen as a co-production agent, offers a much higher manufactured gas capability. It should be labelled Transitional Gas. However, if the carbon-rich exhaust gas from burning Transitional Gas, whatever its origin, is recycled perpetually, this will be genuinely low carbon, sustainable, and properly called Renewable Gas. The same technology that can be used to manufacture Transitional Gas can be kept for use when fossil carbon is replaced with Renewable Carbon to manufacture fully Renewable Gas (Cassidy et al., 2010, Page 154). For this reason, all pathways to Transitional Gas need to be included in discussion of Renewable Gas. Putting it another way, one might quibble that recycling waste gas from a fossil fuel power or industrial chemical plant is not sustainable, but gas that can be recycled once can be recycled many times, and not cause emissions to air from the original fossil fuel source.

Natural Gas can become a true "bridging" fuel to the renewable future, if the flue gases from Natural Gas power stations are recycled into Renewable Gas production. Although the sequestration of carbon dioxide underground is being promoted as a way to keep harmful global warming emissions permanently out of the atmosphere, the development of CCS makes more sense as a temporary repository for carbon-rich gases until Renewable Gas production has reached importance and requires high volumes of these formerly waste gas resources as input.

A range of Renewable Gas-based fuels can potentially be made in a "biorefinery", from feedstocks produced by biological, electrochemical, photoelectrochemical, thermochemical, catalytic and gas recycling routes, and this offers solutions for developing replacements for the liquid fuels currently used in transportation. There is a generally held view that transport fuels will be difficult to replace, as oil refining becomes constrained by peaking crude oil production and increasing chemical complexity in crude petroleum oil resources. It could be found to be more efficient and less polluting to produce liquid fuels from Natural Gas and Renewable Gas resources than manufacture "synthetic" vehicle fuel oils (synfuels) from shales, tar sands and sour or otherwise chemically compromised "unconventional" petroleum fields. Additionally, Renewable Gas production could expand rapidly, bypassing the need to refine complex unconventional gas to make up for progressively widening shortfalls in Natural Gas.

3.2.4 Energy storage – because renewable electricity is sustainable, but variable

Although while the Earth continues to spin ninety-three million miles or so from the central star of our solar system, and it has not become a red giant, the flow of wind and the rain of sunlight will be sustained; in any one location, the sun does not always shine and the wind does not always blow, and so renewable electricity resources are variable. They are also ephemeral, as storage of the actual wind for later re-use is impossible, and sunlight cannot be put in a box to be allowed to shine out later – unless, for example, that reinforced box is filled with molten salts, and the sun's rays are concentrated on it, so that it stores heat that can be used to produce electrical energy when the sun sets. More importantly, renewables, including biomass, are less dense resources of energy than the hydrocarbon "fossil" fuels mined, drilled and pumped out of ancient biomass-based sedimentary strata: oil, gas and coal. Although it is possible to generate electricity in real time from wind, sun and biomass, and supply a power grid with useful, continuous, energy, much more capacity will come from capturing some of the energy from these resources and converting it into a storable form to be used during dips in production. If this stored form of energy is more concentrated, so much the better.

To be practical and affordable, the energy storage technology should be as simple as possible, and as inexpensive as possible, using the least equipment, which, for example, rules out most forms of chemical battery with corrosive fluids that require heavyweight metallic casings. Storage should also ideally be inter-seasonal, something that is not expected to be viable for most actual and proposed storage systems.

Electricity storage options are in development to supply renewable power on a cold, calm night, but for the foreseeable future, regions of the world with extensive power grids are planning to supplement gaps in their renewable provision with Natural Gas- or biomass-fired combustion generation plant. Including Renewable Hydrogen and Renewable Methane in their storage plans would be a solution that not only works for the power grid but also works for grid-independent installations of wind or solar photovoltaic power. The emissions of these options could be far lower than Natural Gas power generation operations. If a power grid cannot use all the renewable power that is output from wind and solar generators, it would be convenient to be able to use this for making gas, thereby maximising the utility.

Gas is straightforward to store, unlike electrical power. However, although we have a gas grid, an increase in methods and facilities for gas storage is essential for the development of Renewable Gas. For example, much Renewable Hydrogen would be made when it is not immediately required. Furthermore, gas needs to be stored in large quantities ahead of weather-related dips in renewable electricity production, if we are not to resort to burning coal or heavy oil on a windless, cloudy evening. Currently, in the UK, for example, there is extensive infrastructure for storing liquid and solid fuels, but not so much for gas – and this needs to be rectified as we move towards a gas energy future. Crucially, Renewable Gas storage will be needed in a distributed arrangement, and not just at centralised facilities. This is because Renewable Gas resources would most optimally be utilised close to where they are produced – particularly for biogas. Renewable Gas can be chemically processed before storage – perhaps into liquid fuels, or into higher density energy gases through processes such as methanation. Maximising the levels of methane in mixtures of carbon-rich and hydrogen-rich gas improves the energy value of the fuel, and decreases the space the same amount of energy takes up, which is a useful storage consideration.

3.3 Why Renewable Gas?

3.3.1 Gas is our flexible friend

One of the key benefits of developing Renewable Gas fuels is that gas is highly flexible. Various compositions of Renewable Gas can be upgraded for use in gas grids, used unmodified in power stations with flexible modern gas turbines or used stand-alone for local independent power generation. And it confers potential global social and economic benefits, as Renewable Gas technologies will be appropriate in economically developing as well as industrialised countries, even for those regions without gas grids. For example, straight biogas is already being exploited in rural India and China to great advantage, and hydrogen-rich or methane-rich Renewable Gas off-grid generator backup will be vital for regions that are developing grid-less solar and wind power systems. Industrialised countries with established gas grids and gas-fired power stations can keep them, if they develop Renewable Gas, whereas they may not be sure of the economics of maintaining their gas grids if they rely solely on Natural Gas. For example, under a supply reduction scenario, Natural Gas would be used in centralised power stations as a priority, meaning gas distribution to the residential parts of the network would be very low, and make operating costs relatively excessive, and very high levels of energy conservation would come to be required for buildings. Alternatively, there could be a shift to electricity for space heating, which could also limit the extent of gas distribution by grid. Very low volumes could make gas distribution to homes unprofitable.

The development of Renewable Gas is not revolutionary – it does not break an energy paradigm, and does not require a complete replacement of technology, or energy processing and distribution infrastructure. It is a gradualist solution – offering change based on what is currently known and done, making use of existing energy sector players and markets and, as such, is likely to be cost-effective. To develop Renewable Gas will not require a large re-balancing shift in the economy, with all the financing risks involved with that, such as the proposed global "Nuclear Renaissance", which would see massive deployment of new atomic energy reactors. The transition to Renewable Gas will also not require a major renovation of fuel-burning appliances, gas pipeline networks and vehicle construction – such as would be required for a completely carbon-free Hydrogen Economy.

3.3.2 Renewable Gas can decarbonise the gas supply

There are three broad categories of options for the management of carbon in the development of sustainable, climate-preserving gas fuels: carbon capture with permanent sequestration, carbon capture and recycling (see Tables 3.1 and 3.2), or the use of "young" carbon – the main resource for "young carbon" being biomass. As well as reducing carbon emissions, carbon recycling could enable lower fuel inputs for power plants. However, whilst needing less input biomass or fossil fuel, carbon recycling power plants would need hydrogen inputs instead, either in the form of water or steam, or in the form of hydrogen gas. It would be prudent to commence the advancement of these options in a short

Table 3.1 Degrees of recycling of carbon dioxide

Type	Purpose	Final destination of carbon dioxide (time to get there)
Re-use	Food and drink industry	Atmosphere (short delay)
	Chemicals and pharmaceuticals	Atmosphere (medium delay)
Relay	Fuel production	
	– Fischer–Tropsch conversion	Atmosphere (medium delay)
	– Fuel additives	Atmosphere (medium delay)
	Agriculture (growth promoter)	Atmosphere (medium delay)
Recycle	Thermal power plant	Atmosphere (long delay – after a number of cycles)
Re-bury	Thermal power plant	Permanent sequestration with Carbon Capture and Storage (CCS)

Table 3.2 Simple medium-term guesstimates for carbon dioxide recycling applications

Type	Curtailed emissions (%)	Technology
Electricity generation	Estimate 5%–50%	Gasification of fossil fuels (thermal recycling)
	Estimate 40%–90%	Gasification with methanation or other chemical recycling
Gas grid	Estimate 3%–30%	Gas from bioenergy (displacing Natural Gas)
Transport	Estimate 5%–20%	Gas from bioenergy (displacing liquid fossil fuels)

time frame, given the risks to the energy sector from continuing high demand for Natural Gas coupled with the potential for failing production or competition for supplies. A transition to low carbon gas does not need to delay, as there are technologies available today that will suffice. Interestingly, some of the technologies required for the processing of conventional and unconventional fossil fuels, both liquid and gas, are also suitable for Renewable Gas production, and so a parallel development can be anticipated – beginning a transition by using the tools we need in future for Renewable Gas to process the fossil fuels we use now.

3.4 Why methanate?

The term "methanation" is here used to encapsulate a range of methods of chemically reacting hydrogen with carbon oxides (carbon monoxide and carbon dioxide) to produce methane (Abbess, 2014, Table 9). The term is also used in process engineering to describe chemically removing trace carbon oxides from an already methane-rich gas. Some use instead the word "methanisation" for reacting hydrogen with carbon oxides to make methane, in an attempt to be more precise. However, this term is also used by others for the process of anaerobic digestion (AD) of biomass and organic waste. In this discussion, the use of the word "methanation" without a qualifier is intended to convey the sense of industrial methanation in chemical process reactors. However, it is also important to note that although there are two broad categories of methanation – chemical-industrial methanation and biological methanation, in some developments, there is a crossover of techniques, so some methanation methods can be a combination.

The development of Renewable Gas will need developments in the production of Renewable Hydrogen. This raises the question of what can be gained from converting the Renewable Hydrogen into Renewable Methane – which would involve a loss in efficiency. Why would we not use the Renewable Hydrogen itself as our Renewable Gas? It turns out that when considered from a systems point of view, Renewable Methane is likely to be vital in the development of Renewable Gas.

3.4.1 Methanation permits a gradual transition in gas energy systems

Step changes in the energy systems of industrialised economies would be significantly more complicated in the present day than the conversion to Natural Gas from Town Gas undertaken in the 1970s and 1980s. Entirely replacing the UK gas grid with hydrogen supplies, for example, would be a major undertaking and could last a very long time. Piecemeal conversions could lead to systemic chaos. The fully Hydrogen Economy therefore appears some distance away. Although small levels of hydrogen could be added to the Natural Gas supply, depending on evolving regulations, in theory there are no technical limits to the amount of sustainably produced methane that could be injected into a Natural Gas grid – provided that any minor deviations in gas quality from network regulations are rectified. This means that the variable supplies of methane from a Renewable Gas production network, particularly at the outset of production when volumes could be anticipated to be low as well as variable, will not greatly impact gas grid control measures, and can be easily integrated. It should be pointed out that keeping the "Methane Economy", with its gas networks, does not prevent an eventual transition to Renewable Hydrogen. There is an energy penalty in converting high value hydrogen into methane, but from a systems perspective, the low carbon transition is more manageable if starting from the current configuration of gas systems and making minor changes in succession.

3.4.2 Methanation can upgrade a mix of gases from a range of sources

There are various gas streams that can be developed – such as recycling waste gases from the use of fossil fuels, or multi-fuel gasification with carbon capture at a centralised power plant. Thermochemical gasification can be a very efficient energy conversion, converting a range of sometimes complex feedstocks into simple gases. Depending on the exact feedstock and the temperature range of the reactor, the output gas can be closer to biogas with high methane content, or closer to syngas (synthesis gas), with both high hydrogen and high carbon monoxide content. Some sources of gas in future, such as those from industrial and waste facilities, including petroleum refineries, will be high in carbon dioxide, as energy processing and conversion plants are anticipated to increasingly use various forms of combustion, including gasification, or partial oxidation, reactors. Gas can also be derived from biological processes, such as AD, and depending on the biomass feedstock and exact microbiological process, be high in methane, and high in carbon dioxide, and potentially have significant levels of hydrogen.

Sources of gas will be very varied and dispersed in a low emissions energy system, and so it would make sense that a range of low carbon gas fuels would be directed to a central storage facility, and one would end up with a gas mix mostly composed of hydrogen, methane, carbon monoxide and carbon dioxide. Carbon monoxide is highly toxic, and so in principle, it should not be stored in large quantities. However, methanation could use up carbon monoxide as it is created, and store it as Renewable Methane, which is far safer. Mixing sustainably produced gases for use as input can assist in the industrial methanation process, as some will be high in hydrogen, and others high in carbon oxides. If there is not sufficient hydrogen, more could be produced from the carbon monoxide and steam via the Water Gas Shift Reaction – or "shift" (Abbess, 2014, Table 9).

The exact composition of this mixed gas in storage could vary significantly over time, but methanation of the mixed gas would both uprate the uniformity and move the majority composition to a higher energy density – thus reducing the space needed for storing the same amount of energy as gas.

Industrial methanation using Renewable Hydrogen could be an option for making biogas suitable for grid injection, as it would limit the amount of "washing" or cleaning that the gas needs before injection into the gas pipeline network.

3.4.3 Methanation can help to resolve problems arising from changing Natural Gas quality

Methanation can solve the problem of compromised supplies of raw Natural Gas in the transition to lower carbon gas. The quality of conventional fossil fuel Natural Gas could be degenerating quite rapidly, in all regions. In some cases the EROEI may be dropping gradually all the time. Worsening quality of raw gas, such as increased production of sour, acid Natural Gas, or Natural Gas with high levels of nitrogen, requires additional processing – which carries an energy and production volume penalty. The first question could be: when does this processing of worsening quality Natural Gas become relatively more expensive than manufacturing Renewable Gas? The second question would then be: when do the economics dictate that Renewable Gas manufacture becomes mandated to support the volumes and qualities of Natural Gas used in gas grids? The third question from that would be: will methanation eventually become an essential treatment for Natural Gas itself to permit continued use of this degraded fossil gas in gas grids?

Technologies for making Renewable Methane can be projected to be useful in processing if a rising proportion of Natural Gas is sour or acid. Sour gas is frequently contaminated with both chemically significant percentages of hydrogen sulphide gas and volumetrically significant percentages of carbon dioxide. Unless solid state electrochemical power generation plant can be developed to run on sour gas (e.g. Cheng et al., 2006), it will remain essential to remove the sulphur from the gas. However, instead of washing the carbon dioxide out of the sour gas, and venting it, or using CCS, it could instead be methanated, if there is also production of Renewable Hydrogen at the same processing plant. If the methanation of sour/acid gas becomes developed, this could be a valuable part of an orderly transition from fossil fuel gas to Renewable Gas, as the same plant equipment could be used for both feedstocks to uprate them for grid quality.

3.4.4 Methanation can enable the recycling of fossil carbon

It can be envisaged to have a power plant that, after start up, requires very small quantities of new carbon input, in the form of coals, say, and recycles all the carbon in its internal system. Renewable Hydrogen would be fed into the plant, or it would produce hydrogen from water using renewable electricity. It would then methanate the carbon – the combustion of an amount of hydrogen would provide the initial thermal energy for the methanation process. The production of further hydrogen by water splitting would be the heat sink from the methanation process – for example, using high temperature solid oxide electrolyser cells (SOEC) (IEA HIA n.d.-b). The Renewable Methane would then be stored and combusted as required to generate power, and the carbon dioxide that results would be captured to make further Renewable Methane. The only reason this plant would require new coal input is that coal is not 100% carbon, and also that some carbon gas losses must be anticipated through inefficiencies and fugitive emissions.

3.4.5 Methanation could avoid exploration for Natural Gas in sensitive regions of the world

Methanation for Renewable Gas could avoid the drilling of Natural Gas in sensitive regions of the world – for example, it could obviate the need to mine the Arctic region. Instead, sour and acid gas could be drawn from extensive reserves in already producing regions, such as the Middle East, and foster a better relationship between gas producer and gas importer regions.

3.4.6 Methanation and methane storage could regulate gas systems

Methanation followed by the storage of the Renewable Methane, for example, by injection into the gas grid, or into new storage facilities, could be useful from a systems regulation perspective. Many microbiological processes that use biomass feedstocks to produce gas have a slow operational cycle, where there are long residence times for the biomass in the reactors. Examples of low gas production rates would be AD of biomass to biogas and low-temperature gasification of biomass. The speed of gas production being low, a differential time lag between the production of the gas and its use could be beneficial, however this necessitates interim storage. If methanation and methane storage are developed for mixed gas streams, these would be useful in justifying increased production of "slow gas" such as biogas, thus optimising the use of this resource.

3.5 Practical measures in an orderly gas transition

Everything must change – but how? A first step will be in the optimisation of the use of Natural Gas. Second, it will be necessary to precisely model the options for the development of Renewable Gas. Third, it will be necessary to manage global gas markets and gas delivery infrastructure to accommodate the development of Renewable Gas.

3.5.1 Efficient management of Natural Gas resources

Notwithstanding global oil and gas industry initiatives, much Natural Gas is still being vented and flared – partly because of the poor return in investing in distribution infrastructure, and partly because the prized aim of most drilling is oil, not gas. This situation is an anomaly, given projections for Natural Gas demand and depletion. In order to ensure that Natural Gas can be a sufficient bridge to a low carbon world, all resources should be best utilised. Regulatory and standards work is ongoing to capture all the value in Natural Gas resources, but it could become necessary to issue not only mandates but premium tariff regimes or subsidies to ensure optimal gas exploitation in some regions. Increased subsidies and other kinds of support for Natural Gas would need to be balanced with the need to concentrate policy on low carbon gas to keep driving the transition to Renewable Gas.

3.5.2 Accurate assessments of Renewable Gas production potential

Gaseous phase fuels are highly versatile and efficient in many applications, not just energy, and there is much plant and other infrastructure invested in Natural Gas, and there is likely to be more with time. The questions are, therefore, can biologically sourced, sustainable and renewable resources of gas fuel displace much fossil gas? Can Renewable Gas offset or take us through Peak Natural Gas? And can Renewable Gas have a low enough emissions profile such that it can be carbon-neutral, to "decarbonise" the gas fuel supply – used for power generation, space heating in buildings, agrochemicals, industrial chemistry and transport fuels? Reliable assessments of these possibilities are essential.

3.5.3 Management of gas market competition

With increasing demand for Natural Gas around the world, the shape of the global markets in Natural Gas can be anticipated to shift. In order to manage the supply and distribution of Natural Gas, it is crucial to monitor trends. This will be even more important to understand if unconventional fossil fuels start to be exploited significantly outside North America, as these resources are often in different physical locations and territories than Natural Gas. Regional and inter-regional disputes can be anticipated with proposals for their exploitation, as unconventional fossil fuel resources pose environmental risks, and could cause complications with transmission and distribution infrastructure. Such fault lines would inevitably affect gas market conditions.

With the repurposing of clean-burning, easy to tap, Natural Gas to a wider range of energy and chemistry needs, demand is set to increase, and this adds to the risk of a peak in the ability to meet demand from production within thirty years. Much Natural Gas is "associated" – co-present with petroleum crude oil – and there are indications that crude oil production will suffer critical depletion within a decade in certain regions, despite the inclusion of what are known as "Natural Gas Liquids" (NGL) into tallies of liquid fuel production by the major oil and gas businesses, databases and consultancies (Abbess, 2014). The global move to diesel vehicle fuel was perhaps considered to be a way to stave off Peak Oil, as diesel grades, fractions or "cuts" of crude petroleum oil could be anticipated to be more copious than those for gasoline going forward, especially when making use of heavier crude oils (Abbess, 2014). Yet even this transition, or the use of other refining by-products as fuel additives, cannot prevent the inevitable global plateau and decline in vehicle oil supplies. Peak Oil will almost certainly impact the production of Natural Gas. If oil wells deplete until they become uneconomic for hydrocarbon liquids production, they will almost certainly be abandoned, and will no longer co-produce gas, and this will have the side effect of skewing the global markets in gas.

As energy systems age, the likelihood of breakdowns, accidents and outages increases. Where energy inputs to one economy are restricted by fault, energy must be sourced from other regions. A recent very telling example is the change in energy imports demanded by Japan following the Fukushima Dai-ichi Multiple Nuclear Reactor Accident. Energy flows in the whole East Asian economy were re-routed, and this had knock-on impacts further afield. Such a significant change could affect market competition, and so could policy trends. Within the EU, for example, governance policies to answer climate change are consistently moving away from coal-fired power generation to Natural Gas, and despite new pipelines and new trade agreements with countries ever further away, it can be anticipated that internal EU competition for Natural Gas could be high, particularly if there were a spectacular failure related to nuclear power in or near Europe – increasingly likely, the older the in-service atomic fleet.

Data intelligence on European Natural Gas imports, demand and consumption is continuing to improve to enable stronger market harmonisation and to limit illegal competition, but clearer data will not remove the risks of scarcity, from both volumetric and market price volatility, should there be a breach in other forms of energy supply, or a country rejects nuclear power, coal, shale gas and other fuels in favour of Natural Gas, and demand for Natural Gas increases. It may even be that countries that produce most of the world's Natural Gas may want to utilise more of their own energy, as they too face demands on emissions and efficiency, and so export supplies to other regions could be negatively impacted. All of these factors need to be considered in the global steer towards gas.

3.6 Pathways to Renewable Gas

The three key areas where gas can be deployed for energy purposes are: building temperature control, transport and electricity generation. The use of gas in the chemicals industry is also important to consider. Each application has its own potential and problems.

3.6.1 Space heating and cooling of buildings

3.6.1.1 Space heating for buildings

A large proportion of Natural Gas is consumed in applications that are not centralised "point sources", that is, they are dispersed, such as for space heating uses. The carbon dioxide in the flue gas from the burning of Natural Gas in millions of distributed heating boilers cannot practically be collected for a CCS system. When the need to sequester any carbon dioxide from the use of Natural Gas becomes an imperative, the usual projection is that either space heating will need to have been eliminated by building insulation, or space heating equipment will need to have been converted to electricity. Failing those options, the gas supply will need to have been decarbonised at source. At the current rates of building insulation, it seems unlikely that space heating will have been eliminated in the medium term. In addition, converting all building temperature control to electrical heating and cooling would require very high levels of new power generation capacity, which is also unlikely, unless wind power deployment is on an unprecedented scale. Furthermore, the option to lower household carbon dioxide emissions through residential scale co-generation, or Combined Heat and Power (CHP, micro-CHP) (Dodds et al., 2015), or CHP at the local area level, would still mean a gas grid remains necessary. It is therefore logical to anticipate a growth in low carbon gas to be used for building temperature control.

3.6.1.2 Space cooling for buildings

It is anticipated that cooling for buildings could become a serious consumer of electrical power, particularly in summer. This is coincident with the expected peaks in solar power production, but if heatwaves become unseasonal as a consequence of climate change, it may be that gas-fired power is needed for space cooling.

3.6.1.3 Urban architecture and biogas

The greatest need for space heating and cooling of buildings is in places where there are high densities of buildings, and these happen to coincide with high levels of food waste, packaging waste and biohazard waste – for example, waste water treatment plants and hospitals. It therefore makes sense in efficiency terms to develop biogas through AD, and biosyngas through gasification – the most appropriate disposal method for biological and hazardous wastes – in urban areas, and use it in CHP (cogeneration), or Combined Cooling Heat and Power (CCHP) trigeneration plant, with either District Heating (DH) or district cooling pipe networks. As urban developments become more efficient and there are higher levels of building insulation, and the global economy starts to reduce once-through product packaging, levels of energy consumption and volumes of waste generated will fall in parallel. So, although the urban feedstocks for manufactured Renewable Gas may decrease, so will the need for it; therefore, Renewable Gas production in towns and cities will still make sense, even if total volumes consumed in the local area are a lot smaller. Household-scale CHP may become practical, if electric transport becomes widely used and home recharging is required. One possible configuration of street-local or domestic-residential CHP would be fuel cells running on grid gas, in which case, the grid injection of Renewable Gas manufactured in centralised facilities would be necessary to decarbonise the system.

3.6.2 The Prime Mover of transportation: the importance of fixing transport options

Transport, like space heating, has many sources of carbon dioxide emissions, making capture impractical. This is perhaps the best entry point for Renewable Gas, as road, sea, rail and air transport is a priority for the global energy transition away from fossil fuels.

3.6.2.1 Trade and transport

Of the key energy applications that need to be transitioned – the most pressing is perhaps that of transportation. The world's economy is now dependent on globalised systems of trade and exchange. To sell across borders and seas and country lines, it is necessary to transport people, goods and products, even if these are merely digital services. All transportation, even of electronic information, requires fuelling. Trade efficiencies depend currently on sustained personal and trade-oriented transport; although personal mobility could become seen as an order of magnitude less essential for global economic capability than the transportation of food, products and goods. The trajectories of the projections for the increase in road, sea and air vehicular use suggest that the global economy is likely to become increasingly dependent on transportation. With almost two billion registered road vehicles in use around the world (WHO, 2013), and nearly three billion passenger airline flights per year (ICAO, 2012), addressing possible problems in supplies of fossil fuels, and minimising their impact on the environment, is critical to international economic security and governance. In a scenario of peak crude petroleum oil supply, changes must be made to vehicle fuel and vehicles in order to keep this system operational. If taking the gas option – transitioning to gaseous phase vehicle fuels – the world is likely to see significant moves towards the use of Renewable Gas. The reasons for this are varied, but mostly have to do with the timescale for the substitution of fossil fuels – something that is being imposed by Nature, rather than energy trading or carbon emissions regulation.

3.6.2.2 A review of the options: efficiency, electricity, biofuels and gas

The three main options for displacing carbon are: fuel use efficiency, electric drive and alternative fuels. One of the key ways overall energy efficiency in transportation can be effected is by reducing the physical size and mass of all new vehicles on the market, although most policy work has focused on engine fuel use efficiency. Early developments in alternative fuels included LPG – Liquid Petroleum Gas – and diesel was introduced as a more efficient alternative to petrol. However, improvements in engine technology are perhaps removing the diesel efficiency advantage. An added consideration is that particulates from diesel combustion are correlated to poor local environmental conditions, high rates of asthma attacks (e.g. McCreanor et al., 2007; Zhang et al., 2009), and the incidence of lung cancer (Silverman et al., 2012), and perhaps even autism (Roberts et al., 2013).

Electric drive and hybrid electric/fossil fuel vehicles can become highly useful in the fight against poor urban air quality; and from a systems viewpoint, overall efficiency of energy use could be higher if vehicles are electrically powered – however, it does depend on the so-called well-to-wheels (or "well-to-tank") pathway – the energy losses and carbon emissions from the initial resource to the final point of use. For example, electricity generation from the combustion of solid bioenergy, such as woody biomass made into pellets, is about the least efficient thing that can be done with it. Some have assumed that most transportation in future will have electric drive train, but the problems associated with much larger numbers of battery manufacture, including concerns about Rare Earth Element scarcity, and the logistics and costs of rolling out several billion new electric vehicles to replace the current fleets, and putting in place the generating and recharging infrastructure, make this energy future seem a distant scenario.

Despite a strong industrial focus on their development, it is not yet clear if liquid biofuels such as biodiesel and bioethanol will be able to contribute more than a small slice of the market in mobility as they are not uniformly carbon-positive or even carbon-neutral; and unexpected hitches have occurred, such as the competition for feedstocks between food and fuel markets, which have impacted European policy (EBTP, 2014). Of the alternatives, several gas fuel options look promising. Until very recently, the production of hydrogen gas was almost entirely from processed fossil fuels, so its use in fuel cells or in combustion engines for transport would be inefficient if fossil fuels could be used directly instead. But if Renewable Hydrogen production becomes prevalent, this could all change. The other main gas alternative is methane.

Compressed Natural Gas (CNG) is rising up the agenda as a vehicle fuel in some parts of the world – for example, in India, where it was introduced in 2003 with the aim of improving air quality (Ravindra et al., 2006), although it cannot eliminate air quality problems entirely (Jacobson and Delucchi, 2009). CNG is beginning to be seriously considered for transport in developed countries as well, for example, in the range of Natural Gas Vehicles (NGV) becoming available. In North America, its popularity is mostly because of cost savings, as the fuel remains relatively cheap compared to liquid fuels. Europe is developing a CNG distribution and filling station network (NGVA, 2014), and Renewable Methane would be highly useful for this, in the form of Compressed Renewable Gas (CRG). Some CNG/CRG could be sourced from biogas from AD of plant and animal waste, upgraded to biomethane, or biosyngas from the gasification of biomass and waste – also upgraded, by methanation, to be closer in composition to biomethane.

Compressed methane-rich gas (CNG or the Renewable Gas equivalent) could prove essential in the long process to transition away from fossil fuels. Combination gas and liquid fuel systems, such as "dual fuel" (where gaseous fuel is used with a small amount of liquid fossil fuel in a diesel engine), or "bi-fuel" (where an engine is run on gaseous fuel, with a liquid fossil fuel tank as a backup), are emerging, and the conversion cost can be cheaper than replacing the whole vehicle, although the success of mass conversion will depend on the roll-out of Natural Gas filling station networks (BCC, 2015; CRD, 2015; EcoDual, 2012; Innovate UK, n.d.; Le Fevre, 2014; Nemeth, 2012; NGV Network, 2014; Truck & Bus Builder, 2011). This is continuing the earlier work done with LPG, and offers cleaner burning fuels and cost efficiencies. This development is equally useful for larger diesel engines as used for stationary and off-grid applications (e.g. ComAp, 2009). It may even be possible to completely decarbonise dual fuel vehicles, if they use biodiesel for their ignition fuel, and Renewable Gas for their main fuel. Besides pure hydrogen gas or methane-rich gas, gaseous phase fuels also include Hydrogen Compressed Natural Gas (HCNG), or Hydrogen-Enriched Natural Gas (HENG) – a compressed mixture of hydrogen gas and methane-rich gas (e.g. Ma et al., 2010; Ortenzi et al., 2008). The technology is improving, but if hydrogen storage units for vehicles remain too heavy to permit the current generation of hydrogen fuel cell vehicles (HFCV) from gaining a wide market, it may be possible to use internally reforming fuel cells that run on methane-rich gas instead (e.g. Priyakorn et al., 2011).

Another route would be GtL – gas to liquids – to utilise biomass gasification and "refine" the biosyngas to synfuels, using such chemical techniques as the Fischer–Tropsch method to synthesise liquid alternative biofuels. A biorefinery that could use Renewable Gas feedstocks to react to produce liquid fuels, such as dimethyl ether (DME), ammonia or methanol, demonstrates the potential for further options. Fossil fuel refineries can use "drop-in" biofuels and even solid biomass in some of their processing, which could decarbonise refined fuels at source (Karatzos et al., 2014). In the long term, however, it may be vital to transition out of using liquid fuels, owing to competition for biomass resources. By contrast, hydrogen-rich and methanated gas phase fuels can be produced using less biomass input.

It can be envisaged that urban private transport, and light duty vehicles (or light commercial vehicles), will tend increasingly towards small-sized electric options, whilst long-distance and heavy duty vehicles will be fuelled with light gases such as hydrogen and methane. Bioethanol will continue to be important in some regions, and in main refinery blending. Algae-based biodiesel could also become useful, in applications that cannot make use of gas phase fuels.

Renewable Gas applications in mobility are likely to hold the highest initial marketable value for more general deployment of the technologies, and can therefore be expected to be developed first.

3.6.2.3 The Hydrogen Economy, the Renewable Gas Economy

Many universities and government agencies researching the potential for a Hydrogen Economy have pinned their hopes to the opportunities to run fuel cell-drive vehicles – and no doubt there will be HFC vehicles (or HFCV), or fuel cell electric vehicles running on hydrogen fuel (FCEV) on the roads in future – perhaps fleets of larger freight vehicles with a widely distributed filling station network to avoid having to carry heavy on-board hydrogen gas containers. This option is possible for Europe, where fuel efficiency regulations could prohibit larger vehicles from running internal combustion engines (ICEs), especially in urban areas. But can hydrogen become the primary energy vector for future transportation needs? Although there have been major achievements in the advancement of hydrogen production, storage, distribution and use, it does not yet seem possible for pure hydrogen gas to capture much of the vehicle fuel market.

By contrast, there are more than ten million compressed methane-rich gas vehicles already in use around the world (Nijboer, 2010), and many models of cars running on diesel fuel can be converted to run mostly on gas instead, and so the market niche for Renewable Methane could be larger than some think is possible. Without replacing all the diesel engine vehicles, it will be possible to perform engine and on-board fuel storage conversion and use alternative methane-rich fuels. Renewable Methane could supply some of this market directly.

If fuel cell vehicles become sought after, those that use Renewable Methane, or mixtures of gas fuels, might be preferred, as this could turn out to be cheaper than Natural Gas. These vehicles would most likely perform on-board reforming of the methane-rich fuel to hydrogen.

In rural areas of the world, especially in regions without electricity or gas grids, Renewable Methane made from agricultural wastes would be optimal for providing local automotive power. It makes sense to begin a transition to fully sustainable gas mobility by continuing with what already exists – the Natural Gas network – as we already have the infrastructure and plant designed to work with it, and existing vehicles can often be adapted.

3.6.2.4 Transition strategies in transport

With every option for a transition in transport, there remains the problem of the large numbers of ICE vehicles already in use. In order to turn over the vehicle fleet to new transport fuels, and to improve fuel use efficiency, to prevent a crisis in mobility, it is necessary to market new vehicles using the alternative fuel choices, or adapt the present ones. Besides partially addressing urban air quality problems, and related public health issues, converting to CNG could also partly address policy to meet carbon emission budgets; and for personal transport vehicles, a conversion from diesel to CNG could be more economically efficient than purchasing an entirely new car. As regards the use of materials, a CNG storage tank fitted to an ICE car could require much less outlay on metal and other materials than a hydrogen storage tank in an HFCV.

In the short term, keeping the cost of transition in transportation down is going to be useful, both because of the risk of continued poor health of the economy, and because cheaper changes could well promote greater, speedier uptake. For example, it could be far faster to arrange to convert a car to CNG than it would be to wait for your order of an entirely new electric-drive vehicle (battery electric vehicle or BEV) to come off the production line (e.g. Hurst and Gartner, 2013). It is possible that conversions to Natural Gas become widespread, particularly as cars can be refuelled at home or in a local residential area with the appropriate compressor pumps. If this trend happens in parallel with higher rates of home insulation, then demand and supply of gas for the residential sector could remain more or less in balance and ensure economic stability. The conversion of vehicles to use gas instead of oil could start to address carbon dioxide emissions from the sector, and this could be enhanced if the gas supply begins to be supplemented with Renewable Methane from biological or industrial sources.

One note of caution is that with high levels of conversion of vehicles to methane-rich gas fuels, it may be necessary to abandon plans for significant levels of hydrogen gas injection into gas grids, owing to issues to do with gas compression ratios, compression equipment and energy density. However, it could be that other conversion options are developed, and vehicles could be fuelled by a range of hydrogen–methane blends, such as Hythane, or partly methanated syngas (known as Synthane), which would permit gas grids to contain mixes of gas, as appropriate. If fuel cell vehicles are developed, they could be able to use a variety of input fuels, and so LMG – Liquefied Mixed Gas – could be distributed to filling stations via Heavy Goods Vehicle tanker truck, or piped as RMG – Renewable Mixed Gas – to be compressed at the filling station. Even if most fuel cell vehicles will only take hydrogen fuel, filling stations could provide it from the high-methane, decarbonised gas in the gas grid – by using internally reforming solid oxide fuel cells to reform it to hydrogen. This is likely to be more efficient than running a fuel reformer in the vehicle itself, which would add extra weight.

3.6.3 Electricity generation

It is in the area of power production that Renewable Gas could have the largest impact on operations – although the total energy provided would likely be greater in the heating or transport sectors. Most power plants in large widespread grids are themselves large, and operations therefore centralised, effectively isolated. As a result, much of the energy of thermal power plants is lost, as heat via cooling chimney stacks, or heat bled to rivers or lakes, as there is no local application for it. Considerations of gas quality and purity for use in these power stations are not as important as they are for grid injection and supply to household end users. A wide range of grades and chemistry of Renewable Gas could therefore be put to use in power generation in centralised power plants, as long as gas storage, gas engines and gas turbines are flexible and adaptable. The waste heat could be used to recycle carbon dioxide to produce new Renewable Gas. Power generation through vanilla combustion could become seen as overly inefficient, and Renewable Gas grades could instead be used in fuel cells, which could support renewable electricity generation at a range of scales. Supplying remote resources of Renewable Gas to centralised power plants is possible if the gas is uprated to a high quality and then injected into the gas grid. For independent power applications, Renewable Gas can be supplied in the same range of ways as Natural Gas is now, including a Renewable Gas-derived replacement for propane/butane LPG cylinders.

The development of Renewable Gas may even enable reliability in the contribution of atomic energy to electricity generation. There are an ever-growing number of nuclear power plants that are over twenty-five years old. This cohort ageing of the nuclear fleet is a risk to power grids, as older reactors and older nuclear power plants are more prone to unplanned outage and other kinds of failure, and so their reliability is thrown into question as providers of "baseload power", the more they age. It can be an option to require certain ageing nuclear power plants to disconnect entirely from the electricity generation grid to ensure supply security. The nuclear reactors would continue to produce heat until they were decommissioned. They could therefore be used to produce Renewable Hydrogen through splitting water – they are usually sited near large bodies of water, to use for cooling water and as heat sinks. Through heat-balanced processing, this Renewable Hydrogen could be uprated to Renewable Methane, recycling locally available carbon oxide gases, should they be available. This Renewable Methane or Renewable Hydrogen could be stored on site (or grid injected if large scale storage of explosive gas near a nuclear reactor were deemed unsafe) and used as required in gas engines or gas turbines to generate electricity, perhaps even re-using the original generation sets of the nuclear power plant – if these are not the most compromised elements. The carbon dioxide flue gas from the combustion of Renewable Methane fuel for power generation could be recycled at these nuclear power plants, as they are centralised and potentially gas-secure. If the widely heralded nuclear power renaissance does take place, rather than being used to provide baseload for power grids, new atomic reactors could perhaps be more usefully deployed for constant hydrogen production, to be either stored for use when the wind dies down, or used for carbon recycling.

3.6.4 Industrial chemistry

There may be several main reasons why Renewable Gas should become more favoured as the basic feedstock in industrial chemical processes as time passes. For example, there may start to be competition between the use of Natural Gas for the production of agrichemicals/agrochemicals and its use for transport. This is where partial or complete methanation of a mix of Renewable Gas streams could provide gas feedstock with a high percentage of methane, vital to the production of fertilisers and pesticides. Bioammonia and other chemicals could also be made on-site at a biomass gasification plant, utilising the biosyngas directly.

3.7 Can we stop digging yet? Technological advance in Renewable Gas

Renewable Gas could contribute to a peak in drilling and mining for energy, and offset any peaks in production capacity of crude petroleum oil and Natural Gas. A greater use of surface-produced, climate-preserving, gas fuels would prevent further destruction of the Earth's crust by hydraulic fracturing, and opencast coal, oil sands and uranium mining.

Making more use of Renewable Gas could encourage efficiency in the use of sustainable fuels. For example, it is likely to be more clean and efficient to gasify straw and wood and combust the biosyngas locally for decentralised power generation, than it is to transport straw and wood to a centralised location for co-firing at a mainly coal-fired carbon-abated power station.

Pathways to energy transition that include Renewable Gas are perhaps a better fit to needs than other energy futures scenarios, purely because of the simplicity of the pathway to deployment. In a new age of financial uncertainty and potential fossil fuel resource scarcity because of strong competition or depletion, pragmatic, low barrier threshold solutions to energy security will always be welcome. A number of the Renewable Gas proposals have prospects for parity life cycle costs with fossil fuels, without excessive capital costs.

A transition to Renewable Methane will be straightforward, but this does not disbar a later transition to the Hydrogen Economy, because the production of Renewable Hydrogen is so important in the development of Renewable Methane.

3.7.1 Renewable Gas is "cuspish" – on the verge of major industrialisation efforts

Basic research is being conducted in every sector of the Renewable Gas matrix, and yielding surprising and encouraging results. Some organisations and companies are not even waiting for these improvements, but are actuating Renewable Gas programmes using current technologies.

Renewable Gas projects (Power to Gas, WindGas, etc.) utilising wind power or other spare electricity are already in development in a number of countries (e.g. dena, 2014a; Electrochaea, 2014). When wind power is high and demand is low, the excess wind power that would otherwise be "curtailed", "constrained" or "shed" is used to make Renewable Gas of one kind or another, and this is then stored. When wind power is low and demand is high, the Renewable Gas is used to generate power to cover the shortage.

3.7.2 Vanguards of research

Renewable Hydrogen production is vital for the development of Renewable Gas. In the US, the federal government's Department of Energy has been running a hydrogen programme for some time (DOE, 2013). In the EU, the Fuel Cells and Hydrogen (FCH) partnership is one amongst a number of initiatives, and the International Energy Agency (IEA) is involved in a number of projects, through the Hydrogen Implementing Agreement (HIA) (IEA, n.d.-b, IEA HIA, n.d.-a), and through a number of other sector tasks, including Bioenergy Tasks 33, 37 and 42 for Biorefinery (IEA, 2014d). A number of universities and companies are directly involved in these working groups, and some have dedicated Bioenergy and/or Hydrogen projects in train. Commercial energy groups are building pilots based on their own research work and their collaboration with academic projects and the international organisations.

3.7.3 The forefronts of knowledge

The critical areas of research, where significant progress is necessary, fall broadly into the following categories.

3.7.3.1 *Physics and chemistry*

Research is ongoing into the improvement of water electrolysis to produce hydrogen, the improvement of hydrogen production through catalysis, the improvement of methanation engineering and the development of solar-hydrogen processes.

Although hydrogen gas can be produced from the electrolysis of water using excess renewable electricity from wind or solar farms, the less power that needs to be used, the better. Catalysts to enable hydrogen "evolution" have been in use in industrial-scale chemistry for decades – and it seems that "Nature's little helpers" do not need to be expensive, rare metals like platinum. In recent years, there have been remarkable developments in the use of special compounds for new catalysts, and tests on nano-structured surfaces on catalysts. Some of the processes to use chemical catalysts to improve "protonation" of hydrogen from water or weak acids make use of sunlight (photocatalysis), and some electrical voltage (electrocatalysis). Most of these improved processes are likely to be possible at around room temperature, and require much less energy input than currently. Some teams are looking at heat-triggered hydrogen production (thermocatalysis).

There is ongoing research, both in academic institutions and in government-sponsored facilities, to improve the selectivity of methanation and other gas improvement processes – with the express purpose of making the most of Renewable Gas fuels. Methane is an ideal replacement for Natural Gas, and so the more methane there is in a supply of Renewable Gas, the more valuable it will be.

3.7.3.2 *Biology and chemistry*

Development of novel inocula for AD and fermentation of biomass to selectively produce gases are being pursued by a number of academic and commercial interests, as are waste water treatment with gas capture, and increasing production volumes from AD reactors by the addition of Renewable Hydrogen (e.g. Reuter, 2013).

3.7.3.3 *Engineering*

Engineering firms are working to provide flexible gas engines and gas turbines; power generation systems that can make use of a variety of gas feedstocks; increases in gas processing efficiency; and successfully scaling up fuel cells that can run on syngas and methane, as well as hydrogen.

Gas fuels of the future are likely to be increasingly decarbonised, and be somewhat less consistent in energy value than at present, whether this is because of additions of Renewable Gas or because of the use of more chemically diverse Natural Gas. These changes will require adaptation. For this reason, all new gas-burning power generation and power-and-heat cogeneration plants (CHP) should be using ultra-efficient flexible gas turbines and gas engines. Engineering firms are working to provide solutions, as required, for example, gas turbines that can use fuel high in hydrogen.

3.7.3.4 Energy storage

Administrations and engineering firms are considering how to develop specialised gas storage and gas separation for a range of Renewable Gas feedstocks, and increasing the amount of gas storage. Interestingly, besides options to store hydrogen (SBC, 2013), methane or carbon dioxide, there are also proposals to store energy underground as compressed air. This suggests that Renewable Gas mixtures, even with a low heating value, could be useful if stored in a compressed form, as power generation capacity from gas turbines relies not only on combustion ability but also on the pressure of the gas feedstock.

3.7.3.5 Uprating and upgrading

Further work is required to establish methods of improving Renewable Gas mixes to be used in specific applications, including setting standards for chemical processes.

3.7.3.6 Gas emission prevention

Continuing work needs to be undertaken to rein in fugitive emissions of gas, particularly methane, from production, supply and storage systems – both fossil gas and Renewable Gas (Liebetrau et al., 2013). Capturing associated gas from oil and condensate wells, and currently uneconomic gas from tight gas and shale gas operations, and the renewal of ageing gas distribution systems are some components of this.

3.7.4 Industrial gas systems design

Encouraging the participation of the current petrochemical, chemical and manufacturing industries in the development of Renewable Gas is key to future progress, as they are the producers of most "waste" gas (e.g. Bermudez et al., 2013), and the players of sufficient engineering know-how to design gas capture, gas recycling and gas storage options to enable the integration of all the elements, besides being best-placed to develop transport fuel options.

3.8 Advances and the way ahead

3.8.1 Market development – volumising

In order for Renewable Gas to be useful in meeting carbon emissions control targets, it needs to be produced in high volumes, to justify displacing fossil gas. It will therefore be important to capitalise on all possible resources. The gasification of unusable waste is becoming more widely deployed in Europe, for example, much Energy from Waste (EfW) or Waste-to-Energy (WtE) in future will be through gasification methods, as an alternative to landfill or incineration, and while the level of absolute waste is set to decrease as product packaging is minimised through legislation, to become more recyclable and compostable, and economic efficiency will reduce the amount of raw materials in product life cycles, there will continue to be a waste water, municipal and business waste stream to employ. Gasification is a useful technique for safely disposing of medical and food waste, as the majority of the solid end products can be safely released to the environment.

The UK Government's National Renewable Energy Action Plan (NREAP) for its part in meeting the EU renewable energy targets was unambitious regarding biogas and biosyngas potential (CCC, 2011; NREAP UK, 2010). Other researchers and industry actors place much higher figures on the potential for Renewable Gas production streams (e.g. E4tech, 2010; Enviros, 2008; National Grid, 2009; NERA, 2009a, 2009b, 2010, 2011; REA, 2009; SKM Enviros, 2011). Gas collection from landfill and sewage sources could be expanded, for example. Importantly, at the time, the UK Government did not appear to consider the potential from Renewable Hydrogen to add volume to sustainable gas fuel production, nor consider what a high availability of Renewable Hydrogen could mean for uprating and upgrading other Renewable Gas resources. The recycling of carbon-rich flue or exhaust gases from the combustion of fuels was not considered in the UK NREAP probably because the implications and opportunities of large volumes of Renewable Hydrogen production were not considered. However, modelling since then has scoped the options (DECC, 2014e, 2015).

Renewable Gas is likely to be a key low carbon energy technology for those countries that choose to generate renewable power. Although Britain has been slow to advance this option, other European countries are actively working on using spare renewable electricity capacity to produce Renewable Hydrogen, mostly by electrolysis of water, done when the wind is blowing strongly, yet there is small demand for electrical power in the grid. This "wind gas" or "Power to Gas" potential is likely to become increasingly important, both in providing more renewable energy, and in balancing wind and solar power outputs.

The example of Germany is a case in point. At its distance from the Equator, there is a large seasonality in solar power output, and wind power in the Northern Hemisphere at this latitude varies seasonally, but also according to the frequent passage of low pressure weather systems. The daily and weekly patterns of electricity demand, by contrast, do not alter significantly between seasons, whereas the gas grid operators have to contend with widely varying demand. Germany is exporting power over and above its own needs (Fraunhofer, 2014), but it can be anticipated that if renewable electricity generation capacity continues to be installed at anything like the same rates as today, depending on the season, there would be considerable excess generation on some days. If the countries around Germany also develop their renewable resources, there could be excess generation in neighbouring countries at the same time, and power exports will not be possible. Excess generation may need to be curtailed, or constrained, from feeding into the power grid at some times. Yet, at other times, perhaps even the following day or week, renewable generation could experience a low and be unable to meet demand. It therefore makes sense to capture excess electricity in some form of storage to use later on when the sun is not shining or the wind is not blowing. The German Energy Agency, dena, has been managing what is being called the Power to Gas (PtG) Strategy Platform (dena, n.d.), to encourage the use of manufactured gas as a means to store energy from excess power until it is needed back in the electricity network.

3.8.2 A summary of the future state of gas

Natural Gas is not a uniform product, and there are questions of quality and composition in future. Resource quality is degrading, as well as being anticipated to be increasingly declining. Peak Natural Gas may not appear for some time, but Peak Sweet Natural Gas may be imminent, and major gas processing plants already need to consider chemistry and transition.

The long-term viability of the gas grid has been questioned. Should we try to retain the use of a majority methane grid, there will be coordination issues over time, because of chemical composition tolerances in pipeline networks of different types and pressures, and because of the potential need to change all the end appliances if the gas characteristics change significantly. As for the backbone of gas supply to power plants, combined cycle gas turbines (CCGT) are particularly susceptible to some contaminants, and are not fully flexible to changing gas composition – although turbines can be designed to operate within a wide range of tolerance windows of gas fuel composition and characteristics – as long as the energy value of the fuel is relatively consistent.

It seems unlikely that the gas grid would be wound down as a policy preference, but if residential connections to the gas grid are eventually abandoned, because gas flows are too low to make economic sense, there may still need to be a national level gas grid backbone. However, instead of Natural Gas, it could be transporting a mix of hydrogen and carbon oxides from a range of sources to places of methanation, and then used in centralised power plants, and in an infrastructure-type gas distribution grid for transport refuelling, similar to the current petrol and diesel station network. New decentralised, more widely distributed power plants could also be built with a Renewable Gas supply, using gas engines or fuel cells for cost-efficient and energy-efficient delivery of local heat and power (CHP) in DH networks. These plants, as well as the production of the fuel they use, could both be useful in helping to load balance power grids transmitting large quantities of renewable electricity supply.

Gas continues to be a good option with which to store large amounts of energy for long periods; however, for the scale of applications envisaged, much more gas storage will become necessary. This is because if a large proportion of gas no longer comes continuously by pipeline or tanker from afar, and daily manufacture of Renewable Gas is low to medium in volume, then to meet surges in demand will require much higher storage capabilities than at present. Options include underground subsea storage in depleted Natural Gas caverns. The other reasons for more gas storage include the wish to decarbonise all aspects of the use of gas, including the recycling of flue/off gas, and because some Renewable Gas feedstocks will have a range of chemical composition, requiring processing. If chemical processing capacity is not available at any one time, then storage of the intermediate gas products will be necessary.

It is unlikely that industrialised economies will give up using gas altogether – it continues to be a good energy fuel, and a vital feedstock for other industrial chemistry, such as the production of agricultural chemicals and fertilisers. Although in the interim, gas is vital for heating purposes, its most important roles in the long-term future are likely to be as fuel for transport and supporting variable renewable electricity in lowering the carbon emissions of the whole energy system. For this potential to be realised, gas needs to be decarbonised, and so Renewable Gas, both Renewable Hydrogen and Renewable Methane, will be essential.

4
A Brief History of Gas

4.1 Eternal flames

Burning vapour flaring out of the ground has been a cause for marvel or even veneration for centuries (Aminzadeh et al., 2001; Berge, 2011, 2013; Dumas, 1859; Etiope, 2010; Etiope et al., 2013a, 2013b; Forbes, 1939, 1958; Galetti, 2005; Hosgormez, 2007; Hosgormez et al., 2008; Ingersoll, 1996; Jackson, 1911; Kamali and Rezaee, 2012; Ker Porter, 1822; Le Strange, 1905; Lockhart, 1939; Marvin, 1884; Pliny, c. 77; Spulber, 2010; Strabo, 7 BC; Thwaite, 1889; Verma et al., 2004; Waples, 2012, Pages 7–9; Yergin, 1991; Ziegler, 1920); however, the widespread use of Natural Gas as an energy fuel only came about after gas manufactured from coal and other feedstocks had become commonplace (Tarr, 2009).

Enterprise scale use of Natural Gas was first developed by the Chinese people of the Han Dynasty, around about 2,000 years ago. Gas vented from wells drilled for subterranean brine, and was collected and burned to evaporate out the salt. Gas was even distributed by bamboo pipeline in possibly the world's first gas pipeline network (Loewe, 1968; Temple, 1986). The Chinese salt industry used advanced drilling equipment for the time, including the spring pole technique; a method also used by the first salt miners in the US (Bowman, 1911; Waples, 2012, Pages 9–13), who used the co-produced Natural Gas in the same way as the Han.

4.2 Coal Gas and Town Gas

Coal mining increased rapidly as part of the Industrial Revolution in Britain. It was commonly known that gas often emerged in large volumes from coal mines, and experimental chemists showed that strongly

Table 4.1 Typical chemistry of various manufactured, natural and biologically sourced gases (% percentage) (see also: Biogas Renewable Energy, 2009; Larminie and Dicks, 2003, Tables 8.4, 8.5)

Gas/species	Methane (CH_4) (%)	Hydrogen (H_2) (%)	Carbon monoxide (CO) (%)	Carbon dioxide (CO_2) (%)	Nitrogen (N_2) (%)	Reference
Early industrial						
Coke oven gas (from coke ovens)	30	50	7	3	7	Thermopedia (2011)
	22	63	6	2	4	ICAE (2012)
Blast furnace gas	0	3	20	24	53	ICAE (2012)
	0	3.9	23.9	26.0	45.9	Thermo Fisher (n.d.)
Water Gas (from coke) (Blue Water Gas)	0.8	48.6	39.2	5.3	6.7	Thomas (2010a, Table 3)
Producer Gas (made from coke)	1	6	25	5	63	Thomas (2010b, Table 1)
Local gas works						
Town Gas (from coal) (Coal Gas)	34	48	5	13	–	Watkins (2011)
	27.5	47	7.75	3.5	10.5	Thomas (2010a, Table 2)
	21	51	15	–	10 ($CO_2 + N_2$)	Rhodes (2009)
Town Gas (made from petroleum refinery naphtha and Natural Gas)	28.2–30.7	46.3–51.8	1.0–3.1	16.3–19.9	0–3.3 ($N_2 + O_2$)	HKCG (2013)
Petroleum refinery						
Refinery Offgas (such as offgas from Fluid Catalytic Cracking (FCC))	30–50	5–35	0.1–0.5	0.1–0.5	3–10	OGJ (2008)
	30–55	10–50	0.1–1	–	2–10	Linde (2012)
Natural Gas						
North Sea	87	0	0	1.2	0.3	Persson et al. (2006)
Groningen (Dutch Natural Gas)	81	–	0	2	14	Persson et al. (2006)
Biological Gases						
Biogas (from anaerobic digestion (AD))	60–75	–	–	19–33	0–1	Naskeo (2009)
	63	0	0	47	0.2	Persson et al. (2006)
Landfill Gas	45	0–3	0	40	15	Persson et al. (2006)

heated coal would give off inflammable gas. Destructive distillation, or carbonisation, of coal – high temperature heating in the absence of air – gave two valuable energy fuels: Coal Gas and coal coke. Coal Gas, or illumination gas, was the primary manufactured gas used for early street lighting, factory lighting and, later, lighting in the home. The by-product, coal coke, as a smokeless fuel, was used in residential environments for heating rooms. Coal coke was also in great demand in industry, to increase the production quantities of iron, and later steel, from blast furnaces. Several different technologies were used to make gas from coal (see Table 4.1), and Town Gas in the UK was a mixture of 70% Coal Gas and 30% Water Gas (Thomas, 2010a) – the Water Gas made by steam gasification of coal coke. With the popular adoption of public and domestic gas lighting in Britain, followed by the promotion of gas fuels for cooking and heating purposes (after electricity began to corner the market for lighting), dedicated localised gas works became prevalent to feed supply, as did large gas holders or gasometers for storing the local Town Gas. It would not have made much sense early on to consider a national gas distribution grid network, as it was more practical to transport the coal as a solid fuel.

4.3 Manufactured gas and the conversion to Natural Gas

In the US, manufactured gas and Natural Gas were developed in parallel. Natural Gas supply was restricted to areas close to gas production wells before the development of an extensive network of pipelines, whilst manufactured gas could be made close to centres of population. However, as the American gas industry took to laying advanced iron pipework, Natural Gas could be delivered over hundreds of kilometres. Early gas fields soon depleted, but a combination of large volumes of new Natural Gas from the Gulf of Mexico, pipelines, compressor stations using gas engines and underground temporary storage of gas in disused wells gave Natural Gas the ability to respond to changing demand at distance. There was a period where manufactured gas and Natural Gas were mixed in supplies before the complete conversion to Natural Gas (Tarr, 2009; Waples, 2005). Natural Gas has a higher heating value than most of the gas manufactured at that time in the US, such as CWG, Carburetted Water Gas (Water Gas with liquid hydrocarbons vapour-sprayed into it), so as gas networks converted to low cost Natural Gas, all consumer equipment needed adapters or replacement for gas burners on end appliances.

Developments in the British gas industry between the 1950s and the 1970s saw naphtha or light distillates, by-products of petroleum oil refining, and also LNG (Liquefied Natural Gas), brought in by ship, being used to thermochemically manufacture Town Gas to meet market demand. Coal production in the UK had been declining since the turn of the century, and the right grades of coal for gas production were becoming scarce. There was concern about energy conversion efficiency, and a consequent recommendation to research the total gasification of coal – which produces syngas – rather than the less complete carbonisation, or "destructive distillation", of coal used to produce Coal Gas (Electricity Council, 1982; Ministry of Power, 1952). Town Gas manufactured from coal was in danger of becoming uncompetitive with electricity, so instead of making gas from coal, there was a transition to making gas from butane, and naphtha – a petrorefinery fraction that lacked a wide market. In the UK in the late 1960s, it became necessary to make a decision about whether to continue to reform hydrocarbons, which included Natural Gas, into Town Gas, or convert to piping unreformed Natural Gas in the gas grid, large volumes of which were anticipated from production from offshore sub-sea fields. Natural Gas became the grid-fed fuel of choice, largely owing to the anticipated scale of discoveries and the differential in price. North Sea drilling for Natural Gas, following the lead of a significant discovery in the Netherlands, showed that there could be a very large volume of gas to be developed, with minimal processing or refining required. Owing to the success of research, it was considered that, should North Sea supply not meet demand, Synthetic Natural Gas or SNG – a substitute for Natural Gas – or gases equivalent in heating value and other chemical characteristics, could be manufactured. This reasoning, and the size of the Dutch gas discovery, increased confidence in a transition, and answered the concern that very high levels of Natural Gas would need to be available to supply the heating for dwellings. In the days of Town Gas made from coal, rooms had been warmed by coke fires, coke being a surplus by-product of the traditional Coal Gas manufacturing process. With the transition to making gas from naphtha, there was a decline in the availability of coke for heating, and so before the wide deployment of Gas Central Heating, the only other modern option was an electric fire in an individual room, which was relatively expensive. However, the transition to Natural Gas heralded cheaper heating for whole homes. Once the national Natural Gas transmission infrastructure was built, bringing the gas to market would turn out to be inexpensive, compared to the cost of coal fuel to manufacture Town Gas. In addition, North Sea gas would provide high levels of tax revenue, important for social and economic development.

In the UK, the transition to Natural Gas was a step change, rather than the patchwork conversion in the US. The difference in the chemical composition and combustion properties between British Town Gas and Natural Gas necessitated a mass transition in end-use appliances and equipment, and a co-ordinated programme of works, but this was considered a one-off investment in cheaper grid gas for the longer term. The conversion of the gas grid, town by town, sector by sector, began only a few years after the first major discovery of Natural Gas in the British North Sea, and ended around about a decade later. Because of the layout of the gas trunk mains, it was necessary to use a substitute for Natural Gas in the south west of England during the changeover, which was a mixture of air and propane (Williams, 1981).

4.4 Research into Natural Gas substitutes

Research into alternative gas manufacture technologies was conducted on both sides of the Atlantic, starting in the 1940s, including methods of enriching lean gas and producing methane-rich substitutes for Natural Gas, commonly known as SNG – Synthetic, Synthesised, Substitute, Simulated or Supplemental Natural Gas – today often referred to, where purified, as synthetic methane.

It was recognised in official British policy that making the best use of Natural Gas required high levels of production, which would inevitably advance the depletion of North Sea gas fields (Ministry of Power, 1967). However, it was considered that this would create an incentive for further exploration and discovery. Serious consideration was given to the long-term sustainability of Natural Gas production and the credible possibility of exhaustion of the whole North Sea province (e.g. Watt Committee, 1982); however, most of the research into synthetic gases (for example, BP, 1972) was scaled back in the late 1970s and early 1980s. Likewise, in the US, where research into gas-to-liquid synthesised fuels began earlier (Schlesinger et al., 1965), the development of alternative gas stalled, owing to the relatively high costs of early liquid synthesised fuels, and the closure of the Synthetic Fuels Agency, and the curtailing of its research budget; however, industrially manufactured gas still had a role to play.

4.5 The parallel lives of syngas during the 20th Century

Research work on the complete gasification of coal, essentially combining other methods of manufacturing gas, and minimising the by-products, such as gas coke, had previously led to a number of developments in the early 20th Century. The making of synthesis gas, or syngas, a mixture of roughly equal parts of hydrogen and carbon monoxide, was accomplished using several key technologies, one of the most significant being the Lurgi gasifier, first developed in Germany in the 1930s. Syngas could be used as a feedstock for the synthesis of gaseous and liquid energy fuels, and also utilised in the production of a wide range of chemicals. Syngas produced in a medium of air was high in nitrogen, and became important in the synthesis of ammonia for agricultural fertiliser. Total gasification of coal required higher reactor chamber temperatures than the earlier carbonisation of coal, which led to less contamination in the resulting gas, but also reduced the amount of methane produced, and increased the amount of carbon dioxide. Methane has a higher heating value than hydrogen by volume – although it is the reverse by weight – and carbon dioxide and nitrogen are non-flammable – consequently, syngas was a lower quality energy fuel. The Lurgi technology was subsequently modified to become one of a range of technologies from several engineering firms to produce a high-methane syngas, which could be processed into SNG, which could directly substitute for Natural Gas in electricity generation plant and in the gas grid pipeline network. Uprating syngas to closely resemble Natural Gas is a process with several stages. Essential chemistry includes the Water Gas Shift Reaction (WGSR), or "shift", reacting the carbon monoxide with steam (Abbess, 2014, Table 9; Young, 2010, Page 167) to raise the ratio of hydrogen to carbon, producing carbon dioxide as a by-product; and secondly, methanation: either through reactions that convert carbon monoxide and hydrogen to methane, or the Sabatier reaction to convert carbon dioxide and hydrogen into methane (Abbess, 2014, Table 9). Some of the original coal feedstocks for syngas production were eventually partly supplanted by Natural Gas: for example, much of the hydrogen that is needed in petroleum refinery comes from the steam reforming of Natural Gas.

4.6 SNG and synthetic liquid fuels from coal and biomass

In the 1940s and 1950s, the production of crude petroleum oil in the US was predicted to peak during the 1970s, and research was conducted into synthetic fuels. Geopolitical instability in the 1970s increased risks for energy supply security, with possible disruptions arising from international trade conditions or geological depletion. Perhaps producer nations could experience stagnation in production capability, perhaps due to infrastructure problems, and act to withhold exports, or find new markets for exports which would create competition for supplies. Two energy crises, "oil shocks", reinforced motivation for renewed research into manufactured and synthetic gas and oil. For the US, most research was into the gasification and liquefaction of coal, of which they still held significant reserves. However, in the 1980s, as concerns about climate change began to come into focus, research into the use of low carbon dioxide emissions technologies emerged, such as the gasification of biomass to syngas. The work with both coal and biomass was essentially a continuation of research and development earlier in the 20th Century, for example, the work done by Fischer and Tropsch, and Fischer and Pichler, on manufacturing liquid fuels from coal, conducted in the 1920s and 1930s. Of note, the use of wood biomass to make gas fuels had an even earlier origin, for example, in Lebon's "Thermolamp" of 1801 (Schivelbusch, 1998; Waples, 2005, Chapter 2), and had a renaissance in 1910 (FAO, 1986, Section 2.5), and continued through the Second World War (e.g. Myllyntaus, 2010).

4.7 Carbon Capture and Storage (CCS) and Enhanced Oil Recovery (EOR)

Although the conversion of syngas to liquid fuels became less essential for research in the 1980s as petroleum markets became more flexible, work continued on environmentally friendly SNG – manufactured gas, enriched by means of chemical process control to have a high level of methane. For SNG derived from coal to qualify as low in net carbon emissions to the atmosphere, the carbon dioxide arising from its combustion would need to be sequestered, ideally underground where it could be permanently contained – methods for this are referred to as Carbon Capture and Sequestration or Carbon Capture and Storage, or CCS.

Several SNG demonstration plants were built in the US, Germany and the UK. A consortia of American companies demonstrated SNG production from coal feedstock at British Gas facilities in Westfield, Scotland, in 1972 (Ghassemi et al., 1979; Higman and van der Burgt, 2003, Section 5.1 "British Gas/Lurgi (BGL) slagging gasifier"; Hutchison, 1987; Kopyscinski, 2010). The Great Plains Synfuels Plant at Beulah in North Dakota, US, using the Lurgi gasification system (NETL n.d.), which was commissioned in 1984 as a commercial operation (Dittus and Johnson, 2001; DOE, 2006), is still operating. The carbon dioxide waste stream that is produced is used for Enhanced Oil Recovery (EOR), by pumping it underground into two Canadian oil fields at Weyburn, to stimulate extra oil production. This project has been monitored for the eventual fate of the carbon dioxide, which shows good permanent sequestration. However, the volumes of oil in this tertiary recovery phase have not been as high as initially projected (Cenovus, 2014a). This throws into question the use of CCS through EOR in permitting the continued use of coal for combustion for electricity generation, as to justify sequestering the exhaust carbon dioxide underground, it would need to create a sustainable revenue stream to finance it.

4.8 SNG made from heavy oils and petrorefinery waste

As demand for diesel and gasoline fuel for transport rose in the 1970s and 1980s, and gave naphtha another market, research on manufactured gas continued by moving on to making gas from progressively heavier fuel oils, for example, the British Gas Fluidised Bed Hydrogenation (FBH) process (Borrill and Easterby, 1985). Despite the fact that most of this research into Natural Gas substitutes for supplying the gas grid was discontinued in the 1980s, the gasification of unwanted crude oil by-products continues to this day, at petrorefineries (e.g. Heaven, 1996), where the gas is frequently used as on-site fuel. At the high temperatures required for efficient gasification of heavy oils, there are a number of issues connected to carbon conversion, contaminants and the control of combustion, suggesting efficiency limits and thermodynamic limits in using heavier feedstock to make gas; however, oil and gas companies are still pursuing this (e.g. Sreedharan, 2012).

Despite much of the chemical industry converting to use syngas made by gasifying Natural Gas and light hydrocarbon fractions from crude oil refinery, much syngas is still produced from coal by gasification, such as by the use of the IGCC gasification process – Integrated Gasification Combined Cycle (Ekbom, 2007). The syngas is used either for power generation or as a chemical base.

4.9 Other gas manufactured in petrorefineries

Refinery Gas (or Refinery Fuel Gas, Refinery Off-gas, or Refinery Offgas (ROG)) is a by-product of several processing stages in petrorefinery (Abbess, 2014), for example, Fluid Catalytic Cracking (FCC), which outputs some hydrogen and methane. Delayed coking, which is often used to process the residual end products of crude oil distillation, besides producing gas oil, produces fuel gas, as well as a considerable quantity of petroleum coke (petcoke). Petcoke can be put through a similar gasification or hydrogasification process to coal to produce further gas.

4.10 Gas for power generation

Although in the US there were restrictions placed in 1978 on burning Natural Gas to generate power, eventually Natural Gas was used on both sides of the Atlantic Ocean for high levels of electricity production. In the UK, privatisation of the gas industry opened the door for the "Dash to Gas" in power generation. Gas is straightforward to store, whereas electricity is not; and so gas engines have been used frequently for providing on-site instant power throughout the history of manufactured gas (Williams, 1981), and now with Natural Gas, independent from the electricity grid network, and so uninterruptible and decentralised.

In recent years, the contribution of gas-fired electricity generation for balancing the demand for grid power has become more crucial, and is acknowledged to be set to become increasingly important as more variable renewable electricity capacity, such as wind power and solar power, is installed.

In the near future, coal-fired power generation needs to be curtailed, unless waste gases are permanently sequestered or recycled, owing to greenhouse gas emissions legislation. With roughly half the carbon dioxide emissions of coal burning, gas-fired power generation will therefore remain important to match supply with demand when other resources are at a low point.

In the longer term, the carbon dioxide emissions associated with the combustion of Natural Gas itself could become problematic for climate change legislators. A response to that concern is that there have been projects to implement more localised, renewably stocked gas production systems, making biologically derived gases – principally from processes related to the decomposition of biomass: anaerobic digestion (AD) and fermentation to create biogas, which can be high in biohydrogen and biomethane (see Table 4.1), although there is also industrial gasification of biomass being developed, and the biosyngas being "uprated" or upgraded to bioSNG. The carbon in these gases would come from recently grown plant organisms, as opposed to coal, so would not be drawing stocks out of the deep geological carbon cycle, and therefore not contributing to net greenhouse gas emissions to air.

4.11 The chemistry of manufactured gas

Essentially, gas from a wide variety of resources will contain significant levels of three flammable gases – methane, hydrogen and carbon monoxide, together with non-flammable carbon dioxide, and with significant levels of nitrogen if the gas has been made by thermal treatments in the presence of air (Abbess, 2014). Each manufactured gas resource will have different contaminants and a range of nitrogen, oxygen and water vapour content which will affect which applications it can be put to. Most gas used for major energy purposes will be processed, mixed, cleaned or upgraded in some way. Some of the processing will be to remove contaminants that would otherwise damage equipment or lower the energy value of the gas. Some of the processing will be to upgrade the gas, either to a higher energy content, or a purer composition.

The direct production of methane, through the decomposition of biomass, via the process of AD by microbial Archaea, has also been extensively studied and developed. Biological routes to high-methane gas may be less costly, but they are slower than industrial thermochemical biomass treatments, and have potentially more complex contamination.

The concept of the "Hydrogen Economy" was proposed several decades ago, and syngas was seen as a way forward in achieving that. However, several things militate against this. Although it is highly valued in industrial chemistry, syngas is highly toxic because of the high carbon monoxide content, and it has a low energy value in combustion, and a high flame speed from having such a high hydrogen content. In addition, hydrogen is composed of very small and light molecules which can easily permeate other materials. All of these phenomena could have costly implications for safe and reliable gas storage, gas distribution networks and gas combustion plant. For these reasons, much gas fuels research has pursued converting syngas into methane-rich SNG, despite the energy penalty of tailoring the reactor for this outcome, and sometimes having a second or more stages of conversion.

Various approaches have been taken to produce SNG in one stage, to increase the amount of methane arising from the processes that produce syngas. These include hydrogasification, the addition of hydrogen gas to the gasification reaction chamber. An alternative approach to a richer fuel is steam hydrogasification, which adds hydrogen gas and steam to the reactor, and produces a syngas far higher in hydrogen, rather than methane – continuing the long history of using steam in manufacturing gas, ongoing since at least 1864 (Howarth, 1864a, 1864b).

4.12 Biomass gasification and carbon recycling

Gasification of biomass, which is more reactive than coal (Higman and van der Burgt, 2003, Section 3.1.3 "Char Gasification"), is performed at lower temperatures (Higman and van der Burgt, 2003, Section 5.5 "Biomass Gasification"), and this tends to produce tars. The formation of tars can be prevented to a certain extent (Abbess, 2014), for example, by recycling carbon dioxide, the waste product from the combustion of many fuels, to the gasification reactor where it performs the function of a gasifying medium (Basu, 2013, Section 6.3.1.2 "Operating Conditions"). Recycling carbon dioxide into the gasifier has been used to improve the performance of IGCC plant. A novel finding is that using carbon dioxide as a gasifying agent in the steam gasification of biomass can increase the amount of syngas produced (Butterman and Castaldi, 2009), opening up the possibility of less biomass feedstock being required. Carbon dioxide can be used to make methane gas by reacting with hydrogen, as demonstrated by Sabatier (Sabatier and Senderens, 1902), eliminating the need for inputs of biomass or coal, if the carbon dioxide from combusting fuels for power generation can be recycled into the gasification unit. There would inevitably be some losses of carbon from such a carbon recycling system, either because of inefficiencies, or because the methane could be drawn off to be used for manufacturing chemicals, as per demand. However, it is now possible to postulate a power generation system where biomass or coal feedstock is only needed from time to time. All the carbon would be recycled repeatedly, and Renewable Hydrogen would be the only primary input.

4.13 A brief future of gas

Sixty years ago, most gas piped to consumers in the industrialised world was manufactured. In Britain, most of what was called Town Gas was manufactured from coal, but rising prices of good quality coal led to partial or total substitution of that feedstock with distilled petroleum oil fractions, such as naphtha. Separately, a market in Natural Gas was created, through the development of LNG, initially done as a way to capture the value of the gas that could be produced in remote locations when drilling for oil. Natural Gas in the form of LNG began to be used as a feedstock to "top up" the calorific value and chemical composition of Town Gas (Arapostathis et al., 2013); subsequently, enough Natural Gas was prospected in British territorial waters to suggest a transition away from Town Gas altogether. The transition of the gas pipeline network to carry Natural Gas answered several major concerns: projections of the continuing high costs of feedstocks used to manufacture gas compared to virtually cost-free Natural Gas, and how to make best use of Natural Gas resources.

To answer questions about the viability of the continued use of fossil fuels, it seems possible that civilisation will return to manufacturing a lot of gas in the future – it is just that this time around it will be low in net carbon dioxide and methane emissions to air. All the essential chemistry for this was done well before the 20th Century transitions to Natural Gas: the carbonisation or destructive distillation of coal, for example, has a history of several hundred years. Apart from studying options for manufacturing gas through the use of fuel cells or membrane reactors (e.g. Minh and Mogensen, 2013; Youngukkasem et al., 2013), there may be little new to learn, although new chemical engineering catalysts and processing equipment will need to be developed. Not that fuel cells themselves are novel: it must be recalled that Nernst developed electrolytic materials in the years leading up to 1900 (Barkan, 1999), and the first fuel cell credited is that of Grove in 1839 (the "gas battery" or "gas voltaic battery") (Perry and Fuller, 2002).

Despite the encouraging signs, there are several problems with the vision for manufactured gas: the transition to Natural Gas in the energy industry means that engineers of manufactured gas have gone into retirement, and technology has not generally been optimised. Also, there is a general belief that manufactured gas is consigned to history, since the world is awash in Natural Gas. Underlying this belief is an unquestioned assumption that Natural Gas will last forever, but that should not be a tenet of energy faith, considering the decline in production from major oil and gas reserves. We are perhaps already seeing the emergence of a plateau in crude oil production, and much Natural Gas is associated with petroleum deposits, or comes from sediments of the same eras as petroleum deposits, so a peak in Natural Gas production might also not be far off. The quality of remaining easy-to-access Natural Gas is sometimes rather poor, or has difficult chemistry, and needs treating. As standards make it necessary to process a large proportion of Natural Gas that will come out of the ground in future, in order to market it, the industry will be essentially reverting to a manufactured product, with a fundamental dependency on chemical process engineering. It appears there is a valid technical and policy debate to be had about reliance on Natural Gas, and how the world could draw on the history of manufactured gas as a contingency strategy. If there is a growing proportion of manufactured gas supplies in future, the reasoning for choices about resources, feedstocks and chemical processing plant is important to engage with now.

If concerns about the security of supply of Natural Gas became pressing, it would be possible to develop gasification and other fuel conversion engineering to meet supply needs, and more efficient systems to use gas could be engineered, such as those that use electrochemical fuel cells. Gasification plant would be industrial in scale, and centralised, whereas fuel cell systems could be of virtually any size.

Natural Gas is commonly associated with petroleum oil in fossil fuel fields, and where it was not explicitly drilled for, was originally seen as an unwanted by-product of crude oil production, frequently vented or flared. Now, however, it has turned from being a waste to being a premium product, in much demand. For future economic stability, it is important to protect the provision of gas, whether from fossil fuel deposits, or its substitutes, manufactured from sustainable resources.

5
Renewable Gas Systems

5.1 The evolution of gas energy systems

A global warming of the troposphere, the lower part of the atmosphere closest to the Earth's surface, has been causally attributed to humankind's burning of fossil fuels, cement production and other industrial manufacture, deforestation and other land use change activities. The thermal inertia of the world's oceans means that even were anthropogenic carbon dioxide emissions to be completely curtailed today, there is enough added heat in the Earth system as a whole, and heightened levels of greenhouse gases in the atmosphere and the oceans, to continue warming the troposphere and land surface of the planet, perhaps for hundreds of years (IPCC, 2013). The risks of the onset of damaging climate change resulting from this "commitment" to further global warming, in addition to that already experienced, are so severe that measures are being implemented worldwide to stem net greenhouse gas emissions to air. Action to curtail what is known as "Black Carbon" – particulate emissions from the burning of wood and fossil fuels, in the very near term, could help to curb short-term atmospheric warming, and provide a window of opportunity to address the more long-term threats from the most significant greenhouse gases: carbon dioxide, methane and nitrous oxide (N_2O) (Boucher and Reddy, 2008; Client Earth, 2012; Highwood and Kinnersley, 2006; JRC, 2014; Minjares et al., 2013). Reinforcing the recycling of wood and pulp industry products, by-products and wastes could assist with efforts to rein back deforestation, which would address up to roughly a fifth of the radiative forcing on the temperature of the Earth's surface caused by excess atmospheric greenhouse gas. Implementing methods to reduce fugitive methane emissions – from the energy industry, including dams – and agriculture and wetland management, is also vital. The scaling back of the use of fossil fuels, and the sequestration or re-utilisation of the carbon dioxide produced by their combustion, and addressing industrial and chemical processes will begin to answer the question of where something like three quarters to four-fifths of man-made net greenhouse gas emissions to air originate.

In the next few decades, it is vital to choose to stop burning coal for energy unless the resulting carbon dioxide emissions are abated. On a life cycle basis, for each megawatt hour (MWh) of electrical power generated from unabated coal, between 781 and 961 kg of carbon dioxide equivalent (CO_2e) greenhouse gas are produced as emissions, compared to only 341–429 kg CO_2e/MWh for unabated Natural Gas in combined cycle (Zhenggang Nie et al., 2013, Figure 5). Natural Gas is clearly far less of a problem in terms of greenhouse gas emissions than coal, but even if all coal-burning is displaced by Natural Gas combustion, the necessary degression in emissions dictated by a safe carbon budget mean that eventually, perhaps within a couple of decades, the waste carbon of combustion from Natural Gas will itself be too high to ensure a "safe" climate. The process to replace coal with Natural Gas will be in the stages of completion when the emissions from Natural Gas combustion itself will need to be addressed. There is clearly an urgent need to consider how to "decarbonise" the use of Natural Gas whilst at the same time deploying it to displace coal. Either the energy systems, or the gas itself, need to be altered so as to prevent net carbon emissions to air.

5.1.1 The existing energy system

The existing energy system in the economically developed countries relies more or less completely on inputs of fossil fuels, and exhausting the resulting carbon dioxide directly to the atmosphere (see Figure 5.1). To avoid the near certainty of dangerous climate change, this situation must be changed.

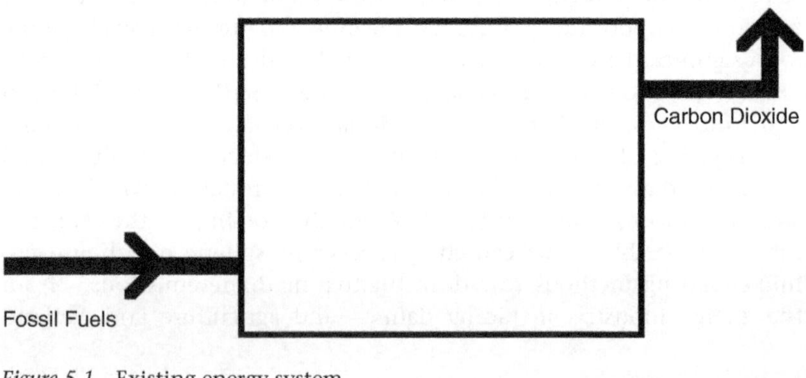

Figure 5.1 Existing energy system

5.1.2 Carbon Capture and Storage (CCS)

Much current dialogue in the energy sector centres on Carbon Capture and Storage (CCS), preventing carbon dioxide from being emitted to the atmosphere, and instead having it pumped deep underground. Most proposals would do this out at sea. Many of the CCS schemes would be unable to capture all of the carbon dioxide, even that produced at centralised fossil fuel combustion utilities, such as power plants. However, it could significantly reduce the burden of excess greenhouse gas emissions (see Figure 5.2). CCS may not prevent the energy industry mining organic carbon that has been held safely underground in geological deposits for millions of years, but it could avoid pumping some of it into the sky.

CCS would not deal with fugitive emissions of methane from the distribution infrastructure and mining operations for coal and fossil gas – principally Natural Gas, and shale gas in North America. However, these should be dealt with regardless of the management of carbon dioxide at the point of use.

Importantly, sequestering carbon dioxide underground in former Natural Gas reservoirs would enable the mining of this carbon dioxide at a later date, for use in making new fuel.

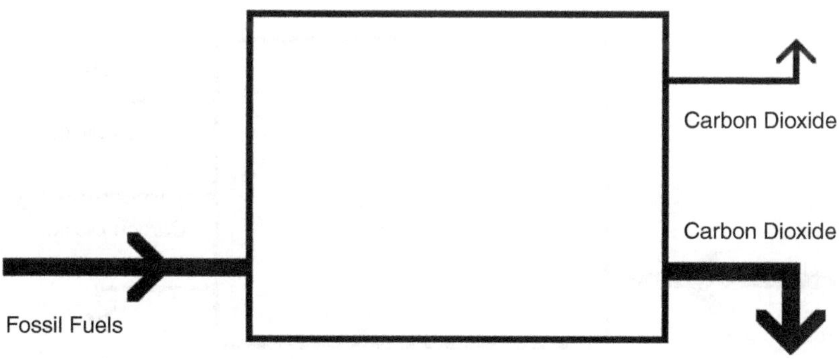

Figure 5.2 Carbon Capture and Storage (CCS)

5.1.3 Carbon Capture and Utilisation (CCU)

Although carbon dioxide is not yet widely used as a chemical or fuel feedstock, it can be envisaged that increasing uses for carbon dioxide will become economic to deploy – Carbon Capture and Utilisation or Re-utilisation (CCU, CCR or CCRU). Currently, most uses for carbon dioxide have the net result that it is eventually emitted to the atmosphere as waste gas after having been used for other purposes. Examples include: using carbon dioxide in the soft drinks industry, or for use in the industrial production of alternative fuels or chemicals. This re-utilisation of carbon dioxide does however potentially avert the use of some fossil fuels, and so could be said to contribute to reducing total carbon dioxide emissions. However, if this type of re-use for carbon dioxide is additional to current fossil fuel use, then it does not curb total greenhouse gas emissions.

Other kinds of uses for carbon dioxide could defer greenhouse gas emissions for a slightly longer period, for example, using carbon dioxide in the growing of food plants in greenhouses. However, when the food is consumed, it may end up contributing to methane emissions, and certainly contributes to carbon dioxide emissions, if human and animal waste is used for biogas and biomethane production.

Some carbon dioxide re-utilisation locks up the carbon dioxide permanently, such as making long-lasting polymers used in manufacture or construction, or making minerals or aggregates for construction or road-building – essentially re-burying the carbon.

If CCU is pursued alongside CCS, it could alleviate the cost burden of widescale CCS, and so lead to much less carbon dioxide being emitted to the atmosphere (see Figure 5.3).

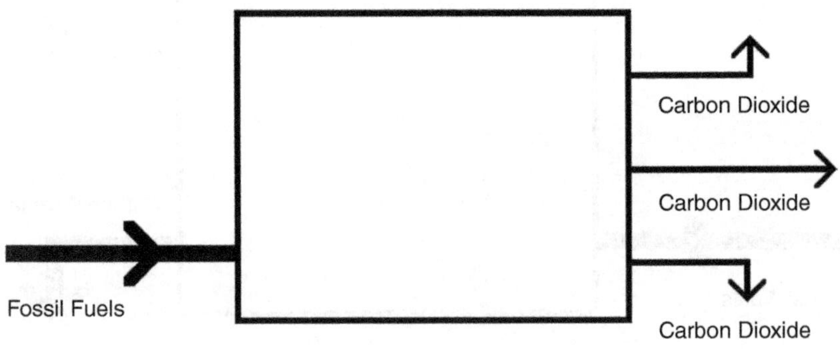

Figure 5.3 Carbon Capture and Utilisation (CCU)

5.1.4 Increased bioenergy

As an alternative to dealing with the "fossil" carbon emerging from the energy system – the carbon dioxide that results from the use of fossil fuels – some of the input fossil fuels could be replaced with bioenergy. This would mean that some of the carbon entering the energy system would be renewable (see Figure 5.4). If this is done widely, new carbon being added to the atmosphere would reduce in total compared to the current situation.

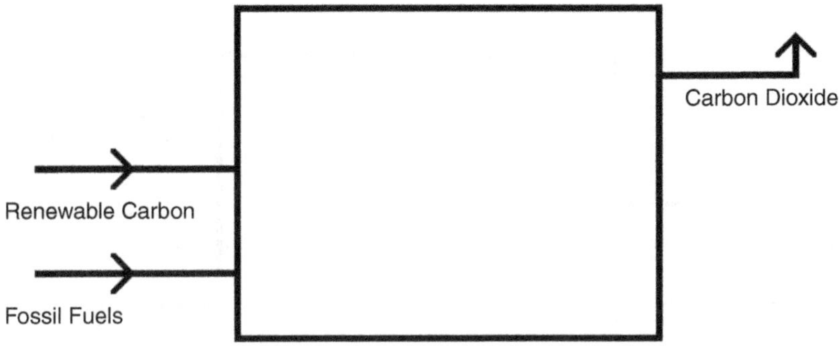

Figure 5.4 Increased bioenergy

5.1.5 Bioenergy with Carbon Capture and Storage (CCS)

If fossil carbon is displaced by Renewable Carbon from bioenergy on the input side of the energy system, and CCS is developed for the output side, then net carbon dioxide emissions to air would be curtailed even more effectively (see Figure 5.5). This concept is referred to as BECCS – Bioenergy with CCS – and theoretically, some implementations could make the total net carbon dioxide emissions to air less than zero.

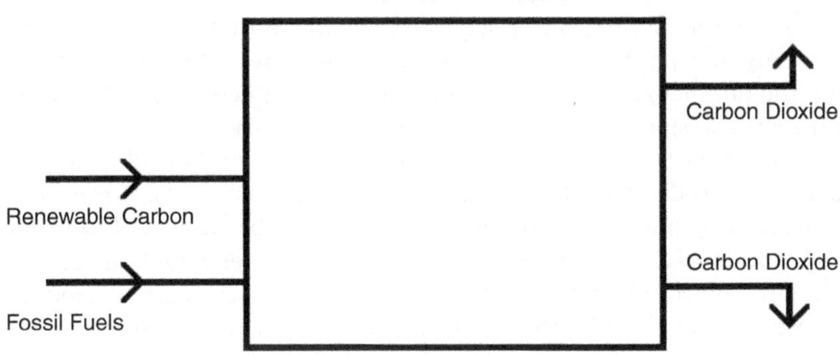

Figure 5.5 Bioenergy with Carbon Capture and Storage (BECCS)

5.1.6 Bioenergy with Carbon Capture and Utilisation (CCU)

The use of bioenergy fed in at the front end of the energy system, coupled with CCU and CCS on the back end, would further reduce the annual atmospheric carbon burden (see Figure 5.6).

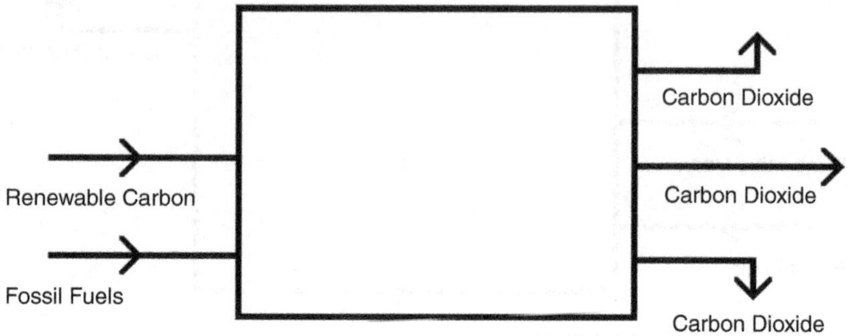

Figure 5.6 Bioenergy with Carbon Capture and Utilisation

5.1.7 Carbon recycling

The major step in addressing carbon dioxide emissions would be to implement carbon recycling – a specialised form of CCU. This would turn carbon dioxide into new fuel for energy use, on a repeated basis. For this to be possible, sources of Renewable Hydrogen would need to be made available at scale, as most chemistry for carbon recycling would require it (see Figure 5.7). Examples could be promoting methanation reactions in thermal reactors – which would synthesise methane to be extra fuel for the reactor, or synthesise methane that becomes part of the outgoing product gases, to be used elsewhere. Alternatively, feeding the product gases from a thermal reactor with Renewable Hydrogen through a separate methanation reactor would make methane fuel. Not all carbon dioxide emissions could be recycled, as this could only be done at centralised, enclosed facilities, such as power plants. However, if most heating and transport applications were to be converted to using electrical energy (renewable electricity, or that generated in carbon recycling power plants), or Renewable Gas, then all the current production of carbon dioxide from heating and transport could be prevented from entering the atmosphere. There could still be carbon dioxide in the economy that cannot be recycled, so some form of carbon capture from air or ocean would be required. The planting of trees could be the most economical and pragmatic approach to this need, especially as this can be done in a staggered fashion.

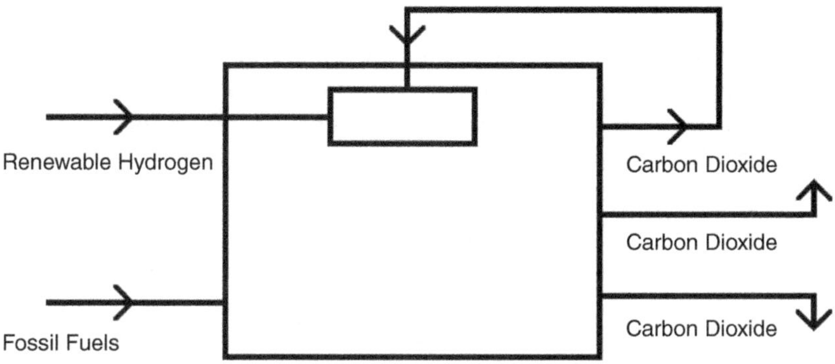

Figure 5.7 Carbon Recycling

The most crucial point to recognise is that with carbon recycling, it is possible to reduce the amount of input carbon – that is, reduce the amount of carbon-based feedstock or fuel fed into the energy system. This depends to a large extent on the source of the Renewable Hydrogen, and the treatment of any carbon in the wider energy system. If the Renewable Hydrogen were sourced from bioenergy, there would be waste carbon, after the hydrogen has been removed. Not all of this carbon could be recycled, and so it would need to be prevented from being used in a way that causes increased net greenhouse gas emissions. Some forms of thermochemical processing of bioenergy result in the carbon remaining in a solid form – biochar – which is possible to dispose of without causing greenhouse gas emissions, and may even be used as an agricultural fertiliser.

The second important result of carbon recycling is that the levels of CCS that would be required in the short to medium term would be reduced. In the long term, it might even obviate the need for CCS to a large extent.

5.1.8 Renewable Gas

If fossil carbon inputs to an energy system were to be replaced by Renewable Carbon – for example from bioenergy – and the carbon recycled, fully Renewable Gas could be produced, and very little carbon dioxide would become greenhouse gas emissions. In addition, where using Renewable Carbon sourced from bioenergy, any output carbon dioxide waste gas would be carbon-neutral – not adding to the extra carbon burden of the atmosphere.

As an example, fully Renewable Gas could be produced by including carbon recycling in a power plant which uses gas fuel manufactured from Renewable Carbon and Renewable Hydrogen (see Figure 5.8). If the Renewable Gas were not needed for immediate power generation, it could be distributed to customers via a gas grid for space heating and transport uses, with no greenhouse gas implications from these dispersed applications.

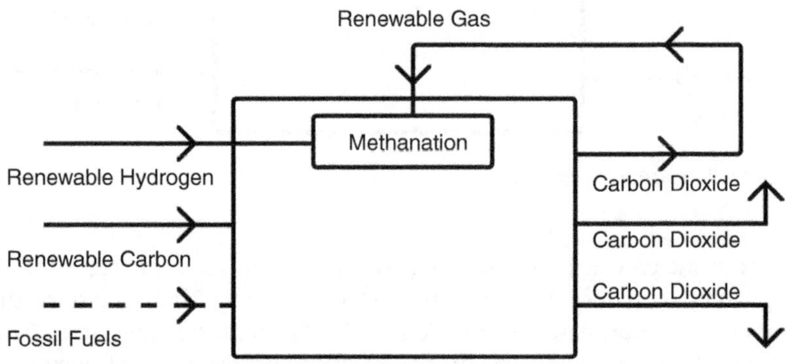

Figure 5.8 Renewable Gas

If the Renewable Hydrogen for the Renewable Gas were to be sourced from bioenergy, then the Renewable Carbon could come from this source as well. However, not all of the carbon from the bioenergy could be utilised. The ratio of hydrogen to carbon (H:C) in biomass is low, and techniques for using water (steam) in bioenergy processing to raise the ratio of hydrogen to carbon in the produced gas are vital. The knock-on effect of the requirement to source "make-up" hydrogen from water is that regions experiencing reductions in rainfall due to climate change will be less capable of manufacturing Renewable Gas. Manufacturing Renewable Gas from bioenergy and water would tip the geographical resources pyramid on its head, as many regions currently producing the most energy – for example, the Middle East – would not be able to produce as much Renewable Gas as Northern Europe and North America. Russia, however, with its large boreal forests and ample river systems, could continue to succeed in energy production.

A possible way to maximise the use of both the hydrogen and carbon in biomass without the addition of extra water (steam) could be to process it using a two-stage system. The first part would use a thermochemical technique to draw off the hydrogen-rich volatiles as a fuel, aiming to leave a carbon-rich residue such as biochar or pure biocarbon, and this carbon-rich by-product would then be used to make electricity via electrochemistry – in a Direct Carbon Fuel Cell (DCFC) (Desclaux et al., 2010). If the concentrated carbon dioxide exit gas were to be sequestered, in total this hybrid system would be carbon-negative.

5.1.9 Hydrogen Economy

Within a 100 years or so, it could become necessary to entirely remove carbon from gas energy systems, and so the logical end point is to use Renewable Hydrogen for the entirety of gas fuel needs (see Figure 5.9). This "Hydrogen Economy" would need to be approached via a transitional route, and Renewable Gas could be viewed as the optimal pathway because of its need to introduce Renewable Hydrogen for carbon recycling.

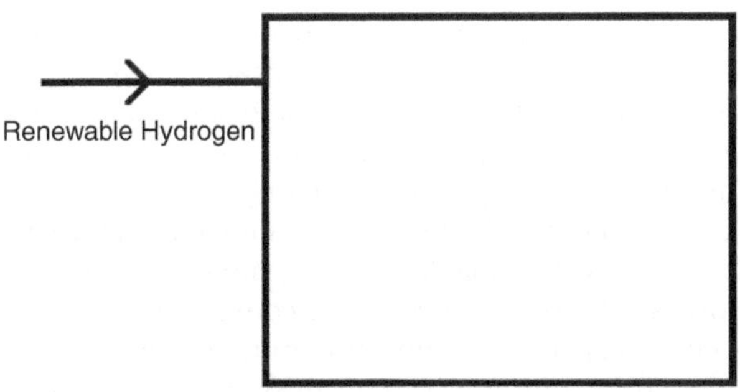

Figure 5.9 Hydrogen Economy

5.2 Key process engineering

The methods for producing Renewable Gas can be thought of as divided into four broad categories of process: techniques for making hydrogen, techniques for making synthesis gas (syngas), techniques for making methane, and techniques for capturing carbon dioxide for use as a gas feedstock in the production of syngas and methane.

5.2.1 The importance of gasification

Gasification is a term normally reserved for the chemical conversion of materials under the action of heat, and not merely vaporisation or atomisation of a liquid or solid fuel. Gasification is often conducted by heat treating of materials in an atmosphere of selected gases, and these, as well as the heated solid or liquid feedstock, can play a part as reactants in the chemical conversion. Gasification can be complete or incomplete, in which case it is normally referred to as pyrolysis, or carbonisation. Gasification is the result of both reduction and oxidation reactions, and is frequently referred to as partial oxidation (POX, POx). If organic material, rich in hydrogen and carbon, is pyrolysed or carbonised, this thermochemical treatment causes the volatile, hydrogen-rich parts of a material to turn to gas, leaving behind the carbon-rich matter, or "char", which can agglomerate into a "coke". Complete gasification by contrast brings the carbon-rich solid matter into chemical reactivity as well as producing hydrogen-rich volatiles, and subsequently there are solid to gas chemical reactions and gas to gas chemical reactions (Abbess, 2014, Table 9). Pyrolysis or carbonisation is conducted without air or oxygen or other additional oxidant; the heat causing liberation of volatile components, and decomposition of the original material, a process also termed "destructive distillation". The carbonisation of coal was used for many decades to produce Coal Gas, used to produce Town Gas; with coke, the carbon, left behind as a by-product. Total gasification by contrast will utilise air or oxygen, and steam or hydrogen, and sometimes even carbon dioxide as well, as gasifying agents, to better control combustion and other chemical reactions, and adjust the ratios of hydrogen, oxygen and carbon, and so convert more of the feedstock material to usable gas. Complete gasification converts flammable materials more effectively to gas fuels, unlike pyrolysis, which is an incomplete conversion.

Although the key chemistry of the combustion of hydrocarbons and other organic matter – the bonding of carbon to oxygen – is part of what happens in gasification, a variety of chemical reactions are sought other than simply oxidation of carbon (Abbess, 2014). At the right temperature, at the right pressure, and under the correct operating conditions, such as selecting the optimum gasifying agent or medium, and using catalysts to promote some reactions rather than others, and perhaps recycling of some of the initial gas product, a range of gas species can finally be produced, in the desired balance, and little residue is left behind. More of the chemical energy in the feedstock is therefore transferred to the product gas, rather than in the case of full combustion, where most of the chemical energy is liberated as heat, and the product gas has no energy value (and is normally referred to as "flue gas"). The use of the term gasification is usually meant to be taken as a reference to complete gasification under POX/POx conditions.

Gasification or fuel cell reforming of fossil fuels by itself does not reduce carbon dioxide emissions; however, it produces a secondary fuel of a form that, when used to provide energy through combustion or other means, provides the opportunity to straightforwardly capture and sequester, or recycle, any resulting carbon dioxide. Pre-processing the fuel in this way can create efficiencies, especially if the syngas produced is cleaned before its use as fuel. In addition, gasification and fuel cell conversion can be done on a large scale, using a variety of raw materials, and give high production volumes, and be continuous. These reasons are why conversion of a range of resources to syngas is likely to be a key part of large-scale Renewable Gas production.

5.2.2 Syngas production

Synthesis gas, or syngas, is essentially a gas mixture with high proportions of (in)flammable hydrogen and (in)flammable carbon monoxide. Depending on the way it is made, it may also have significant amounts of non-flammable nitrogen, (in)flammable methane and non-negligible levels of non-flammable carbon dioxide in it. Syngas with high levels of nitrogen, made by gasifying Natural Gas in a multi-stage process, part of which uses an atmosphere of air, is the usual method used in the process for synthesising ammonia, used in fertiliser production (Abbess, 2014). Syngas could also be made from Natural Gas or Renewable Methane in an IR-SOFC – a multi-generation internally reforming Solid Oxide Fuel Cell (SOFC) producing gas, power and heat (Hemmes, 2010) – but it would make no sense to use a high-methane fuel to produce syngas to be used for the production of methane.

Options for the gasification agent, or oxidant, the medium or "atmosphere" of a gasifier, include using air (Hemmes, 2010, Section 10.3), or pure oxygen, in an SOFC running in Electrochemical Partial Oxidation or EPOx mode (Zhan et al., 2006). Carbon dioxide, which may have been recycled, can also perform the role of oxidant (e.g. Strezov and Evans, 2014, Section 4.4.2), and will certainly moderate a gasification reactor chamber temperature. The operation of a fuel cell can also be altered by feeds of oxidants or other gases.

Syngas made without nitrogen, either by fuel cell conversion of light molecular weight hydrocarbons, or by gasifying biomass or coal in a medium of oxygen rather than air, can be straightforwardly uprated to a gas very similar in characteristics to Natural Gas and with similarly high levels of methane – this is almost universally referred to as SNG. Depending on your point of view, SNG stands for Synthetic, Synthesised, Substitute, Supplemental or Simulated Natural Gas. Although these terms are occasionally used to refer to other things, SNG always means a gas that has high levels of methane, although the percentage can vary depending on the process used to produce it, and the product may need further processing to enhance the methane levels, or reduce the levels of other gases in it, before it can be used to replace Natural Gas (e.g. BP, 1972). Less effort is required to make grid quality SNG if the syngas is produced in such a way that it already has a high methane content (e.g. Graboski and Donath, 1973; Mirmoshtaghi et al., 2013).

Syngas, a "product gas" or "reformate gas", has been conventionally produced either by the gasification of coal, or by the steam reforming of Natural Gas, to be used as a chemical feedstock. However, syngas has a lower heating value than Natural Gas, so making syngas from Natural Gas does not confer any advantage from an energy systems perspective. Making syngas from Natural Gas however is a viable process for the production of liquid hydrocarbon fuels and fuel additives for transport applications, as is currently in production at Shell's Pearl Gas-to-Liquids (GtL, GTL) plant in Qatar.

To make syngas with low net carbon dioxide emissions, the carbon dioxide produced by the use of the syngas or its derivatives would need to be captured and permanently sequestered. Returning this carbon deep underground from where it originally came via CCS would make the syngas low in net carbon dioxide emissions, but not sustainable, as the supply of fossil fuels is non-renewable. The renewable way to make syngas is by gasification (thermochemical) of biomass or electrochemical conversion of a range of biogases, in a much similar way to the gasification or electrochemical conversion of fossil fuels, but needing to take into account the different characteristics of recent living organic matter when compared to fossil fuels.

5.2.3 Methane production

An important method for making Renewable Methane is through the anaerobic digestion (AD) – decomposition and reformation – of biomass to produce biogas. This is already proving to be a very useful small-scale technology, localised to the sources of biomass (e.g. RASE, 2014). Recent developments have aimed to industrialise biogas production (e.g. Bungay, 2009; Bungay and Abdelwahab, 2008; GE, 2014b; Monsal,

2009; NNFCC, 2014), and there are novel forms of Advanced Anaerobic Digestion (AAD) emerging. However, biological treatments are slow relative to industrial chemical plant processes, and usually involve at least one stage of batched or staged reactor treatment with long reaction and residence times, so production volumes of this form of Renewable Methane may not be high enough to displace significant quantities of Natural Gas.

The gasification of a range of coals and heavier molecular weight fossil hydrocarbons (i.e. not Natural Gas), followed by further processing, has been used historically and currently to manufacture methane (SNG) (Abbess, 2014); however, when the methane is burned as an energy fuel, the carbon from the original fossil fuels will be emitted to the atmosphere, unless this carbon were to be sequestered so that it cannot enter the active biosphere. This could be done either by deploying one of the methods for CCS, or by reincorporating the carbon dioxide into new energy fuels. The number of CCS projects needed to capture the carbon from the world's use of syngas, SNG and Natural Gas would be massive, and the infrastructure could be parallel in scale and extent to the entire Natural Gas production industry, and so this route to emissions prevention can be questioned. If instead, the carbon dioxide were to be re-used to make new methane fuel, this would also be a very large scale undertaking if this were only "once-through"; however, repeated recycling of the carbon dioxide into new methane fuel could make capture and re-use more practical in scale, and would also reduce the inputs of original fuel.

Another way to reduce the carbon burden of methane production would be to use biomass as input feedstock. With the right design parameters, the gasification of biomass and post-gasification processing can produce significant levels of methane in the product gas. Methods for recycling carbon dioxide are in development, but the chemistry is not new: Sabatier and Senderens developed nickel catalysts between 1901 and 1905, and found they could synthesise methane from hydrogen and carbon dioxide (Che, 2013; Sabatier, 1912, 1926; Sabatier and Senderens, 1902).

5.2.4 Hydrogen production

Hydrogen, a major component of syngas, is mostly produced at the current time by either the "reforming" steam gasification of Natural Gas, or the gasification of coal. The syngas is passed through a Water Gas Shift Reactor to use steam in converting carbon monoxide to further hydrogen (Abbess, 2014, Table 9). Hydrogen is also produced in the petrorefinery industry, largely for their own use (Abbess, 2014).

Techniques for making Renewable Hydrogen include: the thermochemical treatment of biomass to encourage the production of syngas, which is then passed through a membrane reactor to draw the hydrogen off from the reformed gas; very high temperature thermochemical or plasma treatment of biomass to produce hydrogen, carbon monoxide and elemental carbon, the carbon then being used as the fuel for a DCFC (Steinberg, 2004); the electrolysis of water, for instance, by the use of a Proton Exchange Membrane Fuel Cell (PEFC, PEM FC or PEMFC); a range of catalytic methods, including photoelectrocatalysis (PEC), and what is known as "dark fermentation" of biological material.

5.3 Generic Renewable Gas power plant system design

Employing Renewable Gas is likely to be straightforward for transport applications, but there are likely to be complexities in using Renewable Gas for power generation, especially if carbon recycling is adopted. What follows is a worked example of an idealised and generic design for a medium- to large-scale centralised Renewable Gas system designed to help balance supply against demand in an electricity grid that has a high proportion of variable renewable generators – such as wind power and solar power. All of the elements of the design can come from existing industrial-scale plant, and none of the chemistry is novel.

5.3.1 Mode A: the production of Renewable Hydrogen

Although hydrogen gas can be produced in a variety of ways, the electrolysis of water is perhaps the overall most efficient method available currently, with minimal inputs and no unusable or toxic by-products. With sufficient upfront capital to purchase electrolysers, there can be high production volumes at reasonably low cost of operation, granted that the electricity is available at cheap rates. This could be expected to be the case if power grid operators are in a position where they need to offload excess renewable generation. However, over time, as markets for excess renewable power develop, the higher will be the price that excess renewable power will be able to command. To compensate for this future expected premium placed on renewable power available to make gas, it is going to be important to innovate in the area of hydrogen production, to bring the efficiency up (and the costs therefore down), especially the capital costs of manufacturing plant, and any repair cycle costs (Figure 5.10).

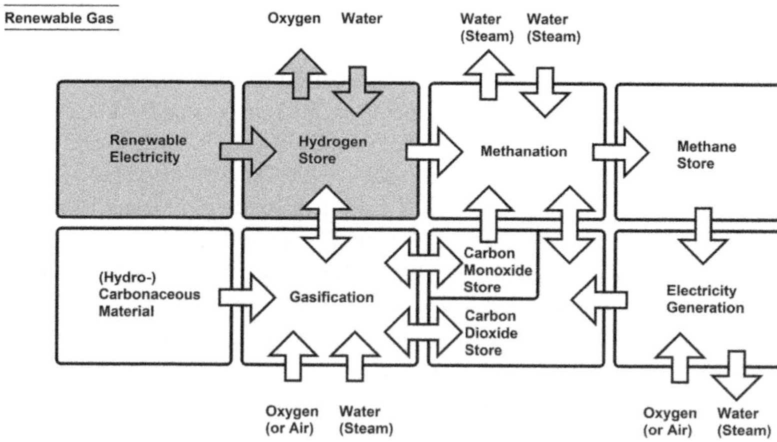

Figure 5.10 Mode A: The Production of Renewable Hydrogen

Two well-developed electrolyser designs are the alkaline electrolyser, and the Proton Exchange Membrane (PEM) electrolyser, based on fuel cell principles. Work on Solid Oxide Electrolysis Cells (SOECs) for hydrogen evolution is still in development. SOECs are effectively SOFCs operating in reverse, which offer prospects for high efficiency. Carbon-Assisted Water Electrolysis (CAWE), if developed, could potentially also offer high efficiencies for hydrogen production (Wang et al., 2014), but would probably require high temperature pre-treatment of the feedstock to prepare suitable particulate or activated (porous) carbon fuel. The use of already-established electrolysers is considered to be costly if they are operated on a periodic basis. In a feasibility study, the British company ITM Power have proposed that electrolysers be operated on a continual basis, therefore making hydrogen production most efficient, and integrating with the energy systems we have today (ITM Power, 2014a). However, if hydrogen were to be intended for the production of methane, and sufficient carbon oxides were not available for methanation, then continual operation of electrolysers would require significant hydrogen storage capability. ITM Power are proposing to inject hydrogen directly into the gas grid, and have recently worked with Thüga Group in Germany to do this, using PEM electrolysis (ITM Power, 2014b). There are currently limitations on the amount of hydrogen permitted in the gas grid, mainly owing to the physical properties of hydrogen (Haines et al., 2003; Melaina et al., 2010), and the legislation on pipeline quality of gridded gas. This is being reviewed, but there are other barriers to hydrogen injection to the gas grid.

The International Energy Agency calculated that injecting hydrogen up to 25% of the gas grid by volume would only result in a carbon dioxide emissions abatement of 4%, mostly owing to the lower heating value of hydrogen on a volume basis (IEA, 2003b). Kiwa GASTEC at CRE assess that to lower the overall carbon dioxide emissions from the use of gas by 8%, around 20% of Natural Gas would need to be replaced by hydrogen (Crowther, 2011). Although gridded Natural Gas could be displaced by hydrogen, which would partly address energy security concerns, carbon dioxide emissions control would not be significant from hydrogen injection into the gas grid. If the gas grid operators choose methane for injection over hydrogen, and continual hydrogen production is economically preferential, but methanation for methane production is not a continual process, then hydrogen storage would become essential, unless the hydrogen were being made in an industrial district where there was hydrogen consumption for other purposes. The suggestion is that continual hydrogen production be shadowed by continual methanation, and methane storage, and therefore a continual supply of carbon oxides or Renewable Carbon would also be needed.

A point of synergy with other industrial sectors is that there is already significant production of hydrogen, mostly made through the thermochemical treatment of fuels to yield syngas (a mix of mostly hydrogen and carbon monoxide), shifted (through the Water Gas Shift Reaction (Abbess, 2014, Table 9)) to a product gas composed mostly of hydrogen and carbon dioxide. Modern petrorefinery makes extensive use of hydrogen, and is expected to need to use more as lower quality crude petroleum oil resources are more regularly used (Abbess, 2014). A natural co-operation between the electrolytic production of hydrogen for Renewable Gas and petrorefinery therefore presents itself. For the oil and gas processing industry to assist in the production of hydrogen for Renewable Gas could be part of the transition to both lower carbon gas and lower carbon vehicle fuels, as crude fossil oil would gradually get displaced by lower carbon alternative liquid or gaseous fuels, such as those made from Renewable Gas.

5.3.2 Mode B: the gasification of (hydro-) carbonaceous material

In a fully developed Renewable Gas system, this plant mode (Mode B) would not be in operation all the time, unless it were the primary source of hydrogen instead of a separate Mode A. The gasification unit of Mode B is the source of the carbon entering the gas system, unless carbon dioxide is being fed in by a piped stream from a power station or a non-energy chemical processing or other manufacturing plant. The primary aim is that all the carbon is recycled in this gas system design; however, some of the gas could be drawn off from the stores to become the feedstocks for the manufacture of synthetic transport fuels, in which case new carbon would need to be input. Despite all carbon in the design being intended to be recycled, there may be some that exits, perhaps as carbon dioxide dissolved in water or steam outlets, or vented during the final stages of methanation when there is a very low level of carbon dioxide remaining in the gas, which would therefore be hard to capture and recycle. Apart from these cases, or venting of carbon dioxide required in exceptional operational circumstances, the gasification unit will not need to be fired up to bring in more carbon to the gas system. To implement this gas energy system, therefore, the most flexible gasification plant should be selected, as it could need to be brought into and out of operation on an irregular basis. The choice of feedstock is likely to be important in enabling this. The most sustainable feedstock to supply into the future would be biomass, and if carbon dioxide were committed to CCS rather than being recycled, this would become a carbon-negative gas system. The usual term for using biomass for energy, and then permanently sequestering the carbon dioxide waste gas, is BECCS.

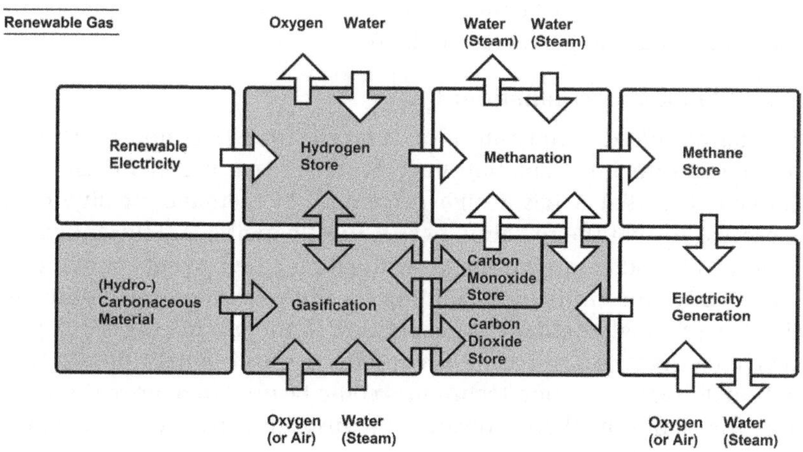

Figure 5.11 Mode B: The Gasification of (Hydro-) Carbonaceous Material

Depending on the characteristics of the feedstock used in gasification, it may or may not be beneficial to pressurise the gasifier, and use another agent or medium to promote the gasification, such as introducing carbon dioxide, hydrogen, or even feeding back in some of the syngas already produced. If oxygen were used as a gasifying agent, this would justify including Mode A in this design, as the electrolysis of water produces not only hydrogen but also oxygen gas. Even if the gasification stage were designed not to be oxygen-blown, oxygen would probably be needed for other process stages, such as in power generation.

The syngas would be composed mostly of hydrogen and carbon monoxide, but could contain a portion of carbon dioxide. The diagram (see Figure 5.11) depicts a hydrogen store and a carbon monoxide (and carbon dioxide) store, which in the case of syngas from gasification may be the same storage unit. If the gasification process is chosen to preferentially produce methane, then most of the gas will end up bypassing the hydrogen and carbon monoxide (and carbon dioxide) stores, and end up in the methane store. If the gasification process is followed by a methanation process, then the syngas can be taken directly through to the methanation stage and the methane store.

An alternative to gasification for the production of carbon for this generic system design would be the production of carbon oxides through the use of DCFC. A variety of fuels could be carbonised, giving off gas high in hydrogen (as Mode A) and leaving carbon-rich by-products. This could be electrochemically converted using a DCFC, producing carbon dioxide and power (as Mode B).

5.3.3 Mode C: Methanation: Type 1

The gas produced from gasification is taken either to temporary storage or directly into a methanation stage, where the gas is either chemically converted so that it has a high percentage of methane, or physically separated so that the remaining gas is mostly methane. Efficient gasification will produce an output gas with either a high syngas content, or a high methane content, depending on the processing conditions. In the case of a high methane concentration in the produced gas, the process of methanation will be more cost-effective and straightforward, and optionally, gas separation techniques could be used to remove the small amount of non-methane constituents, instead of reactional methanation (Figure 5.12).

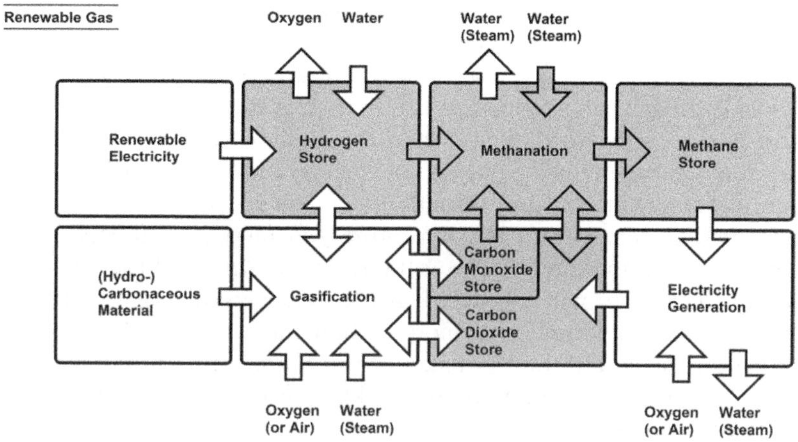

Figure 5.12 Mode C: Methanation: Type 1

In the case that gasification produces mostly syngas – the methanation stage could be conducted in a catalytic reactor, utilising a nickel-based catalyst (IGU, 2012) to directly react the hydrogen and carbon monoxide (Abbess, 2014, Table 9). However, for this to be an efficient process, there is a question of molecular ratios to consider. There will be a number of reactions taking place in the reactor, and the chosen catalyst should promote methanation. Several methanation reaction pathways are possible, but the main methanation reaction in this stage is often known as "CO Methanation" (Abbess, 2014, Table 9, "Methanation 1"), and the balance of the reactions in the combination of hydrogen gas and carbon monoxide should yield mostly methane and water (steam). The balance of hydrogen to carbon monoxide in the syngas should be such that the atomic hydrogen–carbon ratio (H:C) is therefore 6:1, for the efficiency of methane production to be high. Four atoms of hydrogen will chemically bond with a carbon atom, and two hydrogen atoms will bond with an oxygen atom. Since hydrogen comes as a molecule of two hydrogen atoms, the molar ratio of the gases should be 3:1 of hydrogen to carbon monoxide – three hydrogen gas molecules per carbon monoxide molecule.

There are several possible ways to ensure this high hydrogen-to-carbon ratio. One would be to pass the hot syngas through a steam quench to perform what is known in gas processing as the Water Gas

Shift Reaction, reacting the carbon monoxide with the steam to produce carbon dioxide and hydrogen gas (Abbess, 2014, Table 9). A second way to increase the hydrogen-to-carbon (H:C) ratio would be to add additional hydrogen as the syngas passes through to the methanation stage – and this is partly why Mode A is in this design – to provide an extra input of hydrogen for the purpose of efficient methanation of syngas.

Another key methanation reaction is the reverse reaction to the carbon dioxide reforming ("dry reforming") of methane (Abbess, 2014, Table 9, "Methanation 2"). This is highly exothermic, so heat exchange out of the reactor will be vital. This reaction consumes less hydrogen than CO Methanation, and more carbon monoxide, so if it is promoted in the equilibrium or balance of reactions, the carbon monoxide store will be consumed more quickly, but less hydrogen will need to be contributed.

As with gasification, different configurations of methanation technology have been researched, demonstrated and deployed. Varieties of methanation fall into two basic categories: conventional, where shift and methanation are in two distinct stages; and combined or direct, where shift and methanation are in a single stage. Research continues to innovate in this area. For example, it is possible to go one step further than combined methanation, and include carbon dioxide capture in the same step as shift and methanation (Lebarbier et al., 2014) in a process that requires two different sorts of catalyst – or rather, a carbon dioxide sorbent and a carbon monoxide methanation catalyst. This draws on the CO_2 Acceptor Process, developed in the 1960s (e.g. Curran et al., 1967), which uses the same principle of regeneration of the catalyst as Chemical Looping Combustion (CLC) (Nguyen, 2011).

Mode C could also include some methanation of carbon dioxide, if there is any in the store, or is a minor constituent of the syngas. Some catalytic methods can not only selectively methanate carbon monoxide or carbon dioxide, but some catalysts can also methanate both, so if there are trace amounts of carbon dioxide in the syngas, and the process can be done consistently under a range of conditions, there is no reason why all the carbon oxides cannot be uprated to methane in this stage. Were individual reactions to take place in isolation, all the methanation reactions could occur spontaneously in an environmental temperature range of 546–640°C (Abbess, 2014, Table 9), although they could happen at different reactor temperatures with the help of appropriate catalysts. However, in the reactor mix, cascades of reactions will take place, and the reactions that happen more quickly will predominate, and prohibit other reactions by starving them of their reactants. There is certainly room for research into optimising reactor designs for Mode C depending on the input gas composition.

Besides balancing the chemical species for methanation, a further reason why Mode A is included is to allow for hydrogen to be added into the gasification unit (Mode B) as a gasifying agent. At high pressures and high temperatures, this enables hydrogasification to take place – the direct reaction of hydrogen with the carbon in the feedstock – making methane directly – leading to less post-processing required to raise the methane levels of the product gas to SNG standard (Mode C). A different approach would be steam hydrogasification, where both steam and hydrogen are used in the gasifier unit to increase the output of syngas, which when treated by Mode C would give a higher output of methane. In this latter case, oxygen for the gasification could come entirely from the feedstock and the added steam, meaning that an Air Separation Unit (ASU) or other means of supplying oxygen should be unnecessary, and a significant cost reduction could be had.

Methane has a much higher heating value than hydrogen for the same volume – it provides over three times more energy when combusted or burned (Jechura, 2014). It is also roughly ten times more dense at normal temperature and pressure (Alicat, n.d.), so requires less space to store the same energy. This is just one of the reasons why it makes sense to convert syngas and other energy gases up to methane.

There is an interesting synergy between Modes A, B and C, particularly when the gasification of biomass is considered. Most biomass feedstocks will have a considerable level of water in them, and drying is almost universally necessary before most forms of gasification will be efficient. It is common practice to torrefy, or heat dry, biomass prior to gasification. This requires heat energy, which could be provided by the electrolysers in Mode A, or the methanation units in Mode C (Heyne, 2010). An alternative approach would be to make use of supercritical water gasification to gasify wet biomass (e.g. Boukis et al., 2007).

5.3.4 Mode D: power generation

When variable renewable power experiences an extended low of several hours or more, at periods of medium to high demand, the most flexible way for power grid operators to cover the shortfall is to ramp up gas-fired power plant. The generic design proposes that the gas comes from the methane store, but as long as the power plant specification is flexible enough, hydrogen or syngas could be the fuel instead of, or as well as, methane. Some mixes such as Hythane, a blend of hydrogen and methane, could be more efficient for this purpose than the individual gases used separately. Gas blends, and even some Natural Gas Liquids (NGLs) and oxygenates such as ethanol, where they will not break carbon budgets, could become normal fuels for gas turbines used in future, such as aeroderivatives (Jones et al., 2011), or fuel cells that have fuel flexibility (Figure 5.13).

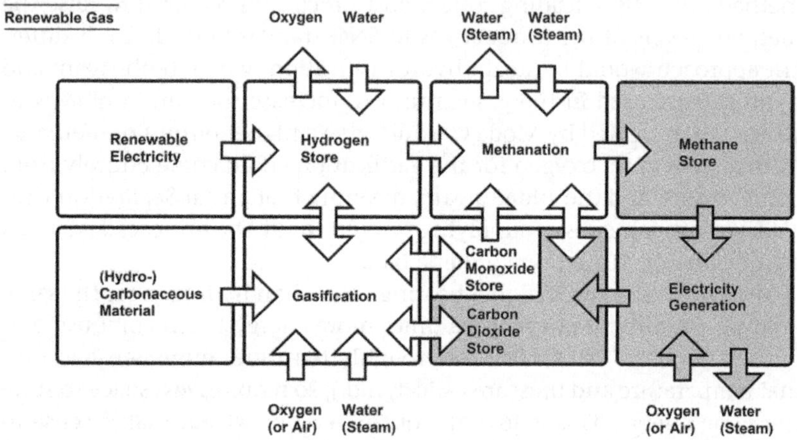

Figure 5.13 Mode D: Power Generation

It is important for this energy system design that the carbon dioxide is captured. Most of the carbon dioxide to be produced in this design is expected to come from Mode D, power generation. The simplest power generation plant would need to use "post-combustion" forms of carbon capture. Other energy system designs that use gasification to enable carbon dioxide separation and capture prior to power generation are generally known as "pre-combustion". However, in this generic Renewable Gas design, the carbon dioxide is put to further use, such as making fuel, before fuel combustion is undertaken; and after combustion, the carbon dioxide is recycled. Estimates of the amount of carbon dioxide capture possible from a conventional power plant, after the fuel has been combusted ("post-combustion"), indicate a maximum of 90%, and in addition, the capture part of the cycle would consume somewhere in the region of 15–25%, or more, extra fuel (AEMO, 2014; Davidson, 2007; Herzog et al., 2009; Rochelle, 2002). However, conventional combustion is not the only answer, as there could be another kind of chemical conversion of the methane deployed in the process of generating electricity in Mode D.

There are advanced combustion techniques for carbon capture in research and demonstration, for example, the White Rose Oxyfuel project in the UK. Oxyfuel combustion takes place in a high-oxygen atmosphere, and flue gas from the combustion of coal is recycled into the combustion boiler to keep the oxygen-aided combustion temperatures from becoming too high (Hong et al., 2009). The net effect is to reduce the amount of combustion exhaust gases, and make sure that they are mostly carbon dioxide and water (steam) (Hong et al., 2009). This is effectively a carbon recycling application in its own right. The final flue gas composition makes the capture of the carbon dioxide straightforward. Currently, the plan for the captured carbon dioxide is subterranean or sub-sea permanent sequestration, but it could be used instead by Renewable Gas plant design to make fresh fuel.

Another approach to carbon capture is the development of CLC. The basic concept was established by patent between 1949 and 1954 (Lewis and Gilliland, 1954) – to use and recycle metal oxides in oxidising the fuel feedstock at normal atmospheric pressure, and releasing the chemical energy as heat, and producing a flue gas that consists largely of carbon dioxide and water. The patent proposed this technique as a source of carbon dioxide for industrial purposes; however, it is clearly useful for carbon capture. Generating electricity from a methane-fuelled CLC unit could operate at around 46% efficiency under normal pressures, which is low compared to the best Combined Cycle Gas Turbine (CCGT) energy conversion of around 58% efficiency (Proll and Hofbauer, 2011); however, the accounting would need to be adjusted after including the steps of recovery of the carbon dioxide and creation of new fuel, to have a more complete picture of overall life cycle energy conversion efficiency.

Power generation by electrochemical fuel cells is also possible, and to date, perhaps the design that could run straightforwardly at megawatt (MW) capacity scale is the SOFC. The SOFCs are designed to operate at high temperatures, and because of the risk of thermal shock to the solid-state ceramic electrolyte membranes or "wafers" from cycling of the fuel cell stacks, it is not thought that SOFC power generation can be made flexible to changing demand patterns. Regular significant changes in stack temperatures would also degrade overall efficiency. However, it is possible to envisage an SOFC being fuelled by methane, and run mostly as a hydrogen production unit to "mop up" excess renewable electricity (as Mode A), and then switched into power production mode to "back up" low wind or solar power production (as Mode D) (Hemmes, 2010, Section 10.4, "The 'Superwind Concept'").

5.3.5 Mode E: Methanation: Type 2

This is catalytic methanation of carbon dioxide and hydrogen to form methane and steam (Abbess, 2014, Table 9, "Methanation 3"). This is also known as the Sabatier reaction, and is exothermic, meaning it gives out heat energy. However, it needs a certain amount of heat energy (reaction activation energy) to raise the temperature of the reactor to the right level to promote the reaction. If Renewable Hydrogen is being produced using electrolysis at the time this mode begins, the electrolyser processing units will generate heat, which could help with starting this version of methanation. If Mode E has been running for some time, and then Renewable Hydrogen production is curtailed, the reactor will not need any other input heat to continue in operation (Figure 5.14).

The Sabatier reaction is much less exothermic than CO Methanation (Abbess, 2014, Table 9, "Methanation 3"), so heat balance is less of a problem.

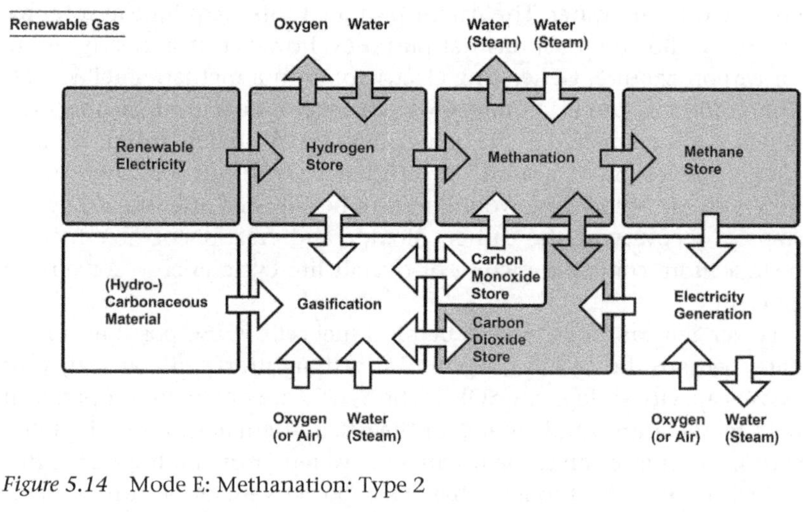

Figure 5.14 Mode E: Methanation: Type 2

5.3.6 Mode F: Methanation: Type 3

Mode F is essentially the same as Mode E, but uses hydrogen from the store instead of making it on-demand. It would need some kind of heat source to the reactor to start the methanation process, and it could only run whilst there is hydrogen left in the store (Figure 5.15).

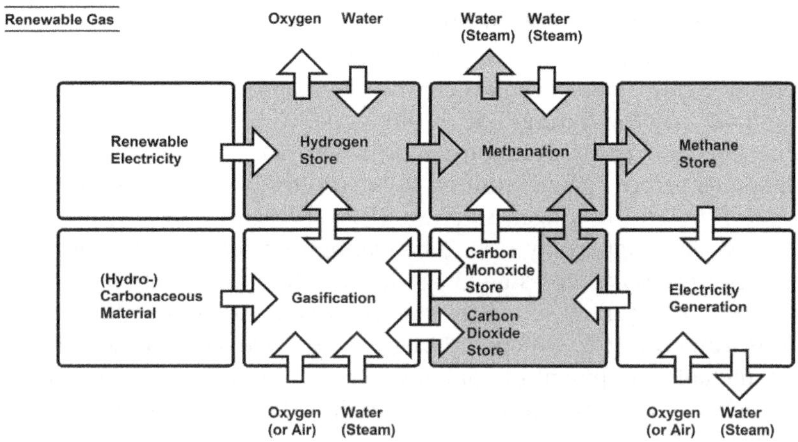

Figure 5.15 Mode F: Methanation: Type 3

5.3.7 Mode G: the production of synthetic fuels

If there is no need to generate power, and the methane store is full, then syngas being produced by the gasification system in Mode B, or methane being produced by methanation in Modes C, E, or F, could be withdrawn to become chemicals, fuels or the feedstocks for the production of synthetic fuels. Hydrogen and carbon monoxide are the key building blocks for a number of chemical processes necessary for producing liquid transport fuels – either directly equivalent or analogous to distilled fractions of fossil fuels (Figure 5.16).

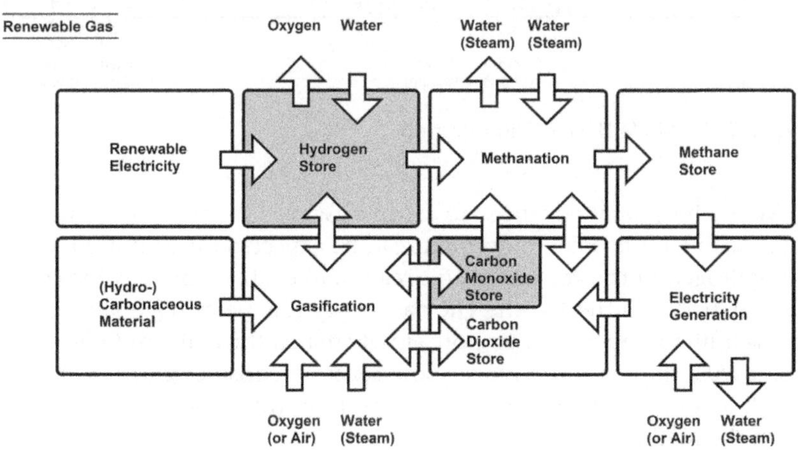

Figure 5.16 Mode G: The Production of Synthetic Fuels

5.3.8 Mode H: gas grid injection

At times when the gas system methane store is full and there is no short-term need anticipated, then the most useful application is to inject the methane into the Natural Gas pipeline grid, where there is a local suitable injection point. Naturally, it will need to meet pipeline quality standards before being permitted to be injected. Of note: carbon monoxide (CO) and hydrogen sulphide (H_2S) levels are strictly controlled as they carry high safety risks. Small amounts of carbon dioxide (CO_2) usually only serve to lessen the heating value of the gas, but in the presence of any kind of water, carbon dioxide goes into solution and forms carbonic acid, which, although a weak acid, could pose a risk to any metallic equipment or appliances (HSE, 2011), which is one of the reasons why permitted levels of carbon dioxide in grid gas are very low (Figure 5.17).

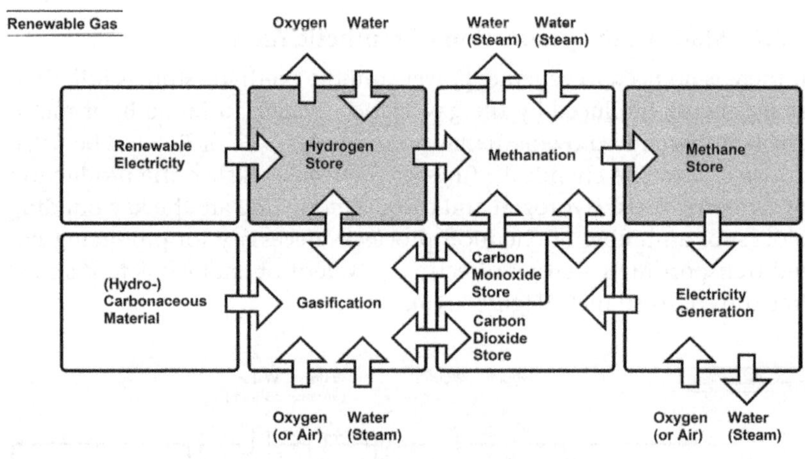

Figure 5.17 Mode H: Gas Grid Injection

With the policy advances to harmonise the gas market across the European Union, some movement can be expected on permitted levels of hydrogen in the gas grid in this region, to enable more flexible cross-border trade. Currently, the UK has a very low percentage allowed. If these limits are increased to more closely match the mainland European standards, then it could make some sense to consider grid injecting a small amount of hydrogen from the store in this generic design along with the methane.

5.4 Alternative routes to syngas besides gasification

The use of Renewable Hydrogen is beginning to be recognised in improving the yield of AD in the microbiological methanation of biomass (Iskov and Rasmussen, 2013). In terms of the generic design, instead of the Gasification Mode B, there would be an AD stage. The AD reactor would produce a biogas, composed mostly of methane and carbon dioxide. Hydrogen (from Mode A) can be either injected into the AD reactor itself, or added into a second reactor that contains a special culture of microorganisms which use the hydrogen to convert the carbon dioxide into methane (Reuter, 2013). Forgoing the creation of Renewable Hydrogen as an intervening stage may become dispensable if biological methanation can be conducted in nutrient-enriched water using purely carbon dioxide and renewable electricity (Gru, 2012).

Although it would make no sense to use Natural Gas as the input carbon-rich fuel for this generic design, in the case that the input fuel is other "light" low-molecular weight hydrocarbons, then internally reforming fuel cells could be used to produce syngas for Mode B.

In the case that the input fuel were to be biocoal or biocoke, high carbon solid by-products from the thermal processing of biomass, then DCFCs could be used to produce carbon dioxide gas for the carbon store.

5.5 Thermal and gas balance

This generic design attempts to demonstrate that it is possible to largely balance water flows in and out of the various stages (apart from inputs of water being significant if Mode A is the electrolysis of water, or Mode B uses steam in the gasification process) which improves its viability for a water-constrained environment. It also shows that oxygen can be drawn from and drawn on in turn in various stages, meaning this too is cycled within the whole gas system. In addition, although not indicated in the diagrams, there are opportunities to recover and re-use heat energy between the various stages, as in the use of process heat to dry biomass prior to gasification. This efficiency in the use of heat is important when Renewable Gas systems are compared to conventional thermal power generation plant, which have virtually one-way heat flow, and heavy inefficiency from heat loss (although this is less with modern developments such as combined cycle or Rankine cycle designs).

5.6 Carbon Capture and Storage (CCS)

This generic design does not include a stage for CCS. This is because the aim here is to recycle carbon. If carbon dioxide were withdrawn at appropriate stages and taken to permanent sequestration, then more feedstock would be required to recharge the carbon in the gas system, which would have cost implications. CCS is itself likely to also have cost implications, so the net impact would be an overall increased cost.

5.7 Activation energies in thermochemical reactions

There may be ways to improve the selectivity for desired product gases in thermochemical reactions, and so improve Renewable Hydrogen production (Mode A), gasification (Mode B) and methanation (Modes C, E and F), for example, by the use of plasma (Hlina et al., 2014; Janajreh et al., 2013) or microwaves (Ferrari et al., 2014; Hunt et al., 2013).

5.8 Renewable Gas for Iran

It could be instructive to consider a concrete opportunity for the transition to Renewable Gas, and a typical case could be found in the gas refinery at Fajr Jam (Fajr-e Jam) in Iran. For manufactured Renewable Gas to be truly low carbon, sustainable and renewable, it must be either used for power generation in a closed loop carbon recycling system, or the carbon that comes into the production system must be "young", which means principally biomass. This has geographical implications for its feasibility, as some of the largest gas refineries are in countries which are semi-arid, such as Iran. A brief look at a satellite image of the province of Bushehr shows nothing much green apart from in the vicinity of towns, such as Jam itself (Google Maps, 2015a). It would not seem practical to engineer the coastline of Iran to increase mangrove wood production significantly. It would seem to be most efficient to make Renewable Gas where the sources of biomass are located, in other parts of the world. If imports of biomass were required by many countries for Renewable Gas purposes, and wood products were selected as most appropriate, it could potentially risk aggravating global deforestation, which already contributes something of the order of 20% of global carbon dioxide emissions to air. This would clearly need to be avoided.

Much of Iran would be suitable for deploying solar power, and even though this is variable in supply, it could be used to make large volumes of Renewable Hydrogen from seawater. With sufficient storage for this Renewable Hydrogen, it could be used when needed to smooth out the energy supply, providing a reliable source of power for industry and society in Iran, and it could also be used for the needs of the oil and gas refineries. Would Renewable Hydrogen be distributed to where it is needed via blending in with Natural Gas in the Iran Gas Trunkline (IGAT) or via a new, parallel hydrogen gas grid? Alternatively, would there be sufficient Renewable Carbon available to make Renewable Methane, to either transport around the country via IGAT, or used to top up supplies for Liquefied Natural Gas (LNG) export? The gas that enters the IGAT pipeline network has its final destination in industry, space heating, cooling and transportation too – as Iran has perhaps the highest number of Natural Gas Vehicles in the world. Yet, the most important question that needs answering is, under what circumstances would a Renewable Gas facility in coastal Iran be manufacturing low carbon gas fuel?

Although Iran has given the world, and can continue to give, a considerable amount of sweet, high calorific value Natural Gas (see Table 5.1), the country is starting to exploit more of its sour and acid gas (IRNA, 2014; POGC, n.d.), as other countries in the region are doing (Hydrocarbons Technology, n.d.; Ravishankar, 2013). Globally, there is something like eighteen trillion cubic metres of reserves of sour/acid Natural Gas with more than 10% acidic carbon dioxide as a primary chemical constituent, and roughly ten trillion cubic metres of sour gas reserves with more than 10% toxic and corrosive hydrogen sulphide (Foster Wheeler, 2008). In total, roughly 40% of remaining global gas reserves are sour, much of this in the Middle East (Foster Wheeler, 2008), and if global demand for Natural Gas continues to be strong, and yet there is a peak within three decades in the amount of sweet Natural Gas that can be produced, as is widely anticipated, considerably more Iranian sour and acid gas could be produced and processed for use. The large quantities of Natural Gas that the Islamic Republic of Iran claims as their territorial mineral reserves could favourably tip the balance in Iran's trading relationships. Although Iran does not currently export much of its very large gas reserves (AOGD, 2006, Section 6, "Gas Exports"), and much new sour gas is destined for Enhanced Oil Recovery (EOR) projects, Iran is developing its LNG export facilities, and it is also developing new gas sweetening plant and pipeline networks. EOR by gas injection into depleting oil wells has shown the potential to increase the total recoverable hydrocarbon liquids, and to raise production volumes by several per cent. However, the number of oil wells in Iran is not going to rise rapidly, and based on experience to date, there is probably a limit to how significantly oil wells can be stimulated by gas injection. Therefore, there is probably only a certain amount of EOR that could be achieved in Iran, meaning there remains the question of what to do with any further volumes of acid and sour gas, concentrated by-products

Table 5.1 Sour and acid gas and nitrogen composition of a selection of Iranian gas fields

	CH_4	N_2	CO_2	H_2S	He	Reference
Agha Jari	66.0%vol	1.0%vol	1.5%vol	–	–	(Moulijn et al., 2013, Table 2.5)
Aghar	89.77%	5.45%	1.298%	0.007%	–	(Rahimpour et al., 2012)
Dalan	87.11%	7.74%	1.627%	0.000%	–	(Rahimpour et al., 2012)
Kabir Kuh	33.64%	64.33%	2%	–	0.03%	(Asre KimyaGari, n.d.)
Kangan	90.04%mol	4.48%mol	0%mol	–	–	(Farzaneh-Gord et al., 2013)
Kangan: Dalan Formation: Well 6	86.9%	8.8%	1.61%	0.02%	–	(Galimov and Rabbani, 2001)
Khangiran	98.6%mol	0.56%mol	0%	–	–	(Farzaneh-Gord et al., 2013)
Khangiran	–	–			~10% ($H_2S + CO_2$)	(Shahsavand, 2008)
Khangiran: Flare Gases	87.42%mol	3.34%mol	0.22%mol	8.34%mol	–	(Zadakbar et al., 2011)
Mazjed-Soleyman (Masjid-i-Sulaiman)	62.668%	–	10.330%	25.535%	–	(Nezhad et al., 2011)
Mozdouran Gas Field	88.901%mol	0.368%mol	6.459%mol	3.588%	–	(OGJ, 2007)
Nar: Kangan Formation: Well 8	86.3%	7.9%	1.34%	–	0.08%	(Galimov and Rabbani, 2001)
Nar: Dalan Formation: Well 16	85.7%	7.5%	2.0%	–	0.07%	(Galimov and Rabbani, 2001)
North Pars	–	–	5%	0.006%	–	(OilGas, n.d.)
Pars	87%mol	3.1%mol	1.85%mol	–	–	(Farzaneh-Gord et al., 2013)
Pars	89.24%mol	1.70%mol	3.28%mol	0.66%mol	–	(Hagoort, 1988)
Sarkhoon Gas Processing Plant	85.73%mol	4.59%mol	0.87%mol	0.01%mol	–	(Bahmannia, 2006) (Sarkhun)
Shanol Gas Field (Shanul)	90.86%mol	3.88%mol	0.52%mol	–	–	(Farzaneh-Gord et al., 2010)
South Pars	82.55%mol	3.38%mol	2%mol	0.54%mol	–	(OGJ, 1996)
South Pars: Well #7: DST #3	85.58%	3.45%	2.02%	0.23%	–	(Aali and Rahmani, 2011)
South Pars: Phase 11: Feed Gas	–	3.5–4.5%	2.3–2.5%	0.4–0.9%	–	(Omidvar, 2011)
Turkman Gas	94.21%mol	1.9%mol	0.14%mol	–	–	(Farzaneh-Gord et al., 2013)

of gas sweetening. Global action to halt gas flaring and venting is going to restrict how much of this sour and acid gas and offgas can be disposed of by combustion and flue. In addition, the rising percentages of carbon dioxide and hydrogen sulphide in Iran's total Natural Gas production will add to its greenhouse gas emissions if the energy industry is permitted to vent to the atmosphere, as both gases (carbon dioxide and hydrogen sulphide) have a strong Global Warming Potential (GWP). Carbon dioxide has a radiative forcing with implications more for the longer term, but Iran could become a land paved with golden recovered sulphur to prevent the strong short-term global warming and toxic biocide from hydrogen sulphide emissions.

A first important step towards Renewable Gas would be to aim for "low carbon gas", using Natural Gas in ways that minimise carbon dioxide emissions. One way to achieve this would be to ensure that all the carbon dioxide from Iran's acid gas and any carbon dioxide generated during gas and oil processing is recycled back into manufacturing gas or liquid fuels. Although this carbon dioxide will eventually find its way into the atmosphere, making fresh fuels with it first will displace the use of other hydrocarbons. Recycling carbon dioxide will necessitate much new gas processing infrastructure. Iran is already producing NGLs from wet gas (IEA, 2010b; PressTV, 2013a, 2013b), and is planning a GtL/GTL facility (ABO, 2014; AOGD, 2006, Section 5, "Gas Processing Plants"). Both of these activities would be good precursors to recycling the carbon dioxide from sour gas into fresh, dry, sweet gas, using a methanation plant, with Renewable Hydrogen production incorporated into its design.

Carbon dioxide may be available via pipeline from oil and gas producing facilities in neighbouring countries, where Natural Gas with very high levels of acid/sour gases is beginning to be exploited, despite the economic inefficiencies arising from the necessary gas processing and cleaning. However, these are new-build facilities, and it would make more sense for these new carbon dioxide-rich sour–acid gas refineries, such as those in the United Arab Emirates, to install the equipment to methanate the carbon dioxide themselves, as part of the project, rather than exporting it to Iran.

To decarbonise Iranian gas beyond "low carbon gas", the country's oil and gas industry could consider using otherwise un-farmed and waste biomass found at home, as a feedstock for gasification, resulting in syngas for use in manufacturing synthetic gas and oil fuels (Guell, 2012; Kaewpanha et al., 2014). There are a range of food, farming and refuse wastes that could be gasified, however, it might be considered that perhaps it would make more sense to anaerobically digest these close to their source, in biogas plants, rather than transport them to the coastal refineries. Alternatively, whilst there may be a dearth of photosynthesis occurring on land in much of the country, offshore of Iran in the Persian Gulf and the Gulf of Oman, there is some potential for cultivating seagrasses and seaweeds – the satellite imagery shows light and deep greens in the close offshore waters, indicating an ideal environment. These macroalgal species, in addition to being used to manufacture Renewable Gas, could play a role in controlling microalgal species that cause Harmful Algal Blooms, or "red tides" (FAO, 2001, Section 2.1, "Seaweed") which are prevalent in the Arabian Sea and Gulf of Oman owing to pollution (Hamzehei et al., 2013) and nutrient upwelling from a "deep mixed layer" of the Monsoon ocean (Honjo, 1997; Honjo and Weller, 1997; Kumar and Narvekar, 2005; Wiggert et al., 2005). Macroalgal farms would compete with microalgal species for nutrients (Al-Hafedh et al., 2012), and with care, algaculture would not disturb the deposition of biological remains on the sea floor, so the sub-sea area would still retain its high productivity (Marra and Barber, 2005) and high carbon "export" to the deeper ocean via the strong "biological pump" of the Arabian Sea. Additionally, onshore, some saline lakes could be suitable for microalgal cultivation, for both gasification, and perhaps aqueous phase reforming or supercritical water gasification (He et al., 2013) and the production of liquid biofuels (Najafi et al., 2011).

Iran does not "over-produce" Natural Gas. It seems that the government of the country wants to capitalise on the social and economic development opportunities afforded by preserving much of its fossil fuel reserves for future use. It seems that they have a view of making optimal rather than maximal use of their fossil fuels, and so for them, the case for the transition to low carbon gas should perhaps be made less on preventing net carbon dioxide emissions, and more on making the best use of the carbon flowing through their economy. Renewable Gas systems, implemented alongside their Natural Gas systems, would ensure that they wring the most energy out of the carbon dioxide that comes out of the ground or is created in high temperature gas processing.

Apart from direct capture of carbon dioxide from the air and the oceans, Iran may find it hard to source all the carbon it would need for a large volume of Renewable Gas production, but even on a small to medium scale, it could provide an economic benefit. Rather than gasify biomass to obtain hydrogen, the best configuration of Renewable Gas for Iran would probably be the wide deployment of solar power for the production of Renewable Hydrogen, as the country has excellent sunlight resources and lots of available land. This Renewable Hydrogen can be used to produce Renewable Methane, with the Renewable Carbon coming from biomass. Renewable Methane can be injected into gas grid pipelines, to complement Natural Gas, and applications would include space heating, and space and equipment cooling in urban environments, with steam capture at the point of use, to provide a source of water in peoples' homes. Water in Iran is perhaps going to become a more valued commodity than Natural Gas as climate change begins to exacerbate surface air temperature extremes and evaporation rates in the Middle East, and seems likely to disturb patterns of rainfall (Amiri and Eslamian, 2010). Changes in rainfall would have knock-on effects on agricultural irrigation demand and, consequently, affect the integrity of inland water resources, such as lakes, which could shelve plans for algal farming. Any disturbance to the Monsoon in the Arabian Sea resulting from climate change could affect its biological productivity. Climate change is likely therefore to affect the prospects for sourcing biomass for use in Renewable Gas production, so the solar power route for producing hydrogen appears optimal: if Renewable Methane can no longer be made because biomass for carbon becomes scarce, the country could turn to using the Renewable Hydrogen as it is, instead.

5.9 The carbon problem: an ocean solution?

By considering the scenario in Iran, it can be seen that the transition to Renewable Gas depends crucially on identifying the sustainable source or sources of low-emissions carbon that can be fed into the Renewable Gas system. For the meantime, using the fossil carbon dioxide associated with sour–acid gas, and the fossil carbon dioxide from advanced oil and gas processing plant, for recycling into fresh gas, may be an adequate low emissions solution. In the longer term, although elsewhere biomass would be an ideal source of "young" carbon, biomass productivity in Iran, both onshore and offshore, may be compromised by climate change. American Navy researchers may have a solution for sourcing "young" carbon directly from the oceans, by proposing electrochemical acidification technology to degas carbon dioxide from natural seawater (Willauer et al., 2013), incidentally producing some hydrogen. The Earth's atmosphere has become increasingly laden with carbon dioxide over the last 300 years, owing to increasing use of fossil fuels, and deforestation and other land use change. Air and water circulate carbon dioxide between them at their boundaries, with the ocean taking up a net amount as a carbon sink, and as a result, the surface waters of the oceans are becoming increasingly acidified by carbon dioxide. It takes anywhere from decades to centuries for the aquatic biochemical carbon cycle processes (the "biological pump") to translate carbon to the deeper waters, and thence eventually deposit it on the sea floor, whence it is subducted by Tectonic processes, eventually becoming geologically sequestrated. The organic carbon produced in the surface ocean at the end of life of organisms that grow by photosynthesis from sunlight and carbon dioxide is approximately 100 gigatonnes annually, of which somewhere between 5% and 15% is exported downwards to the deeper ocean (Giering et al., 2014). The deeper parts of the ocean form a carbon sink of roughly 38,000 gigatonnes of carbon, of which more than 95% is in the form of dissolved bicarbonate ions, and the rest carbonate ions (Willauer et al., 2014). At an acidity of around pH 6 or less, carbon dioxide gas will re-form and leave the water, and this technique for harvesting carbon dioxide is being developed with a view to creating carbon feedstocks for synthetic fuel production.

Despite the fact that producing carbon dioxide this way will involve the processing of large quantities of seawater, and quite some electricity, the costs of doing so could be significantly less than Direct Air Capture (DAC) – recovery of carbon dioxide from the air. In addition, recovery of carbon dioxide directly from the ocean would be far more effective in mitigating global warming. Any carbon dioxide removed via DAC from the atmosphere would need to be removed again within a period of around eighty years – because removing carbon dioxide from the air will cause carbon dioxide to come back out of the ocean, undoing the benefits of air capture (Cao and Caldeira, 2010).

Harvesting carbon dioxide from the ocean to make combustible fuel would displace the use of fossil hydrocarbons. However, the same carbon would become carbon dioxide emissions from the combustion of "ocean fuels" and this would end up in the atmosphere, so at best it can only be a "carbon-neutral" option. This can only be achieved by removing equivalent amounts of carbon dioxide from the atmosphere in parallel with the production of ocean fuels. Removing carbon dioxide from the atmosphere could perhaps best be achieved by sequestering ocean carbon dioxide (Rau et al., 2013) permitting the ocean to draw down carbon dioxide from the atmosphere.

5.10 Transitions and choices: major branching choices: alternative narratives

The development of Renewable Gas will most likely involve the engagement of the oil and gas industry – primarily because of the need for increasing amounts of hydrogen for petrorefinery (Abbess, 2014).

Important choices to be made over the next twenty or so years (see Figure 5.18) will largely determine the growth rate of Renewable Gas.

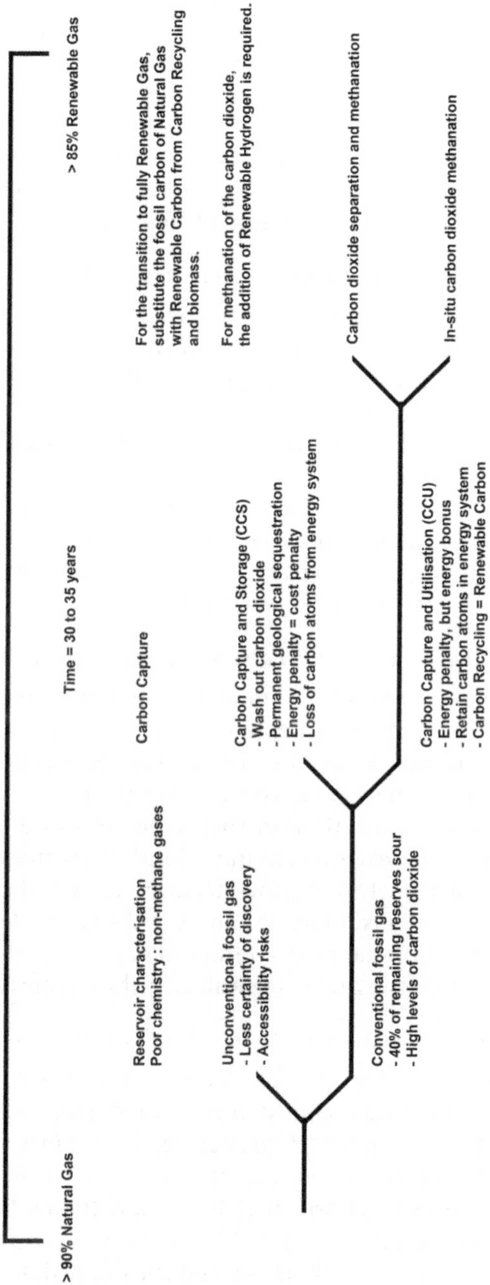

Figure 5.18 Renewable Gas: The Transition to Low Emissions Energy Gas and Major Choices

5.10.1 Option A: conventional or unconventional Natural Gas?

The first major choice along the timeline to 2050 is whether the fossil fuel-producing companies and countries prefer to pursue extraction from large "conventional" Natural Gas fields, or develop novel "unconventional" resources, such as shale gas, Arctic gas, coal seam gas (coalbed methane) and methane hydrates. In the short term, it may seem optimal to pursue both conventional and unconventional fossil fuels, but there could be hard limits to the continuation of this strategy.

5.10.1.1 Unconventional gas could be sweeter gas, but riskier gas

Unconventional resources in some cases offer the prospects of sweeter Natural Gas, whereas approximately 40% of conventional resources are sour – containing significant percentages of hydrogen sulphide and/or carbon dioxide (note: some consider sour gas to be unconventional, or "non-conventional"). However, compared to conventional gas fields, producing unconventional gas resources entails a number of new risks, such as less certainty of the discovery of economically viable hydrocarbons; new challenges in terms of accessing the resources; and the higher potential for drilling "empty wells" – owing to the inability of seismic prospecting to accurately size or scope the gas in place.

5.10.1.2 Conventional gas depletion and remaining large gas fields

In currently exploited reserves, a number of major conventional gas provinces are showing signs of depletion, and discoveries in conventional oil and gas in the last few decades are tending towards dispersed, minor fields. However, there are a large number of known "giants" with reasonable reservoir characterisation that have yet to be produced in any great quantities – for example, in Iran. Considering the size of these, it might be more practical to improve technological and trading relationships between gas-consuming countries and Iran (and other countries with large, accessible gas fields), rather than going to the trouble of mining regions such as the Arctic, regardless of the chemistry.

5.10.1.3 The obfuscating re-categorisation of both oil and gas resources

There is a transition underway in the kinds and qualities of fossil fuel resources that are becoming included in accounted resources, considered viable for economic exploitation, a quantity enshrined in a figure known as "ERR", or economically recoverable resources. Advances in technology and engineering such as horizontal drilling and hydraulic fracturing, and more extensive refining, now provide the means to exploit complex and inaccessible resources, such as gas and oil from under the seabed at great depth; tight oil (crude petroleum oil found in shale sediments and shalestones); heavy oils (generally considered to be degraded crude petroleum fields); extra-heavy oils such as bitumen (a by-product of kerogen expelling lighter hydrocarbons), heavy oils and bitumen in sands and sandstones (tar sands); and immature kerogen from incomplete catagenesis, such as found in oil shale. Many of these unconventional resources are best mined by processing in place – shales are hydraulically fractured, heavy oils and bitumen perhaps steam-melted or oxygen-fired underground (SAGD – Steam-Assisted Gravity Drainage, and THAI – Toe-to-Heel Air Injection), deep coal is gasified (UCG – underground coal gasification), and methane hydrates would be pressurised.

Because of the confidence placed in new technology and engineering, revisions of the TRR – technically recoverable resources – have been made. Resources that would otherwise previously have been considered uneconomic or unmineable are now included in tallies of a region's reserves. For example, Canadian oil production from tar sands was included for the first time in BP's Statistical Review of World Energy in the June 2004 edition (BP, 2004), having previously been explicitly excluded (BP, 2003). Another important example is Venezuela's quoted oil reserves, which were almost doubled from one year to the next, to include Orinoco Belt heavy oil that had been included in the OPEC Annual Statistical Bulletin for 2008 (BP, 2010; OPEC, 2009, Page 22).

Unconventional gas reserves are also gradually being added to reports (GEA, 2012, Section 7.3, "Natural Gas"), such as for North America, Australia and China (Lecarpentier, 2012, Section "Continued growth of gas reserves") as shale gas and coalbed methane are developed. The reasons for re-classifying unconventional resources as reserves are probably partly policy, such as the several-decades-old stated aim of the US's pursuit of "energy independence", which stimulated indigenous shale gas development, for example, through the sale of public lands and the rescinding of environmental legislation. The evidence suggests that treaties and agreements with energy trading partners in North and South America, such as the North American Free Trade Agreement (NAFTA) with Canada, and the diplomatic focus on the unstable relationship with Venezuela, are partly governed by the aim to lessen dependence on "foreign oil" from the Middle East. Again, the expansion in the exploitation of Canadian oil sands and heavy oils can be seen as a response to the US's demand for closer supplies of oil, underwritten by a commitment to the development of a variety of pipeline and other oil transportation networks. The American drive for "energy independence" could be viewed as a tacit admission of crude oil depletion in key fields in the

Persian Gulf – a medium-term scenario would then have continued high levels of American oil imports, in the context of competition from world markets for diminishing supplies from the Middle East, unless alternatives were brought to production.

5.10.1.4 The inequivalence of conventional and unconventional resources

The inclusion of unconventional resources in statements of oil and gas reserves masks problems of inequivalence – it is not possible to substitute in a like-for-like fashion – producing these reserves has different risks and rewards. This is an "apples and oranges" problem that has several cross-cutting elements, which broadly condense into potential financial and investment risk for both capital expenditure and operational expenditure.

Although some unconventional hydrocarbon plays may have the advantage of being large and concentrated, they have several disbenefits, such as in their reservoir characteristics. An unconventional oil deposit may be too young or too old, or only part of it has passed through the "oil window", the ideal subterranean pressure, temperature and isolation conditions for producing light, sweet, crude petroleum oil (Abbess, 2014). So although conventional fossil fuel deposits exhibit a range in their chemistry, unconventional fossil fuels are from a much broader spectrum, and are widely divergent in a number of respects. The plant used to refine and process them is inherently more complex, and in addition, the level of complex or unusable by-products is potentially higher than with conventional resources, risking lower overall production volumes. This possible variability in output will almost certainly impact on operational costs.

The need for oil and gas majors to shore up their Reserves-to-Production (R/P) ratios) – an indicator of company viability and investment-worthiness – is probably leading to the nudge to include increasingly complex resources into their listed reserves. However, unconventional resources need to be "proven" through a certain level of market development before this can be acceptable to auditors as genuine "reserves growth".

5.10.1.5 "Easy" or "edgy" oil and gas

The line between conventional and unconventional resources is being blurred somewhat by engineering advances and trends, and a new relativity is perhaps whether a resource is "easy" or "edgy" to produce. A resource may not only be on the borderline of economic favourability but also be on the edge of acceptable chemistry. Whichever route is taken, adaptation is necessary.

5.10.1.6 Undesirable by-products

Unwanted polluting by-products are set to increase from energy fuel and petrochemical refinery, whichever kinds of resources an oil and gas company wishes to exploit, and this will inevitably give financial risks. For example, using heavier crude oils in petrorefinery will lead to an imbalance in the plant hydrogen-to-carbon ratio, and require higher levels of carbon rejection, which will inevitably lead to higher carbon dioxide emissions (Abbess, 2014). Another example would be that unconventional Natural Gas may have unacceptable levels of certain organic compounds or toxic minerals and require extra care, cost and purifying processing plant. What to do with the extra carbon arising from high levels of carbon dioxide in sour/acid unconventional gas is perhaps the most important question. Analysts would suggest that processing that results in by-products with firm, sustainable markets would be the best option.

5.10.1.7 The impact of climate change

Also, in an era of climate change, with changing rainfall patterns, melting ice caps, rising seas, increased storm erosion and possibly mining restrictions (quotas) or charges for carbon dioxide emissions to air, optimal choices for fossil fuel production and processing would be in the scenarios of choices that adulterate or consume less freshwater overall, are not situated on the coast, result in less atmospheric global warming-causing particulate or greenhouse gas emissions and do not remove or disturb forests or soil. The cost of or fines for environmental protection and remediation requirements of any kind are not the key indicators here. Rather, it is the risk to the margin of profit per barrel of oil equivalent, pushing some unconventionals developments over the edge of the financeability cliff. In economic profitability, as in thermodynamics, prevention of entropy is better than cure, and so the development of remaining "easy" global gas reserves, even with chemistry issues, but with uniform characteristics and good accessibility in already-infrastructured fields, is likely to be more preferential than borderline profit margin, "edgy" wilderness or very deep provinces.

5.10.2 Option B: utilise or vent carbon dioxide?

If the choice is made to place more confidence in sour–acid conventional (easily accessible and large volume) Natural Gas resource, then another choice appears as a result of the need to reduce net greenhouse gas emissions: should the carbon dioxide in the gas be rejected or chemically incorporated into the finished pipeline product? If the carbon dioxide is to be discarded, it should be geosequestered ideally, injecting it underground. Depleted gas caverns or oil domes, man-made salt caverns or saline aquifers may be appropriate places to permanently dispose of carbon dioxide (EIA, 2004), although some may wish to use it as a stimulant for crude oil or Natural Gas production – especially in declining fields. If the carbon dioxide is to be chemically reacted to become a viable energy fuel, methanating it would seem to be a sensible option. This would not lock the carbon dioxide away, but would prevent other fuel being burned, so would preclude other carbon dioxide emissions.

Despite the current good press for complex and paradigm-breaking unconventional resources, it could be projected that the oil and gas industry will begin to re-focus on more conventional, but sour–acid, gas and oil in major fields. It is hard to imagine how this would not affect the capex and opex costs of the current suite of plant for chemical processing and refining of oil and gas. Added to which, if there were an evolution of chargeable liabilities attached to "unburnable carbon" and fugitive emissions, climate protection measures, such as capturing carbon dioxide and methane from refining facilities, could become legal requirements. In these circumstances, extra value would need to be scoured from refining. The current favoured proposal for dealing with captured carbon dioxide is permanent sequestration underground; however, this would bulk up processing plants, and both their capital and operational expenditure costs, unless refining and purifying processes were significantly altered to incorporate emissions mitigation chemistry. The alternative is the re-utilisation of carbon dioxide in the processing of fossil fuels – Carbon Capture and (Re-)Utilisation (CCU) – through the use of methanation. Crude oil refineries and gas and oil processing plant are highly centralised, and so CCU could be designed in relatively straightforwardly. This ease in adding CCU would also be true for centralised gas-fired electricity generation or power plants, and to a lesser extent for other major industrial processes, especially in petrochemicals, cement, steel-making and pharmaceuticals.

5.10.3 Option C: separation or in situ methanation?

If methanation is adopted, the third key branching decision is whether the carbon dioxide should be separated from the rest of the Natural Gas and reacted in a parallel process, or should it be integrated: methanated in situ with the rest of the gas? Because of the energy penalty of washing the carbon dioxide out of the gas, it would seem pragmatic to develop ways to methanate it in-stream with the Natural Gas (Abu Bakar et al., 2012).

The decision will be based on the exact chemistry of the gas, as the efficiency of methanation in place could be affected by the balance of oxidants and temperatures. Methane auto-ignites at 537–539°C at low concentrations in air, because of the oxygen present (PTT, n.d.), and methanation is generally conducted at 200–550°C (Schaaf et al., 2014), although this can be reduced to 200–400°C under pressure between 20 and 70 bar (Er-rbib and Bouallou, 2013). In the case that the processing plant needs to be flexible in the sources of gas it refines, the optimal choice might be to separate the carbon dioxide from the rest of the methane-rich gas for its methanation. In the case that the main gas stream is supplemented from industrial feedstocks before processing, it could have significant quantities of carbon oxides, which would affect the choice of methanation catalysts, or affect the efficiency of conversion to methane. In the case of wet Natural Gas, there will still be a certain amount of higher carbon chain compounds yet to be distilled off for liquid fuels (Natural Gas Plant Liquids). In this case, separation of the carbon dioxide for its methanation may be elected to prevent the possibility of complex chemistry with the higher carbon chained compounds.

From a systems perspective, as greenhouse gas emissions become increasingly restricted, the carbon dioxide management choice in Natural Gas processing could need to be largely dictated by the final intended use of the gas products. If the methane produced from carbon dioxide is used for fuel in Natural Gas-burning vehicles (NGV), or for space heating in buildings, then the carbon dioxide originally present in the Natural Gas will end up in the atmosphere. To comply with carbon dioxide emissions regulations, these refineries will probably opt for CCS, unless they can divert some of the carbon dioxide in their mined Natural Gas to other petrochemical or industrial chemical uses, where the final products have long decomposition times – essentially locking the carbon up. In this case, the refineries may still choose to methanate – chemically react their carbon dioxide to be included in a methane-rich gas, if the destination will be durable, recyclable plastics used in manufacturing (e.g. cars and bicycles) or construction. In processing plant where the gas products are most likely to be used for power generation, in situ methanation CCU is likely to be the optimal choice, and carbon capture or carbon recycling requirements delegated downstream to the electricity generation plant operations.

Gas processing plant and biorefineries of the future could have solar power panel banks and wind turbines to provide the power needed to produce the Renewable Hydrogen for methanation, and be located near a plentiful source of water.

5.10.4 Option D: fossil gas or Renewable Gas?

For the gas refineries to opt to integrate carbon dioxide into their products via methanation to a Transitional Gas would be significant medium-term strategic planning. As scientific evidence accumulates on the severity of climate change, greenhouse gas emissions regulations can be anticipated to become more stringent and enforced. Those in the gas industry who are methanating the carbon dioxide in Natural Gas into Transitional Gas will be in prime position to offer gas products lower in fossil carbon by substituting Renewable Carbon for inputs of Natural Gas, so manufacturing a truly Renewable Gas. They will have constructed virtually all the processing plant they need to make this switch. They will already have sourced Renewable Hydrogen, and so they can increase this supply to their processing whilst exchanging Natural Gas for renewable sources of carbon – such as biomass or carbon dioxide harvested from renewable resources. The mitigation of carbon dioxide emissions by a transition to using Renewable Carbon and Renewable Hydrogen to make Renewable Methane – an almost direct substitute for Natural Gas – will create a production capability of gas with very low net atmospheric carbon emissions. If this transition is anticipated, then the oil and gas industry can be expected to choose CCU methanation as an optimal pathway and move in that direction.

6
The Technology of Renewable Gas

6.1 Key areas of research and development

In the development of a Renewable Gas system using the generic design elements previously described, there are a number of "pinch points" – process steps which could benefit from further research and development where the current articulations of the technologies have snags or pitfalls.

6.1.1 Mode A: the production of Renewable Hydrogen

6.1.1.1 Pinch Point: the changing profile of "excess" renewable electricity

Splitting water into hydrogen gas and oxygen gas via the action of electrical current is environmentally generally benign, if the power is sourced in a renewable fashion. Those who advocate Power-to-Gas (P2G, Power to Gas, PtG) have as one of their baseline assumptions that otherwise unused renewable electricity, for example, from wind power or solar power, where peaks in production will not match peaks in demand, can be diverted to the production of Renewable Gas. This could be either hydrogen produced via electrolysis, or methane produced from carbon oxides reacted with hydrogen produced via electrolysis. Whilst the deployment of renewable electricity is accelerating, there should be "spare" renewable electricity, particularly in summer in the middle of the day where solar energy is collected. This situation will continue until the markets develop to make good use of the provenance of virtually free electricity at these times of "excess" generation. If the rates of deployment of renewable electricity technologies eventually start to flatten out, or renewable electricity uptake reaches the same rate as renewable electricity deployment, there may be competition from a variety of users to purchase any unexportable power, and so it may not all be available to manufacture gas with. If energy storage of renewable electricity as gas became seen as a vital function of the power grid to balance supply against demand, Renewable Hydrogen and Renewable Methane operations could be granted preferential access to "spare" solar, wind and hydro- or marine power, but this would depend on energy regulator policy, such as obligations for green gas and green power to be produced, which could undermine to a certain extent cost and deployment efficiencies derived from free market conditions.

One way to approach this problem could be to set a premium price on Renewable Gas, thereby providing a step up for green gas producers in their ability to settle suitable long-term contracts with renewable electricity providers. If renewable electricity providers were guaranteed to be getting healthy business from Renewable Gas manufacturers, it would encourage the relationship between the two sectors, and encourage increased provision of both green power and green gas. An alternative would be a market intervention without a price attached – such as requiring renewable electricity generators to offer a fixed percentage of their power to Renewable Gas manufacturers, determined by a calculation that takes into account the nameplate capacity compared to the actual generation of green power plant – requiring green power producers to cater in part for their own variability in production.

Technological advance may well mean that the amount of electrical power required to manufacture Renewable Hydrogen via electrolysis and electrochemical techniques could reduce significantly, and Renewable Gas producers may then become the winning purchasers of excess renewable electricity, outcompeting other purchasers of green power, due to the premium utility of the Renewable Gas portfolio.

6.1.1.2 Pinch Point: getting beyond initial development and deployment issues

Renewable Hydrogen is likely to become increasingly in demand owing to the need to reduce carbon dioxide emissions from a range of energy and chemistry applications. Due to this cross-sector nature of demand for hydrogen, it is important to build Renewable Hydrogen capacity starting from today. A key point is that this should not be prevented by a lack of understanding about the exact levels of future excess green power generation or its availability to produce green gas.

Although it would be expedient to begin major Renewable Hydrogen production projects, it is not yet clear which technology would be optimal. There is still a considerable level of research work going into finding suitable catalysts for hydrogen production to be used by a range of methods (e.g. Karunadasa et al., 2010). It is possible that weak acids, such as seawater (Abdel-Aal et al., 2010) and effluent from water treatment facilities, or ionic liquids (de Souza et al., 2008), could be used for very inexpensive forms of electrolysis. And there are risks, such as economic risks from the perhaps continued need to use expensive noble metals, or chemical risks from the co-evolution of hydrogen and oxygen in the same reactor. A two-step process to firstly oxidise water, and then subsequently to evolve hydrogen in a separate process stage could vastly improve safety (Symes and Cronin, 2013). Hydrogen production using a form of chemical looping could even split water by thermal dissociation using the power of the sun (Muhich et al., 2013).

The ideal source of hydrogen is water, and water is important in virtually all thermochemical, biochemical and electrochemical techniques of hydrogen production. For the scale considered, whether using electrolysis of water or some other method to produce hydrogen, a large amount of water is expected to be necessary (although it could partly be recycled in the generic design described), and this would mean siting a Renewable Gas plant either at a coast or near a large body of inland water. This might conflict with other needs for the Renewable Gas plant, such as being close to sources of biomass suitable for gasification.

If chosen for commonplace hydrogen production, electrolysers may not be efficient if there is frequent ramping up and ramping down in operation. For example, regular cycling might interfere with efficient heat and water management in Proton Exchange Membrane Fuel Cells (PEMFC), so electrolytic hydrogen production could need to be an always-on capacity, rather than an on-demand one. This could impact the economics of Renewable Hydrogen production, and also its contribution to carbon dioxide emissions control, and also dictate more complex storage arrangements.

6.1.2 Mode B: gasification

6.1.2.1 Pinch Points: coal gasification

If coal is the feedstock, it is best gasified at very high temperatures, 900°C or above (Higman and van der Burgt, 2003, Section 5.5), which has implications for the engineering of the reactor. The quality of the coal feedstock is important in the performance of the various designs of coal gasifier developed. Mostly, it should not have a high initial moisture content, or high ash content, as some trace elements deter gasification, and coal particles should not adhere together too easily at high temperatures, as then they become unreactive. Some designs of gasifier are therefore precluded for some grades of coal. There may be high levels of char and ash to dispose of owing to the very severe conditions of gasification, and it may not be possible to recycle them into a gasifier owing to the high mineral content.

6.1.2.2 Biomass gasification

There are several problem areas in biomass gasification, and most of them relate to the reactor temperature.

6.1.2.2.1 Pinch Point: pyrolysis conditions versus gasification conditions. The optimal conditions for the partial oxidation of biomass to useful gas products depend on a wide range of factors, and can differ with each type of feedstock. Owing to the high reactivity of biomass (Higman and van der Burgt, 2003, Section 5.5, "Biomass gasification"), gasification of biomass can generally be conducted at or below 900°C, meaning that in some configurations, the reactors are operating more under pyrolysis conditions than full gasification. Pyrolysis of biomass results in some gas products, but also the formation of complex tars and biochar products, which would themselves be susceptible to gasification (or partial oxidation – to produce syngas), or combustion (if the product gas is to be used for heating applications). Therefore, it is not surprising that many designs of biomass gasifier systems have dual reactors operating at different temperatures (Higman and van der Burgt, 2003, Section 5.5). In the MILENA process, pyrolysis and oxidation are performed in different but integrated parts of the same reactor (ECN, 2007a; van der Meijden, 2012). Heat exchange between the exothermic and endothermic parts of these systems is critical to the overall energy efficiency of the system, and frequently, hot sand or char is circulated to achieve this (e.g. Rasmussen, 2012). In addition, to create uniform conditions for gasification across the reactor's active zone – the "bed" – the feedstock may be put into motion – "fluidised", or "circulating". The appearance of heavy hydrocarbons (tars) in the output gas, despite these measures and designs, is the key hazard from biomass gasification. In the second reactor of the Carbona design, for example, high temperatures and special filters are used to reform the tars (Patel and Salo, 2007). Other designs take alternative approaches – such as increasing the gasification reactor temperature to prevent the formation of tars to a large extent (Abbess, 2014).

6.1.2.2.2 Pinch Point: non-uniform and high moisture feedstock. Biomass is often not uniformly structured as a feedstock, and may contain a lot of fibre that cannot be gasified (or pyrolysed). Pre-treatment of biomass is the rule, and from a single source, such as wood by-products from trees, or a particular energy grass. The naturally high levels of water in most biomass demand that most pre-treatments should also include a drying phase.

6.1.2.3 *Pinch Point: stop–start nature of gasification*

Choosing the thermochemical treatment of gasification means that the reactor environment must be at a high temperature. If a gasifier is asked to start and stop several times over the course of a day, this will exact a toll in terms of thermal efficiency, as re-starting the gasifier will have implications for the need to combust some fuel in a secondary or support "booster" heater to revitalise the high heat levels in the reactor until the core reactions are once more established and the reactor can more or less maintain its own temperatures.

If the gasifier could be switched from medium-high temperatures (where it is producing syngas to help replenish the gas stores), to higher temperatures (using syngas as a combustion fuel) to produce heat for power generation, then it could be operated on a constant basis. However, this would have implications for its construction as the metallurgy considerations could be problematic. If continual operation were not specified, not only gasifiers, but some fuel cell designs as well, would find this problematic. A fuel cell stack might have advantages in this case however, as a simple system of gas flow valves could switch the function of the unit from majority gas production to majority power production (Hemmes, 2010, Section 10.4).

If it is considered that a thermochemical gasifier should be operated on a continual basis to avoid inefficiencies from heat loss during shutdown periods, then a suitable application would be the production of methane for injection into the gas grid, constantly consuming biomass fuel and methanating it with Renewable Hydrogen, intended for interseasonal storage. In the UK, this application would raise questions about the levels of gas storage in the grid, as these might need to be expanded to apply the strategy of saving Renewable Gas for cold weather periods.

6.1.2.4 *Pinch Point: the slow-moving development of electrochemical fuel cell systems*

If the choice is made to pursue electrochemical hydrogen or syngas evolution, rather than the thermochemical treatments of fuels, then there remain a number of technical and operational issues with fuel cells across the whole spectrum of designs (Abbess, 2014). Electrolyte lifetimes, catalyst degradation and cell sealing are particular concerns, as are issues of corrosion or deposition in oxidising or reducing hotspots, calcification in Molten Carbonate Fuel Cell (MCFC) stacks and carbon deposition when using complex carbon-rich fuels. A solid-state option, the Solid Oxide Fuel Cell (SOFC), operating on gas fuels, in principle reduces the complexity of fuel cell operation by minimising the risks of phase-boundary chaotic events. However, to guarantee reliability, the systems may be set up such that overall efficiencies may be lower than the theoretical maximum. This is partly answered by having a hybrid system, such as an SOFC-GT or SOFC/GT – an SOFC producing gas to be used as fuel in a gas turbine. Demonstration and production projects of SOFC-GT have been conducted by large established engineering companies such as Siemens Westinghouse and GE (General Electric Power Systems). There are also straight SOFC systems available, such as from newcomers Bloom Energy and Redox Power Systems (Behling, 2012; Bloom, 2014; Microsoft, 2014).

6.1.3 Mode C: Methanation

6.1.3.1 Pinch Point: heat management

All the methanation reactions are strongly exothermic, and this heat needs to be extracted or recovered to maintain the optimum reactor temperatures, otherwise the heat generated could risk the integrity of the reactor and the reactants and products. Common industry practice is to introduce additional gas and/or steam to control the temperature (e.g. Haldor Topsoe, 2009). It is possible to use the recovered heat in other process stages, such as in the drying of biomass feedstock prior to gasification.

6.1.3.2 Pinch Point: catalyst sensitivity

Catalysts for the Sabatier reaction can be sensitive to operating conditions; for example, they can be "poisoned" or passivated by trace amounts of sulphur; or in the presence of other species such as water or oxygen, the catalysts can be deactivated (Muscatello and Santiago-Maldonado, 2012). This means that, unless a specially designed catalyst is employed, desulphurisation and desiccation of the gas stream would be advised before methanation.

6.1.3.3 Pinch Points: multiple passes and gas purity

The balance of chemical reactions in methanation reactors and in gasification reactors does not automatically imply a product gas of pure methane. There is usually a need for gas clean-up of one kind or another. Two methanation units back-to-back are sometimes used – one at high temperature and one at low temperature, to maximise the methane content of the gas. A final stage to remove traces of remaining carbon oxides could also be required. This increase in the number of conversion stages has implications on the efficiency of the overall process. If the methane-rich product gas is destined for use in power generation local to its production, then cleaning the gas would not be so vital, apart from the removal of components that would damage the gas turbines, gas engines or fuel cells, whichever are utilised. If the methane-rich gas is intended for the gas grid, then clean-up has to be thorough.

6.1.3.4 Pinch Point: catalyst development

As with the production of Renewable Hydrogen, there is much research and development in the area of methanation catalysts. It may be possible to successfully develop inexpensive, long-lasting catalysts that act on both carbon monoxide and carbon dioxide at the same time, but in all likelihood, catalytic methanation will need two stages, with catalysts treating one or other carbon oxide as inert in turn. This process configuration could be integrated with the need to have multiple passes of methanation to raise the final methane content.

6.1.4 Mode D: power generation

6.1.4.1 Pinch Point: carbon capture

Carbon dioxide capture and storage (CCS), or its alternative, Carbon Capture and (Re)-utilisation (CCU), are a collection of measures and installations proposed to reduce carbon emissions to the atmosphere. In order to make the Renewable Gas system make carbon sense when using a fossil fuel feedstock, carbon dioxide must be either continuously recycled (CCU) or permanently sequestered (CCS) – both of which require capturing the carbon dioxide out of the gas stream. Techniques include capturing carbon dioxide from the flue gas of power stations, and then compressing it and piping it offshore, to bury it under the sea in now emptied former gas caverns. Although there are many operational and demonstration CCS projects, as yet it is not widespread. This is partly owing to the fact that methods for capturing and compressing carbon dioxide are energy-intensive, and mean that more fuel must be burned. Parasitic energy demand accounts for the high projected costs of CCS (Global CCS Institute, 2012; Herzog et al., 2009; NETL, 2007), and cost analyses indicate that carbon dioxide prices or taxation would have to rise significantly in order for CCS to become cost-efficient. Energy system designs that use the combustion of fossil fuels and capture carbon dioxide for storage or re-utilisation are split into three main categories by their methods of capture: pre-combustion, post-combustion and oxy-fuel (or oxyfuel) carbon recycling and separation.

6.1.4.1.1 Pre-Combustion Carbon Capture. Pre-combustion carbon dioxide capture is made possible by separating the carbon-bearing material from other energy-rich materials in the fuel before combustion. This essentially means thermochemical (de-)volatilisation or chemical liberation of hydrogen gas from the fuel feedstock, and using this for energy generation instead of the carbon-rich constituents.

The pre-combustion technologies have largely been developed for coal, such as IGCC – Integrated Gasification Combined Cycle (or BIGCC, if using biomass feedstock). In IGCC, the first oxygen-blown gasification stage produces syngas, high in hydrogen and carbon monoxide. This is then shifted with the Water Gas Shift Reaction (Abbess, 2014, Table 9), so that the carbon monoxide is reacted with steam to produce hydrogen and carbon dioxide. The major output gases are hydrogen and carbon dioxide, which are separated. The hydrogen goes on to the combustion chamber, whilst the carbon dioxide is prepared for burial. Heat management is important to make this conversion process efficient, as although the shift reaction would take place at the same temperature range as in the gasification reactor, the processes to remove carbon dioxide would in most cases take place at much lower temperatures. It may be possible to have the carbon dioxide remain in the gas under some circumstances, for example, very high temperature combustion, where the carbon dioxide would help control reactor temperatures.

IGCC of coal has been in development since the 1970s, and has been given the label "Clean Coal". The best examples of IGCC are perhaps Polk Power Station and the Wabash River Generating Station (SourceWatch, 2012), both in the US. In Europe, there were plans originally to upgrade the Nuon Magnum Natural Gas Combined Cycle plant at Eemshaven in The Netherlands to coal IGCC, but these have been postponed (Gas Turbine World, 2011; MIT, 2013). The initial plan for FutureGen in the US was based on the IGCC design, but it was abandoned for "FutureGen 2.0" oxy-combustion, and this project has now been shelved.

The important point to note about designs of this kind is that the hydrogen from the original feedstock becomes a portion of the product syngas, and more hydrogen is added from steam if the syngas is shifted. Also, since the gas is also then washed of carbon dioxide, what takes part in the combustion is simple hydrogen gas, whereas in the original feedstock, it is in the form of more complex hydrogen-rich or hydrocarbon volatiles. Firing the hydrogen with pure oxygen means that the flue gas from the combustion reactor is less likely to contain contaminants, and be mostly composed of steam (water). Supplying pure oxygen rather than air to the hydrogen combustion reactor is a parallel to gasification, where oxygen is supplied to avoid inert nitrogen becoming part of the syngas, whilst at the same time increasing the temperatures, and therefore fuel conversion rates, through encouraging carbon–oxygen reactions. It can be seen from this simple generality that the use of oxygen in thermochemical treatments is useful in ending up with cleaner outputs that are more easy to process.

6.1.4.1.2 Post-Combustion Carbon Capture. Post-combustion carbon dioxide capture is thermodynamically disadvantaged in most designs, which frequently depend on extensive flue gas cleaning (Bellona, n.d.; Global CCS Institute, n.d.), separating carbon dioxide from flue and other exhaust gases, where the carbon dioxide is at low concentrations and low pressures in most plant designs – although it may be at high temperatures at the outlet of the system, which can assist in capture. In addition, it takes energy to then compress the carbon dioxide for pumping into the storage location – and in its entirety, this method of carbon capture requires the use of a lot of extra primary fuel.

Post-combustion CCS is an expensive adaptation, or retrofit, to existing power plants, and is a method that can only be employed at "point sources" of carbon dioxide, such as power plants, as it would be impractical and highly energy-inefficient to collect dispersed emissions from combustion vehicles and appliances such as cars or domestic gas boilers. Pre-combustion capture, by contrast, could in theory be used as a source of hydrogen gas for distribution to vehicle filling stations or delivery via the gas grid, as long as consumers had the correct appliances to burn it.

The parasitic load on a conventional thermal power plant to do post-combustion capture of carbon dioxide and compress it for sequestration is estimated at something like 30% (Global CCS Institute, 2012, Page 7). In addition, most designs do not seem to offer the possibility of capturing more than around 90% of the carbon dioxide in a post-combustion configuration – mostly due to thermodynamic constraints.

Capturing carbon dioxide from a pressurised SOFC-GT is not expected to damage the efficiency of operation (Singhal and Kendall, 2003, Page 373), a great contrast to post-combustion CCS in a conventional power plant.

If a fuel combustion reactor or electrochemical fuel cell stack is pressurised, this would make carbon capture more efficient. One of a number of methods of Chemical Looping Combustion (CLC) could also enable less energy-intensive carbon dioxide separation, by dividing the combustion process into reduction and oxidation phases.

6.1.4.1.3 Oxyfuel Carbon Capture. Oxy-fuel combustion (or oxyfuel combustion or oxy-combustion) is another centralised capture technique, where pure oxygen is used in the oxidation/combustion of the fuel at a power plant, and where the carbon dioxide-rich flue gas is recycled and used to moderate or control reactor temperatures, as it is non-flammable. This changes the final output gas profile to be almost pure carbon dioxide in steam (water). Simply cooling the flue gas will condense the water out, and the carbon dioxide can then be compressed for piping to a sequestration facility, or re-used in other chemical or energy applications. Recycling the gas exiting from the combustion reactor operating at high temperatures ensures that as much carbon as possible is either fully oxidised to carbon dioxide, or is rejected as carbon ash.

Oxy-fuel (oxyfuel) carbon capture is essentially a post-combustion method with a high temperature combustion reactor and carbon dioxide recycling.

6.1.4.2 *Pinch Point: air separation for oxygen*

One of the main costs in high-efficiency gasification and combustion is the need to use almost pure oxygen for the high temperature oxidation processes. This is normally provided from an Air Separation Unit (ASU); however, in the design outlined, oxygen would be a product of the production of Renewable Hydrogen in Mode A, promoting higher cost efficiency.

6.2 Transition pathways: technology adoption

For Renewable Gas to become widely deployed, there has to be a pathway for its introduction. Most of the elements of the generic Renewable Gas design previously described are already in use in various parts of the petroleum oil refinery, Natural Gas processing and chemical industries. It would be possible, therefore, that if a major fossil fuel refining enterprise were to consider its corporate development over the next couple of decades, that it might elect to set up a Renewable Gas system, producing either Renewable Hydrogen or Renewable Methane, or a combination, amongst its other activities.

6.2.1 Mode A: the production of Renewable Hydrogen

The production of Renewable Hydrogen is a process element unlikely to be seen at the current time at oil and gas refinery, however, that could be due to change in the near future. Increasing amounts of hydrogen are required in petrorefinery. There are several reasons for this, including improvements made in process efficiency, and the evolution in reducing waste by-products (Abbess, 2014). With a more divergent range of quality in the oil and gas feedstocks being refined, and with an increasing need for additives, low sulphur fuels, and a wider range of fuels, more chemistry must be done in refinery. An example of this is the focus on processing heavier crude petroleum oils, which yield lower levels of light and medium distillates, and leave residua more complex to manage (Ancheyta and Rana, n.d.; Rana et al., 2007). To process heavy oils into the usual hydrocarbon fractions requires such steps as hydrotreating and hydrocracking. The hydrogen that is needed for this petrorefinery is currently produced from fossil fuels; for example, the steam reforming (SR, SMR, SRM) of Natural Gas, or the separation of hydrogen from gasified refinery "bottoms" (ExxonMobil, 2011). The volumes of hydrogen required are so large that it would make sense to consider the electrolysis of water to source the hydrogen, using renewable electricity to do so. It could be envisaged that petroleum and gas refineries erect wind turbines or solar farms to be sure of their own renewable power. This perhaps could be a way forward to revitalise the quiescent BP Hydrogen programme.

6.2.2 Mode B: gasification

The gasification process element, often referred to as "partial oxidation" in the oil refinery context, is already being used where difficult to treat heavy oils and residua are left behind after fractionation in the distillation of fossil hydrocarbons (Higman and van der Burgt, 2003, Sections 5.3 and 5.4).

Gasification is also used in the processing of Canadian tar sands into syncrude – the residue from the primary conversion is gasified to provide hydrogen for the secondary conversion (Higman and van der Burgt, 2003, Section 4.2.2; Rettger et al., 2004, 2006).

There is already the ability to feed in various biomass feedstocks (Johansson et al., 2012) to gasifiers, and otherwise waste refinery gases are also suitable for gasification for the purposes of producing hydrogen and syngas (e.g. Breault, 2010; Orhan et al., 2014). If a large-scale slagging gasifier were chosen with a view to consuming all refinery wastes, this would embody the principle of zero waste – or rather near-zero-waste, as long as there were a market for the vitrified final aggregates. It would however have carbon emissions implications.

A fuel cell stack might be an appropriate way to produce hydrogen from lighter refinery offgases, but not for complex heavy oils and residua, unless it were very carbon-rich and pre-processed into fine particles.

6.2.3 Mode C: Methanation Type 1

The first type of methanation, "CO Methanation", is already part of ammonia production, as carbon monoxide appearing in the feed needs to be removed or it could compromise the ammonia synthesis catalyst (Abbess, 2014). Methanation could also be used in conjunction with the Shell Gasification Process (SGP) (e.g. Sep and Wolff, 2004), or considered as part of a Shell Gasification Hydrogen Process (SGHP) (Higman and Eppinger, 1994). It could be used as a post-treatment for any thermal cracking unit where carbon monoxide is undesired in the output chemical stream.

(*Note*: The second type of methanation, "Methanation 2" (Abbess, 2014, Table 9) – the reverse of carbon dioxide reforming of methane – tends to occur under the same conditions in which the "CO Methanation" reaction is promoted. It is probably the source of much of the carbon dioxide produced in a methanation reactor.)

6.2.4 Mode D: power generation

Heat demand at modern petroleum refineries is high, and there is also power demand. The benefits of on-site co-generation (Combined Heat and Power (CHP)) would be appreciable. Currently, power and heat are mostly provided by the combustion of raw feedstock, light condensates or distillates, such as unwanted refinery gases ("tops") or raw Natural Gas. However, gasification or "partial oxidation" units have made an appearance, to produce power as well as hydrogen and syngas, so heat could come from there instead.

As refineries are likely to be creating more unwanted carbon-rich by-products in future, from the use of heavier hydrocarbon resources, if these were to be thermochemically processed into a fine particulate state, they could be suitable for DCFC power generation. Power and heat could be generated by operating a fuel cell stack that internally reformed light refinery offgases to produce hydrogen – which might be more efficient overall than conventional steam methane reforming.

6.2.5 Mode E: Methanation Type 2 and Mode F: Methanation Type 3

Carbon dioxide is usually disposed of via ventilation to the atmosphere of flue gases and unwanted refinery offgases at Natural Gas processing plant and petroleum refineries. It is generally a minor, unwanted by-product of a number of processing units. However, as it becomes important to industry to capture carbon dioxide and either sequester it or recycle it, and particularly as the carbon dioxide content of crude oils, condensates and raw Natural Gases increases, the Sabatier reaction could help. The economic case for it will probably rest on the volume of carbon dioxide that needs to be captured for sequestration (CCS), and whether the carbon dioxide gas streams in the plant are high pressure and pure, or mixed and diffuse. Pressure Swing Adsorption (PSA) is generally used to remove carbon dioxide under pressure from "shifted" syngas to yield hydrogen for use in petrorefinery, and is cost-efficient. In gas sweetening plants, carbon dioxide is frequently washed out of other gases by the action of amine gas treating units at reasonably low pressure, which means the carbon dioxide is not suitable for CCS unless the gas is then compressed, for which there is a significant energy penalty: Sabatier methanation for recycling the carbon might be more cost-efficient.

6.2.6 Mode G: the production of synthetic fuels

Fischer–Tropsch reactions are already being used in, or being planned for, petroleum refinery – for example, where heavy oil fractions have been gasified, and the plant management want to synthesise light oils from the synthesis gas (syngas) produced by gasification (de Klerk, 2011). Synfuel manufacture methods can be BtL (Biomass to Liquids), CtL (Coal to Liquids), GtL (Gas to Liquids), or collectively and non-specifically, XTL or XtL.

6.2.7 Mode H: gas grid injection

In Natural Gas processing plants, sweetened, dry gas is the main product, and is normally taken away by pipeline to customers.

In summary, it can be seen that bringing gasification and methanation into gas and oil processing, and including Renewable Hydrogen gas production, and with some conceptual adjustment about feedstock, and some gas storage, a Renewable Gas system could be implemented. Where Natural Gas processing is conducted in the same facilities as crude petroleum oil is refined, all of the units required, and the engineering skills to operate them, would already be present and closely located. Gas and oil refinery plant take up a large amount of land, so it could be envisaged that a Renewable Gas facility could be constructed within the grounds of a current installation with minimal intervention required. It could be a problem to bring renewable power to the plant for Renewable Hydrogen production, if there were no solar or wind resources in the area, or the refinery were to be some distance from a power grid connection, but there would be no problem with sourcing water, as oil and gas refinery are normally located near the seaports for deliveries, and use water in many of the processing units.

The main reason why Renewable Gas system elements will be added to modern petroleum refineries is because the future of fossil fuels is more diverse, requiring more intervention by increasingly technical chemistry. As the era of fossil fuels begins to wind down, the industry will turn increasingly to heavier oils, more immature oils, and more sour crude oil and sour Natural Gas, and the amount of hydrogen required for refinery will rise sharply (Abbess, 2014). A change in direction could come about if it was recognised that refinery is sufficiently advanced to replace much of the use of heavy hydrocarbon feedstocks with gas and liquid components created purely by industrial chemistry, carbon recycling and hydrogen. Although this may seem revolutionary at the current time, all the technology is available, and there are suggestions that the economic balance may eventually tip away from making complex hydrocarbons that substitute for gasoline and diesel, to making high technology Renewable Gas for gas-drive vehicles.

6.3 Coal routes to low carbon gas (CtG)

Although there has been significant growth in the use of renewable energies, coal is still a major fuel for power generation around the world. Until now, in some senses, the development of green power has enabled the continuing use of coal, as countries and companies can point to their support for renewables as a strong indicator that they are taking steps to address carbon dioxide emissions. However, this truce between coal and renewable energy cannot continue indefinitely, as international treaties and regional regulation stipulate that greenhouse gas emissions from coal-burning must begin to turn down. This is probably not possible without some disruptive technologies that can remove coal carbon from energy systems.

Regional air pollution regulations, such as the European Union Large Combustion Plant Directive (EURLEX, 2001, 2001/80/EC), and its successor, the Industrial Emissions Directive (EURLEX, 2010, 2010/75/EU), will almost certainly prevent some new and replacement coal-fired power stations being constructed, and it seems likely that gas-fired power stations will be built instead. However, there will still be a number of large coal-fired power plants across the region.

Coal remains a valuable fuel, despite its environmental impacts on the areas around mines and post-combustion coal ash heaps, and on air quality and greenhouse gas emissions in the areas where it is burned. If carbon dioxide and other chimney stack emissions can be curtailed, it is likely that coal will continue to be used for energy applications. It has to be borne in mind that economically recoverable coal resources are of the order of 100 years or less at current consumption and consumption growth rates, according to some projections (GEA, 2012, Section 7.4.9; Hook et al., 2010; Mohr and Evans, 2009; Patzek and Croft, 2010; Rutledge, 2011), which means that new infrastructure might only be guaranteed usable for one plant life cycle, unless flexibility in fuelling is included in the design.

6.3.1 Carbonisation, gasification and carbon capture

The carbonisation, or destructive distillation, of coal was one of the earliest methods of producing gas, and the total gasification of coal is a field with several decades of study and operation. Variations of the earliest total gasification reactor designs are worthy of attention as they hold out the promise of efficient energy conversion, pollution mitigation and the opportunity of reducing coal consumption through the use of co-firing with biomass, or with agricultural, municipal (Koukouzas et al., 2008) or commercial waste, for example, Shell's Nuon Magnum (TREC-STEP, 2011; van der Ploeg et al., 2007; Wang and Massoudi, 2013). Innovations could lead to high quality, high pressure carbon dioxide streams being capture-able (e.g., Day et al., 2013), which would enable coal to produce lower carbon gas and power.

6.3.2 Co-firing

An important short-term step for lowering the carbon emissions from coal-fired power generation is to enable co-firing with either biomass or a variety of wastes. From research done so far, there are some synergies in terms of emissions reduction and energy conversion in co-firing. In effect, biomass and/or waste-derived fuels could have a positive catalytic effect on the gasification of coal. Coal will probably offer useful energy for some time to come, but in the medium term, however, without significant means to prevent carbon dioxide emissions, through carbon capture or otherwise, coal burning should be ended.

6.3.3 Fuel cell conversion

It should in theory be possible to efficiently and inexpensively operate Direct Carbon Fuel Cells (DCFC) with coal chars or particles as feedstock – from lower and higher rank coals; however, the emissions advantages would be far greater if the fuels were biomass, so this is likely to be a superior aim for research and development.

6.3.4 Hydrogen, steam, catalysts and sorption

All measures that introduce more Renewable Hydrogen or Renewable Carbon into a gasifier could increase the total available energy from the resulting gas fuel, and lower the net carbon dioxide emissions.

The development of hydrogasification has been seen for coal (and biomass) – gasification using a gasifying agent high in hydrogen.

Research is being conducted on an ongoing basis into catalytically promoted gasification of coal, and coal char, and sorption-enhanced steam hydrogasification (SE-SHR).

6.4 Waste routes to low carbon gas energy (WtE, EfW)

The use of waste materials, whether from agroforestry, local authority refuse, clinical waste, agricultural or industrial waste, or from other sources, is variously known as Energy-from-Waste (EfW) or Waste-to-Energy (WtE), and has significant potential in the short term (e.g. Defra, 2014; Welfle et al., 2014).

The use of waste materials in energy systems is not strictly carbon-neutral, as much waste is originally derived from petrochemicals or other carbon dioxide-forming mined minerals, or from wood or other biomass that used to form a carbon store prior to extraction, that may not have been replaced with new growth. However, the use of biologically derived waste to derive energy is highly strategic, as energy system conversion to carbon dioxide is preferable to its eventual natural decomposition, perhaps to methane, which has a much higher Global Warming Potential (GWP) than carbon dioxide. It is important to make a distinction between biologically derived waste (biomass) that can be anaerobically digested (composted) to produce methane and hydrogen, and that biologically derived material that is non-compostable in the short term and so is rightly included in the category of waste.

Countries with managed forests can hope to have higher levels of wood waste available in future for energy purposes; however, there is likely to be increasingly strong competition between countries for this resource, so there should not be an over-reliance on this feedstock (Defra, 2012, Figure 1).

The good thing about the prospects for developing energy systems that use waste as feedstocks is that there are large amounts of waste in landfill globally that could be used to produce energy, offering the potential to remediate the sites, and prevent fossil fuel use at the same time. Several countries have blocked further landfill, and federal regions are regulating against it (EURLEX, 1999, 1999/31/EC), and yet it will be some time before manufacture and commerce cease producing non-compostable materials which travel to consumers through economic activity and end up as waste, so WtE is a useful policy to adopt.

There has been a parallel development between the rise in the use of fossil fuels for energy, and a rise in their by-products being used in the manufacture of non-compostable plastics. Reducing fossil fuel use and a transition to biodegradable plastics and bioplastics will almost inevitably mean a reduction in the amount of petrochemical-derived plastics in the waste stream (SDC, 2010), which is a limiting factor on strategies for EfW, as most of them are probably relying on the high combustion energy value of plastics for their economics (Defra, 2011, Figure 2; Komilis et al., 2012, Table 5; Themelis and Mussche, 2014). Under the European Waste Framework Directive (EURLEX, 2008, 2008/98/EC; WRAP, 2010), the total volume of waste is anticipated to decline (Defra, 2011, Figure A.1; 2014, Chapter 1, Section 24), with more rigour in the application of waste hierarchy strategies, which implies a limitation to the scope for growth in WtE plant.

Because of the high level of inert or complex composite materials in waste, most proposed energy schemes are based around high temperature gasification. One example is the use of high energy plasma, used as a torch to flash gasify very complex materials into useful carbon-rich and hydrogen-rich gas products. The types of refuse that were acceptable feed for an early Monsanto pyrolysis plant included household appliances, presumably including refrigerators, but not automobiles (Klass, 1998, Chapter 9, Section V).

The use of fuel cells, for example, DCFC, for the processing of waste could be envisaged where the feedstocks are high in carbon, such as plastics (Desclaux et al., 2010), and are pre-processed into fine grains to permit efficient reaction contact surface area.

6.5 Biological routes to low carbon gas (BtG)

Anaerobic digestion (AD) – the decomposition of recently living matter, acted on by microbiological organisms which produce methane and carbon dioxide – is likely to provide many communities around the world with useful gas and power, particularly in urban areas where human and food waste are in abundance, and in agricultural centres of activity.

The fermentation of biomass to alcohols and hydrogen is another biological route to gas, and liquid, fuels.

Some countries and regions are much farther along the road of making full use of their biomass resources than others. For example, although there is an issue about the conflict between food and energy crops, Germany has over 5,000 local biogas production plants (Siemens, 2011), and Nepal will have over 30,000 (GPOBA, 2007). In India and China, biogas production is an essential part of national energy provision, and this could eventually also become the case for the countries that industrialised earlier.

Climate change is likely to alter policies for land use, to enable continued high production of food, so the production of energy crops might be unable to accelerate. However, many biomass waste streams will continue to exist, so biogas and the upgraded biomethane are important to develop.

Research is continuing into improving AD biogas production and consumption, such as combining biogas networks with District Heating (DH) (and cooling) networks, and using CHP or co-generation (or tri-generation) of electricity and heat (and coolth).

Fuel cells, in particular high temperature SOFCs and MCFCs, are likely to be able to make better use of biogases than traditional gas engines, as they are tolerant to higher levels of inert diluents such as carbon dioxide and nitrogen (Larminie and Dicks, 2003, Page 237).

Routes to deriving useful energy fuels from the novel uses of organisms include: the gasification or liquefaction of microalgae, the use of unusual Archaea in AD, and the use of pressure in biogas reactors to increase methane production.

Biological methanation of renewable hydrogen in reactors fed with biogas is possible with suitable microorganisms, as biogas has a high carbon dioxide content (Bensmann et al., 2014).

6.6 Hydrobiological (hydrogen-biological) routes to low carbon gas (H+BtG)

The chemistry of biologically derived anaerobically digested gas produces mostly methane and carbon dioxide, whereas high temperature thermochemical treatments produce mostly hydrogen and carbon monoxide. However, the general biological process whereby biogas is formed has a stage which is high in hydrogen and carbon dioxide. At this point, if more hydrogen is added, and a specific culture of microorganisms is present, the total production of methane will be greater (Reuter, 2013).

Another way to produce more methane, if the production of pipeline-quality biomethane is required from biogas, is to either filter out the carbon dioxide, or pass it into an industrial methanation unit with additional hydrogen gas.

If pipeline-quality gas is not required, simple chemical mixing of hydrogen with biogas will produce a gas with a higher heating value than the biogas alone.

6.7 System integration

6.7.1 Candidates for process configuration for carbon capture

To enable the carbon dioxide recycling of the generic Renewable Gas design, carbon dioxide must be captured. This is most easily done where a gas stream – whether input or output to a process stage – is at high temperature or pressure, or a simple mix, such as flue gas from complete combustion, which is mostly carbon dioxide and water in the form of steam.

There are four process configurations that could be considered the best candidates for this – although only the oxyfuel/oxy-fuel and SOFC-GT designs are currently viewed as having potential for large-scale and centralised plant.

6.7.1.1 Chemical Looping Combustion (CLC), Absorption-Enhanced Reforming (AER) and Calcium Looping (CaL)

The CLC process design, first patented in 1954 (Lewis and Gilliland, 1954) for the production of pure carbon dioxide, and named in 1987 (Proll and Hofbauer, 2011), uses a metal oxide as an oxygen carrier. In the first reaction chamber (the air reactor) metal is oxygen-loaded. Then in the second chamber (the fuel reactor), the metal oxide mediates oxidation of the fuel, giving up its oxygen to become metal once more. Then the metal passes back to the first chamber to take on a fresh load of oxygen. The main usefulness of this process is that it permits the use of air for oxidation, as it separates nitrogen from carbon dioxide in two final output flue gases, making it straightforward to create a pure stream of carbon dioxide, when the water (steam) has been condensed out. The fuel can be in gas, liquid or solid form, but the more complex the feedstock, the more the gas will need cleaning after oxidation. A CLC design could be implemented as a Dual Circulating Fluidised Bed (DCFB) gasifier, a twin reactor system with differing control conditions in each (TU Wien, n.d.).

Absorption-Enhanced Reforming (AER) (e.g. Heffels et al., 2014; Specht et al., 2009) and Calcium Looping (CaL) or SEWGS (Sorption-Enhanced Water Gas Shift) (e.g. Cormos, 2014) are other forms of chemical looping that could progress gasification – the latter calculated to be capable of 95% carbon dioxide capture.

6.7.1.2 "G Twin": indirect pyrolysing gasifier and reformer (indirect gasification)

Many recent designs for gasification configured for biomass feedstock have a combustion chamber alongside the gasifier, to provide heat to the gasifier. This is known as indirect gasification. Feedstock is first gasified at medium temperatures, then the combustion chamber provides the heat indirectly for the gasification chamber by the combustion of the syngas, the biotars and biochar produced by the gasifier. If the desired output is electricity, all the syngas can be combusted; alternatively, if the desired output is syngas, only part of the syngas is combusted. Part of the flue gas from the combustor can be recycled to the gasifier to control the temperature and the balance of the thermochemical reactions. The simplified chemistry of dividing the process into two stages can result in increased efficiency of conversion (e.g. Materazzi et al., 2013). It may also be possible to make use of sorbent-enhanced gasification to treat deleterious chemistry and combine both reactors into one (e.g. Heidenreich and Foscolo, 2015, "UNIQUE" reactor concept). The final flue gas from the combustion chamber should be principally carbon dioxide and water if the temperatures are sufficiently controlled in both reactors.

6.7.1.3 Oxy-fuel/oxyfuel combustion – Allam Cycle and White Rose

There are two forms for the NET Power process, designed by Rodney Allam (Allam, 2013). The first uses gasification of coal, and then combusts the resulting syngas. The gasification is probably intended to be done at a high temperature to maximise the conversion to gas. In the second form of the design, there is no gasification stage, as Natural Gas is the fuel directly used in the combustor. The combustion in both cases produces a gas high in carbon dioxide, and this is recycled into the gasification reactor (in the case that the fuel is coal), as a temperature moderator or gasifying agent. Carbon dioxide is also recycled into the combustion chamber in both forms of the design. The net produced gas is high in carbon dioxide and water in the form of steam, and it is straightforward to separate these.

The White Rose CCS project is being developed by a consortium of Alstom, Drax and BOC in the UK. It is designed to use the oxyfuel process for the combustion of coal, and capture the carbon dioxide from the flue/exhaust gases from the oxy-boiler, recycling a part back into the combustion chamber.

6.7.1.4 Hybrid Solid Oxide Fuel Cell with gas turbine (SOFC-GT)

In design terms, this is similar to indirect gasification, having parallel reactors; however, the gasifier is replaced by a fuel cell stack, utilising electrochemical liberation of electricity and gas by-products from the fuel, in preference to the thermochemical reactions in a gasifier. Several fuel cell designs could be developed, but the most appropriate could be the SOFC. A design hybrid, using methane in an internally reforming fuel cell to produce power, and then combusting the syngas output using a high-efficiency gas turbine, appears to offer optimum energy conversion and is already in use (Bullis, 2013; GE, 2014a; Kuramochi et al., 2009; Maurstad et al., 2006; NFCRC, 2002, 2004; Siemens, 2002; Singhal and Kendall, 2003, Chapter 13). With appropriate flue gas recycling and gas species adjustment before or after it is fed into the fuel cell stack, the final output gas could be "clean" – with high levels of just two gases – carbon dioxide and steam (water). Fuel input flexibility could make hybrids of this kind suitable for use with a range of Renewable Gas feedstocks – whether rich in methane or not (Singhal and Kendall, 2003, Section 12.1). Suitable gas phase fuels would include (bio)digester gas (biogas from AD), biomethane (uprated or upgraded biogas), sewage gas, landfill gas, industrial waste gas, and mixes of biogases and industrial or petrorefinery offgases. Pre-treatment gasification of the feedstocks may be necessary to eliminate some long carbon chain hydrocarbons; and if pre-treatment becomes standard, the use of small amounts of coal could be justified, as known as the "integrated gasification fuel cell cycle" (e.g. Keairns and Newby, 2010); however, the overall destination of the carbon would need to be carefully accounted for.

The method of fabrication of the ceramic electrolytes (Singhal and Kendall, 2003, Chapter 1) in an SOFC stack has been the determining factor in their longevity if the stack undergoes frequent thermal cycling. So large-scale SOFCs, ramping up and down for power regulation, such as would be required for grid load-balancing, could seem unachievable. In addition, the high temperatures inside the SOFC stack, typically between 700 and 1,000°C (Singhal and Kendall, 2003, Chapter 12), would need to be maintained for optimum fuel use efficiency

(Singhal and Kendall, 2003, Chapter 13), another factor precluding the frequent turning up and down of an SOFC power generation stack. Furthermore, gas flow regime changes could lead to over-oxidation or under-oxidation of parts of a fuel cell which could lead to unfavourable "redox" chemistry and equipment damage.

With the generic design being considered, power generation is not required all the time. It could be that an SOFC stack could be designed with two modes so that it can operate continuously: power mode – when generating power, and gas mode – when assisting with Renewable Gas production. It is now thought possible to design and build solid-state fuel cells which could switch between several efficient modes of operation that prevent redox damage and electrolyte cracking, particularly as fabrication methods are improving (e.g. Basu, 2007). One option that could enable SOFC to be used in the generic design considered here would be to switch hydrogen-fuelled SOFC stacks to SOEC or solid oxide electrolysis cell mode (Brisse et al., 2010, Section 9.3.2; Klotz et al., 2014). In SOFC mode, the stack would be producing power, and in the SOEC mode, the stack would be electrolysing water to produce hydrogen – so flipping between Mode D and Mode A.

Another switchable SOFC design could be an internally reforming SOFC that takes Natural Gas or methane as fuel. When not in power generation mode, it would increase its production of hydrogen through maximising the production of syngas – another way of switching between Mode D and Mode B (Hemmes, 2010, Section 10.4), or between Mode D and Mode A. It would need to be determined if a hybrid SOFC-GT design could be adapted to support these mode switches (Singhal and Kendall, 2003, Section 11.6).

Combining hydrogen production with fuel cells for power plants would be a larger-scale version of the regenerative fuel cell design used by the National Aeronautics and Space Administration (NASA) for space applications (NASA, 1989), and related to Francis Thomas Bacon's original design for the Alkaline Fuel Cell (AFC) (Larminie and Dicks, 2003, Section 5.6).

6.7.2 Doubling down on doubling up: peaker plant design

In a Renewable Gas "peaker" power plant, used to meet power demand peaks, and load balance variable renewable electricity, all the carbon is intended to be recycled internally, and no carbon is intended to leave the plant in the form of syngas for producing fuel, or methane for gas grid injection. In this case, Mode B for gasification of the input fuel is only needed to bring carbon into the system if any is lost from gas "leakage" – where carbon dioxide needs to be rejected in small quantities from some units, and the most efficient way to do that is as flue gas vented to air. Most of the carbon is recycled by carbon dioxide capture and fresh gas-making through methanation with new Renewable Hydrogen.

In this version of the Renewable Gas design, Mode B, gasification, will not be operating at the same time as Mode D, power generation. Therefore, there is no need in principle to have separate oxidation equipment for the Mode B gasification and Mode D power generation. This would be in the case that the combustion–gasification reactor is designed to take either biomass or methane-rich fuel. This could theoretically be achieved with an Allam Cycle design, as the methane-rich fuel mode would match the Natural Gas mode of the NET Power design, and the biomass fuel mode would match the coal mode of the NET Power design.

Using the "G Twin" design, in gasification mode (Mode B), air or oxygen (a by-product from the production of Renewable Hydrogen in Mode A) would be introduced into the gasifier with biomass, perhaps also with steam, as gasifying agent and fluidising medium. In power generation (Mode D) with methane from the store for fuel, only oxygen would be added, and the reactor can be expected to run at a higher temperature, so maybe some of the flue gas rich in carbon dioxide would be recycled for temperature control – heat would be drawn to produce the power.

For a chemical looping implementation in the gasification mode (Mode B), some types of biomass in the fuel reactor might not be suitable, and in addition, recycling carbon dioxide into the fuel combustion side might detract from the oxidation. In power generation mode (Mode D), methane from the store would be the fuel, and that could be expected to work optimally.

In general, for gasification mode (Mode B), followed by methanation (Mode C), process heat from methanation (Mode C) could be used to pretreat the biomass fuel for gasification (Mode B), and compress or liquefy the methane for storage. In this case, there would be no need for a carbon monoxide or syngas store, as biosyngas would be methanated directly upon its production (Mode C) when in gasification mode (Mode B), and taken to the methane store.

An SOFC could be used as both an internally reforming methane-to-syngas unit (Mode B) or an electrical power production unit (Mode D), thereby reducing the number of separate processing units required. Another option would be to alternate operation of a fuel cell stack between SOFC (Mode D) and SOEC (Mode A) as required.

6.7.3 Doubling down on doubling up: Sabatier reaction versus syngas methanation

In Mode D in the generic system design, power generation is via the reaction heat of the combustion of methane from the store, and this produces carbon dioxide which is either stored or methanated with Renewable Hydrogen (Mode E or Mode F), and destined for the methane store once more. Mode E and Mode F make use of the Sabatier reaction, as they methanate carbon dioxide, but in system integration terms, this is not strictly necessary, as there is already a process arrangement for making methane from carbon monoxide, in Mode C.

If instead of a CLC arrangement, Mode D is conducted as Chemical-Looping Reforming (CLR) (Proll and Hofbauer, 2009), then with the feedstock of methane from the store, reformer gas, or syngas, is produced, much like the steam reforming of Natural Gas. The hydrogen can be separated for combustion for power production, and the carbon monoxide can be methanated using Mode C.

In CLR, combustion of the fuel is intentionally incomplete to avoid the production of carbon dioxide as much as possible; however, some will probably end up in the gas output. This would be inert in the methanation of carbon monoxide, unless the catalyst were appropriate for carbon dioxide methanation as well (Habazaki et al., 1998).

The kinds of biomass that may be used as fuel in CLR might be restricted; however, a number of bio-oils have been tested (Dupont, 2011), so this could be a solution where biomass has been pyrolysed elsewhere, then transported as bio-oil to a place of power generation.

There are two configurations for CLR – autothermal (CLR(a) or CLRa) and steam (CLR(s) or CLRs) (Lyngfelt, 2007). The precise arrangement would be dependent on the levels of carbon dioxide and water in the syngas that would not take part in the methanation and would need to be filtered, condensed or washed out.

6.7.4 Doubling down on doubling up: heat management and heat integration

The exact choice of power generation and methanation route would need to be decided by heat-balancing and integration, as well as consideration of side effects, the efficiency of catalytic activity, and by-products. The elements with the highest need for thermal management are below.

6.7.4.1 Combustion in power generation (Mode D)

For power generation in Mode D, using a conventional gas turbine, energy input for ignition of the methane–oxygen combustion/oxidation reaction is fairly insignificant, but internal temperatures can be high, for example, 1,200–2,000°C, when there is high oxygen availability. CLC and "G Twin" can be tailored to bring the fuel reactor temperatures down to a more manageable level, whilst keeping it high enough for power generation: in CLC by metal oxide flow rate control and in "G Twin" by carbon dioxide recycling. By contrast, the NET Power designs could be interpreted as relying on the highest possible combustion temperatures.

In power generation Mode D, conducted as CLR, the steam reforming of methane (SRM or SMR) needs activation energy from a reactor in a temperature range above around 1,040°C (Kim et al., 2010, Figure 4) and is essentially two reactions end to end that both produce hydrogen – the reaction of methane with steam, which is endothermic, and the Water Gas Shift Reaction – the reaction of carbon monoxide with steam, which is exothermic. In total, this reaction is endothermic (Abbess, 2014, Table 9), as it is essentially the Sabatier reaction in reverse. The hydrogen produced is combusted to produce power. Some of the heat of that combustion will presumably need to be directed to keeping the CLR unit in operation.

If power generation (Mode D) is performed by a fuel cell stack, this will be electrochemical "combustion", and internal temperatures can be finely tuned. An SOFC or a MCFC will have a high internal temperature, in the region of 700–1,000°C, and expected to be operated continuously to manage heat efficiently. However, if a fuel cell stack were cycled between two modes – one for producing power (Mode D) and one for reforming or synthesising gases (Mode B), the internal temperature and the temperature of exit gases could swing, and so heat-balancing would become an issue. Excess heat for the particular mode would need to be used in other parts of the design.

6.7.4.2 Methanation (Mode C, Mode E, Mode F)

In Mode E and Mode F, using the carbon dioxide produced by Mode D and hydrogen from the store, the Sabatier reaction methanation commences with an activation energy derived from a reactor temperature in the range of 148–527°C (Verostko and Forsythe, 1974); higher than about 546°C, the thermal cracking of methane becomes spontaneous, mitigating any further methanation (Abbess, 2014, Table 9). The reaction proceeds in both directions, and needs to be maintained in the direction of methane formation by the use of catalysts. The reaction is exothermic, meaning that net heat is produced during the reaction, and this heat energy needs to be removed from the reactor to prevent cessation of methane formation. It may be possible to use some of the reactor heat from Mode D to initially activate the Sabatier reaction, but thereafter, the heat generated should make it self-sustaining.

Mode C, the methanation of syngas, will operate favourably in a reactor at a temperature of 280–400°C (Kopyscinski, 2010), and involve several reactions which are all exothermic. Some carbon dioxide will be produced, so theoretically, the Sabatier reaction (the methanation of carbon dioxide) could also take place.

6.7.4.3 *Other thermodynamic questions*

There may be some advantage to combusting some of the methane from the store in Mode D in the CLC configuration and some in the CLR configuration in parallel, or running CLC and CLR in series, in order to balance out carbon oxide species, or heat recovery and exchange, but this might lead to duplication of equipment.

Steam reforming of methane is used in the petrochemical industry for the production of hydrogen, but that is not required in the generic system design under consideration, as hydrogen can be produced by electrolysis. However, if there were thermodynamic advantages of using CLR over CLC, then it might be useful to use part of the methane store to top up the hydrogen store.

6.7.5 Further heat integration

Methanation is exothermic, and heat is required to prepare biomass for gasification, and raise the temperature of the gasification reactor, so naturally there is a synergy between gasification and methane gas-making.

Using the heat in this way does not necessarily reduce the efficiency of the whole energy conversion system, as methanation is essentially a way to lock the energy value of Renewable Hydrogen into the gas.

6.8 Pinch Points

Whichever energy conversion is considered, there will be bottlenecks in chemical reactions or heat flow. The "pinch points" of the generic design are considered below.

6.8.1 Pinch Point: carbon slip

In some of the pathways to methanation, in particular the methanation of syngas, not all the carbon can be converted, or at least, not in the main process or first pass. Early operational plant, such as those designed to produce Synthetic Natural Gas from coal, had several catalytic methanation stages in series (BP, 1972). Residual carbon dioxide will almost certainly appear, and from past experience, it seems likely that something like 3% of the carbon, in the form of carbon dioxide, will end up having to be separated from the other gas products and discarded, through venting to air, as it will not be of sufficient density, pressure or temperature to be usefully captured and recycled. Other inevitable carbon slip could come from the important gas cleaning stages in the chemical conversion of biomass to useful energy fuel gases. In a centralised Renewable Gas power plant, there would be no losses due to gas pipeline transmission, but there may well be some loss of gas during storage, either long-term storage of carbon dioxide or methane. Carbon could also be lost in any parasitic energy loads requiring combustion of one or other fuels for auxiliary purposes, such as gas cleaning, washing and filtering – unless hydrogen can be used for fuel.

Unless the input feedstock is biomass, there is therefore an inevitable net emission of carbon dioxide to the atmosphere, but on a small scale, maybe less than 5% of the total carbon input. This makes a Renewable Gas power plant as carbon-abated as a thermal power plant with 95% post-combustion carbon capture, which compares favourably with post-combustion CCS, where, for reasons of thermodynamics, most proposals offer the prospects of a maximum of 90% capture. The Renewable Gas power plant can also achieve this significant carbon emissions abatement without requiring a system for permanent underground sequestration of carbon dioxide. In addition, it would avoid the parasitic fuel demand for CCS.

The step change in this proposal is that a Renewable Gas plant would not be introducing carbon-rich fuel on a continual basis, so the 5% carbon slip would be from the use of much less carbon-rich fuel than in a once-through thermal power plant. The total carbon dioxide emissions could be as low as 2–3% of a conventional thermal plant, so somewhere between 10 and 30 g CO_2/kWh. The exact figures could only be determined by extensive modelling and demonstration plants.

6.8.2 Pinch Points: the production of Renewable Hydrogen

The management of carbon through recycling in the generic design for Renewable Gas depends on an input of hydrogen gas. At the present time, hydrogen is produced mostly from fossil fuels, which results in carbon dioxide emissions. In addition, Natural Gas, which is mostly methane, is often the source of hydrogen used in industry and for petrochemical refinery needs. However, it would not make any carbon sense to use hydrogen produced from fossil fuels to make methane.

6.8.2.1 Making hydrogen from excess renewable generation

A number of PtG proposals have been put forward that make use of "spare" renewable electricity. The output of wind power and solar power installations can vary quite significantly, and there are some days, and even some hours within various days, where there is too much electricity production for local demand. In the UK, there is a steadily rising amount of "constrained" wind power – generation that is turned down as the grid network does not have the capacity to transmit it, or there is not sufficient load to utilise it. How much excess renewable electricity generation that could in future be made available for hydrogen production and other uses depends on a number of factors, including how much new "nameplate" or rated capacity is installed.

Germany is a prime example of what is possible. As of 2 April 2014, the country had 33.668 GW (gigawatts) of installed wind power capacity and 36.858 GW of installed solar power capacity (Fraunhofer, 2014, "Electricity production from solar and wind in Germany in 2014"). As of 29 October 2014, the country had 35.678 GW of installed wind power capacity, and 38.124 GW of installed solar power capacity. This can be compared to 16 October 2013 when the country had 32.513 GW of wind and 35.651 GW of solar (Fraunhofer, 2013).

Rated capacity does not always translate into power generation, as renewable resources such as wind and solar energy are variable. However, the figures show that Germany's renewable energy policy is achieving its aims. In 2013, wind and solar power generated 76.9 TWh (terawatt hours), and the country exported 31.4 (32.3) TWh (Fraunhofer, 2013, Slide 8) from their national power network, which was more than 20 TWh above the average export of 2003–2005 (Fraunhofer, 2014, Slide 7). In the first eleven months of 2014, German wind and solar plant generated 75 TWh of electrical power.

The increase in renewable electricity generation has clearly had an impact on levels of power exports. For the time being, there is an export market for this power, but that situation could change in the short to medium term. Germany is one of several neighbouring countries growing its renewable electricity generation, roughly in line with its National Renewable Energy Action Plan, a schedule that each country in the European Union has devised to meet the Renewable Energy Directive of 2009. The continued implementation of these plans suggests that the rate of growth in "excess" renewable electricity generation could be of a similar order of magnitude in other countries. In the period from January to May 2014, Germany exported 16.1 TWh of power, so the final figure for exports in 2014 could well be higher than in 2013.

Between January and May 2014, wind and solar generated 37.7 TWh, 8.7 TWh more than in the same period in 2013.

Since the rates of power consumption across Europe could rise, but coal-fired generation is scheduled to be terminated, it would need to be replaced by gas-fired power and renewable generation. Therefore, although they are currently very varied, there could be similar rates of growth in renewable electricity deployment across European countries in future, which would deny Germany the opportunity of exporting its excess power. A first-order projection based on this growth data in renewable electricity suggests that Germany could be unable to export somewhere in the region of 40–60 TWh of power annually by 2030. The question would then be – what can be done with all that essentially zero-cost electricity, roughly 10% of national power demand, and an energy equivalent of roughly 5% of national gas demand for all purposes? The German Department of Energy (dena, n.d.) is seeking to implement a Power-to-Gas strategy to answer that question. The plan is to make gas from spare power and inject it into the gas grid, where it can be stored from season to season. It could also be stored underground in depleted gas fields. The strategy is aiming for 10% of German gas to be renewably sourced in future – 2% from Renewable Hydrogen, and 8% from Renewable Methane and biomethane.

Although Germany is planning for renewable electricity capacity that will enable them to displace 10% of their Natural Gas consumption within ten years, warnings about slowdowns in renewable energy investment in the UK suggest that "spare" wind and solar power there could be far less than in Germany. A rough estimate based on a range of assumptions about onshore and offshore wind power, and solar power, in the UK, could put the figure at around 5 TWh of excess renewable electricity generation annually in the 2025–2030 time frame. Therefore, for the British, although Renewable Gas could initially be developed using Renewable Hydrogen made by the electrolysis of water, its expansion would depend on other routes to hydrogen production.

6.8.2.2 Renewable Hydrogen from biomass: biomass to hydrogen (BtH)

If it can be accepted that carbon dioxide emissions from the energy use of biomass are technically carbon-neutral as far as net loading of the atmosphere goes, then Renewable Hydrogen can be made from biomass by simple thermochemical or more complex biological techniques – and the carbon dioxide vented. However, this has several disadvantages. Notably, there are questions about the long-term reliability in the availability of some types of biomass, and also there is likely to be strong competition for the most energy-dense and uniform kinds of biomass. Second, there are continuing questions about the net carbon balance of the production and use of some kinds of biomass, in particular, tree wood. Third, it is inefficient to obtain hydrogen gas from biomass and then discard the carbon-containing by-products, especially as this could make hydrogen production from biomass more expensive than other methods, owing to the quantities of raw materials required. On this point, it should be noted that there are ongoing projects aiming to raise the percentage of hydrogen make from the thermochemical treatment of biomass, and these rely on the use of steam as an input, thereby incorporating the hydrogen from the water (Ni et al., 2006, Section 3). In a sense, in these methods, the biomass is the promoter or catalyst in the thermolysis of water, and this means that regions of the world with water stress, remote from the oceans, would not be able to apply these techniques (Gandia et al., 2013, Section 1.3.1.2, Page 14).

6.8.2.3 Renewable Hydrogen from thermochemical processes

The thermochemical treatment of biomass, principally gasification, fast high temperature pyrolysis coupled with combustion, and the steam reforming (SR) of the products of pyrolysis and gasification, could become economically competitive with the steam reforming of methane (SMR or SRM), which is the most prevalent way that hydrogen is currently produced. In much the same way that water is used beneficially as a gasifying agent for coal feedstocks, steam gasification (SG) of biomass has been widely researched for higher hydrogen gas yields.

Fast high temperature pyrolysis of appropriate biomass feedstocks produces bio-oil of a sufficient composition for gasification. The benefits of this approach are that fast high temperature pyrolysis can be done on the medium scale and does not require an oxygen supply, for example, from an ASU: despite advances in air separation, air separation could remain relatively costly. Bio-oil is more compact than raw biomass, so having bio-oil produced in a decentralised fashion, and having bio-oil transported to centralised gasification facilities, would enable a more efficient biohydrogen production system than mobilising raw biomass. In addition, with this two-stage approach, it can be possible to minimise undesirable compounds reaching the gasification stage.

6.8.2.4 Renewable Hydrogen from catalysis, sorption and chemical looping

The ash content of biomass can act as a catalyst for gasification, and so can biomass char. The addition of a range of other catalysts has been researched. Biomass ash is composed of inorganic compounds, and by extension, the use of inorganic salts has been trialled for catalytic effects, including molten salts. Dolomite and olivine rock are often used for catalytic effects. Sands have been used as an aid to gasification – with a utility not only as a catalyst, but as a functional device to ensure heat transport in fluidised bed designs, and good contact between solids and gas. Some process designs have several types of catalysts all in the same reactor. Using the carbon dioxide acceptor calcium oxide (quicklime or burnt lime) as a sorption enhancer can raise the production of hydrogen in gasification, and largely prevent tar formation (He et al., 2013, Section 6.5.7). The limestone cycle has also been deployed in Chemical Looping Gasification (CLG) (Huang et al., 2014). Sorption, as in the established CO_2 Acceptor Process, permits the capture of carbon dioxide during the thermochemical process, which can then be efficiently sequestered or re-used, as a relatively pure stream of carbon dioxide is produced when the sorbent is regenerated.

6.8.2.5 Renewable Hydrogen from Aqueous Phase Reforming (APR) and Supercritical Water Gasification (SCW, SCWG)

Hydrogen production relies principally on the use of water, either in the liquid or vapour phase, as the hydrogen-to-carbon ratio of many fuels can be low. From one perspective, the fuel is acting more as a catalyst for hydrogen evolution than the source of the final hydrogen gas. Taking this concept of carbon-rich-fuel-as-catalyst, water-as-hydrogen-source further, very low grade carbohydrate and hydrocarbon feedstocks have been trialled in gasification and other thermal treatments, such as in Aqueous Phase Reforming (APR) and Supercritical Water techniques (SC, SCW, SCWG) (He et al., 2013).

6.8.2.6 Renewable Hydrogen from microbiological processes

Anaerobic dark fermentation and photofermentation have been usefully researched (Martinez-Merino et al., 2013). Photofermentation needs a large land area, and has low conversion efficiency, but dark fermentation appears to be promising, and does not need large open spaces. Direct biophotolysis, to split water through the use of photosynthetic organisms, and indirect biophotolysis, with several stages, could be relatively more costly, considering the relative time taken to produce the hydrogen. The direct use of enzymes on a range of sugars is in development (e.g. Lewis, 2014b).

6.8.2.7 Renewable Hydrogen from WtH: waste to hydrogen

Although post-consumer and industrial waste is not carbon-neutral, the use of waste to produce hydrogen is a good use of otherwise unusable material. Organic waste should preferentially be composted instead, so the remaining inorganic or inert waste would have a low hydrogen ratio, unless it is high in plastics or stable biologically derived polymers or solids.

6.8.2.8 Renewable Hydrogen from the sun: solar hydrogen

This is a suite of techniques that aims to directly split water using concentrated solar energy and oxygen carriers (Roeb et al., 2013). For example, there has been recent research showing that this can be done in a single stage, deploying metal oxides in a similar way to chemical looping – known as isothermal redox (Muhich et al., 2013).

6.8.2.9 Renewable Hydrogen methods: by-product carbon

In the biological and thermochemical treatments of biomass for hydrogen production, there is generally some carbon oxide, or biochar output. Some of this resource could be used for the carbon required as input to the Renewable Gas power plant design. This would be perhaps more efficient than using fossil fuels or fresh biomass as the source of carbon.

6.8.2.10 Renewable Gas dependency on Renewable Hydrogen

There are many research projects into the production of Renewable Hydrogen, and some of these are likely to bear fruit within ten to fifteen years. For example, the American Department of Energy have ambitious plans for cutting the cost of industrial-scale production of clean hydrogen. It must be kept in mind, however, that deciding to embark on a programme to develop Renewable Gas requires taking a calculated risk that Renewable Hydrogen production can be scaled up. Like all new energy systems, there must be a certain amount of speculation in initial investments, and a good deal of focus on priorities. Although Renewable Gas produced from biomass could be the basis of low carbon transport fuels and home heating, concentrating on the carbon recycling "peaker" power plant concept is possibly the best initial option for demonstration, as this would be a relatively compact and independent project.

6.8.2.11 Electrochemical methods for the production of Renewable Hydrogen

Renewable Hydrogen production from water by fuel cell electrolysis, such as PEMFC, is an efficient technique – the only proviso being that the electricity consumed needs to be from renewable resources. To produce significant quantities of Renewable Hydrogen therefore depends on strong growth in renewable power generation. Partially decarbonised hydrogen can be economically produced by PEMFC electrolysis using grid power in industrialised countries where the contribution from renewable electricity is rising (ITM Power, 2014a, "Grid Connected").

6.8.3 Pinch Point: methanation efficiency

The main problem with taking syngas up to methane is that the reaction stages are not perfect – there is generally significant residual carbon oxide gas remaining in the methane. This means that there is usually a separation or further reaction stage of one kind or another after a methanation stage, or series of methanation stages. In the first development of SNG from coal through gasification, several methanation reactor stages were used (e.g. BP, 1972), and it can be effective to pass the products of gasification through methanators at different temperatures with differing catalysts, to push the balance of the chemical reactions along a desired pathway.

However, any extra process stage implies a drop in overall energy conversion efficiency, and so the ideal scenario would be accenting methane production in the gasifier itself. The integration of methanation into the main gas production process in the gasifier is the opposite choice to directing the process equilibrium – the net result of the combination of chemical reactions that take place in the gasifier – towards hydrogen in the form of syngas. Catalysts are being developed to enable the methanation pathways in a variety of process designs, tilting the gas balance towards lighter hydrocarbons and away from hydrogen. The implications are that longer chain hydrocarbons than methane may be produced, or that the overall efficiency is lowered; however, the heating value of the products could be higher than a hydrogen-rich output.

6.8.4 Pinch Point: economics

Owing to democratic pressure resisting underground carbon dioxide storage, and the projected costs, coal power emissions in Germany, as just one example, are unlikely to be abated in the near term by CCS. The imperative to displace coal-fired power generation must be met in the short term therefore by low carbon electricity generation, but variable renewable electricity sources cannot do that on their own, because the resources are variable in output, whereas coal- and gas-firing can be constant, providing "baseload", always on, generation. Despite the fact that Natural Gas remains relatively low cost, the economics are poor for maintaining and operating a gas-fired power station only on an on-demand basis, for when renewable generation is low. Several governments have considered "capacity" payments or similar subsidies to keep gas-fired power plants out of "mothballs", on standby for "backup" to wind and solar power. These subsidies could be justified, if they are made on the basis that a prerequisite for qualification for the subsidy is to take slightly more expensive Renewable Gas fuel rather than Natural Gas. If gas-fired power plant operators were to take this option, it would drastically lower the net carbon dioxide emissions of the plant, and this could stand them in good stead in the case of an imposed carbon tax, levy or fuel quota. This is one way in which Renewable Gas production could be supported in its early stages.

7
The Energy Policy Context for Renewable Gas

7.1 The policy waymarker

The British Government, in language similar to that used by other governance bodies, have identified an energy "trilemma": of having to design policy that meets the needs of the combined criteria of energy security, climate security and economic stability (DECC, 2014b). Energy policy should build in safeguards for energy supply and energy demand control; meet the Carbon Budgets set out by the Climate Change Committee under the articles of the Climate Change Act of 2008 (CCC, 2013), whilst ensuring affordable energy provision, and sustaining energy markets, without destabilising the wider economy. It may seem that the way forward for policy points in several different directions at once, like a city centre waymarker; however, a common framework for policy can be constructed, provided that counter-productive requirements are removed and enabling measures adopted.

Counter-productive conditions include the declaration of a level playing field for all players in the "energy mix". As a very first step, it is important to recognise that not all energy technologies are equal, and that energy system considerations can outweigh individual technology choices (Platt et al., 2014). The governance language that asserts neutrality in energy technology choice has no utility in meeting the energy trilemma – governments cannot be "technology-blind", or "technology agnostic", in policy.

Secondly, there has been an unhelpful assumption that the private sector can leverage the finance needed for the construction of profit centres in the energy system, such as power plants and gas production facilities. When setting policy, it is necessary to consider global and national economic health indicators, and how these might impact the capital commitments and cash flow condition of the energy companies. It might be difficult for private energy companies to embark on significant investment at this time, and so it might be that the optimal way forward is to consider some parts of energy systems as infrastructure assets rather than private enterprises, to permit state-led financing and ownership, especially since market economies are naturally resistant to large-scale projects.

Thirdly, it is absolutely essential to recognise that effective energy demand control measures must be implemented and enforced by policy, as the transition to a low carbon energy economy cannot be effected quickly without strong energy conservation. The growth in renewable electricity, although rapid, will not necessarily be sufficient of itself to meet the carbon budgets in the time allowed, and so the deployment of renewable power, low carbon gas and energy conservation must be advanced in parallel.

Fourthly, the carbon balance of any alterations in the energy system must be carefully considered, both through risk analysis and carbon flows analysis. For example, it is important to avoid high levels of governance intervention where there are risks of low outcomes in measures intended to curtail carbon dioxide and methane emissions.

Fifthly, the energy sector should not be considered in isolation from the rest of the economy, which is highly dependent on the energy sector being successful. The energy sector needs new investment and it needs to be decarbonised, but this needs to be a smooth transition, and so piecemeal or fragmentary responses are not sufficient.

Finally, it is necessary to consider the economic impact of Renewable Gas.

7.1.1 Why we cannot be "technology-neutral" or "technology agnostic" in transforming and re-investing in energy systems

7.1.1.1 Climate change demands a rapid response

The Fifth Assessment report from the United Nations Intergovernmental Panel on Climate Change (IPCC, 2014b) confirms earlier analysis, and indicates that the time frame for significant emissions reduction is at best two decades, in order to safeguard a stable climate. Although there are opportunities for Research and Development, it is clear that the technology toolkit that we have now is the one from which we need to deploy solutions. In addition, the energy installations that are quick-to-build should be used by preference. Given the two decade window, it is not acceptable to base policy for electricity generation on power plants that cannot be built and made operational within ten years. This rules out most new designs for fission nuclear power plants, and definitely rules out fusion nuclear power plants. Besides de-selecting a preference for Generation III or Generation IV nuclear power plants, and unproven Small Modular Reactors, it also rules out starting to build any new nuclear reactors where the lead time to generation could be subject to delays. This eliminates entirely greenfield nuclear reactor sites from consideration. It might be possible to build new nuclear reactors at existing installations within two decades, and there are somewhere between sixty-three and seventy-two new nuclear reactors currently under construction worldwide (IEA, 2015; Kee, 2015), but policymakers should always look to the state of ongoing nuclear power projects globally to assess the risks of this strategy failing through overruns.

The requirement for energy plant to be "quick-to-build" rules out Carbon Capture and Storage (CCS) projects for the most part, where continuing high cost implications, including higher operating costs from increased fuel use for post-combustion configurations, have meant continued resistance to the application of CCS. Carbon markets do not appear to be creating an actionable charge on carbon dioxide emissions, so without a radical reform of the price of carbon, which would be politically unpopular, economically damaging, and complex to enact, CCS might well remain cost-effective in only a small number of circumstances, such as carbon dioxide re-injection into Natural Gas fields where the carbon dioxide content of the Natural Gas is high (Enhanced Gas Recovery), and where carbon dioxide injection into oil wells enhances oil recovery (Enhanced Oil Recovery). Trading mechanisms for carbon could keep the cost of emissions low enough to pass the costs on to end consumers without altering the levels of emissions. It will be near-nigh impossible to implement effective absolute carbon quotas on fossil fuel production, and there are many economic forces working against the imposition of an effective carbon tax or carbon (floor) price. The risk is that the price of carbon cannot be lifted sufficiently in the next decade or so to stimulate CCS investment.

The development of some new energy resources, such as shale gas, Underground Coal Gasification, coalbed methane (CBM, coal seam gas), shale oil, tar sands oil and heavy oil, is ongoing, and in some regions such as the US has been successful in the short term. However, the carbon dioxide and methane emissions implications of bringing on these new resources have not yet been fully appreciated, particularly as regards fugitive emissions in production, abandoned (Davies et al., 2014; Kang et al., 2014) and depleted wells, the emissions resulting from their production and refining, and the final destination of the residual by-products of producing and refining. It is not clear if the production of shale gas, in particular, can be significantly developed within two decades in the UK (Howell, 2014; Pöyry, 2014) and other regions, as geological, political, social and economic conditions are unique to each shale deposit, and so it will inevitably take time to develop.

Marine, wave, lagoon, estuary and other tidal power projects are potentially very significant, but some views are that they still have several decades to mature, so their development should not be central to energy or climate change policy. Many of these are large projects, and although they are highly technically feasible, it is unreasonable to expect markets to unlock major capital when there is so much lingering economic uncertainty.

The only technologies that are "quick-to-build" and that can definitely respect the carbon budget are renewable electricity from wind power, solar power, small hydropower, geothermal resources and gas. In the very short term, this gas can be Natural Gas and still meet emissions requirements, but in the medium to long term, gas energy should be transitioned to be Renewable Gas. Natural Gas is therefore a "bridge" to a low carbon future, although it is not a final destination.

A full risk analysis of a country's energy sector and the steps needed to be taken to transform it to low carbon status might suggest that should wind or solar power fail to live up to expectations, development of new nuclear power, CCS and indigenous unconventional fossil fuels may be necessary as a stopgap plan, but these technologies should not be or become the primary focus of policy, as they do not meet immediate carbon budgetary requirements, neither for the most part could they contribute to energy security in a short time frame. Work to bring these technologies forward should not in any way hinder attention given to energy conservation and the deployment of renewable energies, although not all renewables can be deployed rapidly. Of note, grand marine and tidal energy projects are likely to be too long in duration to meet carbon budgets in a timely fashion. No subsidies or grants or other investment financing mechanisms for longer-to-build technologies should be permitted to undermine the continuing progress in planning for large wind power projects, nor must they be allowed to prevent considerable additional solar power generating capacity. Since Renewable Gas development is going to take a couple of decades to be widely established, it could seem consistent to dismiss focus on low carbon gas. However, a way to transition out of Natural Gas must be prepared, as countries and regions will still be relying on gas energy for at least the next six decades to a century, if not longer, yet not be able to afford the expenditure from their carbon budgets for Natural Gas.

The combination of renewable power and gas is a solution that can work for all types of economies and enterprise, and at all scales. Industrialised countries with already-established gas and power grids, plant and other infrastructure are well placed to ramp up the deployment of renewable electricity technologies and displace coal-fired power plant with gas. Developing countries with partial grids can deploy on-grid and off-grid renewable power, and biogas, if not Natural Gas, for off-grid applications as well as on-grid. China and India are examples of nations that have a well-developed biogas capability and can therefore rely on gas to support their renewable power systems development.

Strategies and incentives to reduce the impacts of coal-burning by co-firing, or substitute firing with biomass, may not address local environmental legislation concerning air (EC, 2013a), soil and water quality and, in addition, may not render the power plants truly low carbon in a full life cycle analysis (Clark, 2014; Harrabin, 2013; Stephenson and MacKay, 2014). Proper carbon accounting may yet conclude that the optimal course of action will be to relegate the large-scale combustion of solid fuels for the generation of power to the history books, and instead transition to a gas-to-power regime, with a view to decarbonising the gas supply at source.

Policy recommendations would be: legislate against coal combustion for power generation using a variety of rationales, increase the flow of capital and finance to renewable electricity, continue developing the gas supply market, promote gas-fired power generation, and incentivise Renewable Gas production asset-building.

For the UK context, the European Union (EU) Industrial Emissions Directive (IED), which follows on from the Large Combustion Plant Directive (LCPD), is one driver that could displace coal combustion from the electricity generation mix, although there are compliance "loopholes" that could keep coal plants open (Jones and Worthington, 2014). Legislation was originally introduced to control sulphur dioxide, nitrogen oxides (NOx), metal and particulate emissions, but if existing coal power plants close, the side effect would be to also influence carbon dioxide emissions. In the UK, the Energy Act of 2013 introduced an Emissions Performance Standard (EPS), which will effectively bar new unabated coal-fired power generation (Brown, 2014a, 2014b), as the permitted carbon dioxide emissions will degress in stages over the next few decades. Since it is unclear when CCS could be available to abate the carbon dioxide emissions of coal-fired power, the outcome of instituting the EPS is that, in the very short term, investment decisions for new generation plant should be shifted from coal technologies to gas, or the EPS cannot be implemented by the dates set. The EU is also implementing a suite of measures and instruments to harmonise and grow the gas market (EC, 2013b, 2014a), in parallel with the electricity market, that should percolate down to the UK. This, in combination with continued international diplomatic efforts to secure access to Natural Gas (FCO, 2014; Yee, 2011), should address the short- to medium-term gas supply needs for the UK (Ofgem, 2012), even as it switches from coal to gas.

What is currently missing from British policy is a mechanism to grow and accelerate renewable electricity generation, consistently, accountably and predictably (Harvey, 2013), and a mechanism to promote the decarbonising of gas supplies, beyond the preliminary work that has been done with biomethane injection into the gas grid (DECC, 2014c). These two policy strands are perhaps the most important, and yet the most underdeveloped, and it is to be hoped that this will not remain the case.

7.1.1.2 Energy system efficiency and integration

Gas and power are highly complementary, and the integration of gas and power in developed countries enables high system efficiency. For example, a great deal of energy can be distributed by a relatively small gas pipeline, and so handling rapid heating or cooling demand swings by using gas energy obviates the need to build a far greater number of power generation plants, and reinforce electricity transmission networks. Decentralised electricity generation in Combined Heat and Power (CHP) plant, whether using gas engines or gas-fed fuel cells, can reduce electricity grid transmission losses (Basu, 2007, Page 3). Gas is especially important as a backup to variable or intermittent renewable electricity generation, as gas-fired power plants are highly flexible and responsive. Without gas-fired generation, a far higher number of renewable electricity installations would be needed, to compensate for the sawtooth wind and undulating solar generation.

Gas can also act as a seasonal storage mechanism, which is crucial as cold months heating demand can be nearly an order of magnitude higher than in warmer months. Even with a comprehensive insulation plan for all buildings, energy demand in winter weeks could still be far higher than power grid capacity, which would also need to provide home charging for electric vehicles, as these gradually arrive off the production lines. Electrical heating should therefore be considered as less desirable than gas central heating, until air, ground and water source, or gas-absorption, heat pumps are much improved, and used in very well-insulated buildings.

The integrated gas and power networks are a key feature of the North American and European energy sectors, with companies such as National Grid, which maintains gas and power grids and load balancing services in the US and the UK, occupying a sphere of action that is to all intents and purposes in the public sector. There is considerable government policy support for gas and power, for example, within the EU, which will apply common standards of compliance and service to all levels of the energy networks, including fuels, as well as supporting trade in energy, which requires a strong interface with transmission and distribution operators.

Power and gas can be partners at all scales of operation. For example, the National Grid's Short-Term Operating Reserve (STOR), which frequently calls on gas-fired power plants (National Grid, 2014a), is designed to enable the balancing of supply and demand in the case of emergency generation shortfalls over the whole power grid in the UK. By contrast in scale, manufactured gas can also act as a good backup for remote off-grid island renewable energy systems (Carter, 2013; IPHE, 2011; PURE, 2008), and in Germany, biogas has been researched for its capacity to supplement and balance local renewable power (German Renewable Energy Agency, 2013). With the anticipated steep climb in the adoption of renewable electricity, which is variable and intermittent over all timescales, flexible gas generation will be increasingly sought after to provide backup. Ultimately, it will be considered that a region can never have too much wind power capacity (WWEA, 2014) – as excess can always be stored in the form of gas. Power plants that are "always on" and providing "baseload" generation will become increasingly inappropriate in this context. For businesses, hospitals and other organisations that wish to have independent power for energy security, the option of gas-fired power and heat, and on-site gas storage systems are going to offer a far lower emissions profile than coal, diesel or oil versions, and offer improved local environmental safety.

Although electrical power has a high utility in every setting, gas is normally considered to be restricted mostly to power generation and use in buildings for heating purposes. However, there is no reason why gas cannot come to have much more utility outside buildings, such as in the area of transportation, where it will complement electric drive. There are already a number of initiatives to develop Compressed Natural Gas (CNG) and hydrogen distribution and filling station networks for use by adapted internal combustion engine (ICE) and fuel cell vehicles.

As fuel cell development continues to be strong, it can be anticipated that gas refill facilities could be installed at docks and train stations for use by seagoing and railway fuel cell-powered electric vehicles (DNV, 2012). The use of gas in fuel cell-powered electric trains could eventually permit the eradication of costly and dangerous overhead cables and electrified lines, which would be highly useful in areas with higher risks of storminess, owing to climate change. The use of gas in fuel cell-powered electric ships, or gas in dual fuel diesel engines, would help meet the need to reduce ocean and atmospheric pollution from marine-based transportation, as demanded by the Marine Pollution (MARPOL) treaty (IMO, 2014). The use of gas in transport is likely to become key in addressing a range of public health and safety concerns, as well as providing a way to prevent urban and global warming from the combustion of diesel, petrol (gasoline) and heavy oil bunker fuel. Gas-powered aircraft have a precedent in the form of space-going rockets, although these relied on the combustion of gas, rather than the electrochemical liberation of energy used in fuel cells. A transition to fuel cell electric propeller aeroplanes can be easily envisaged for smaller aircraft, and could be invaluable in increasing efficiency in flight, thereby reducing emissions.

Gas is highly flexible as an energy vector, and there are good prospects for the increased use of gaseous fuels for transportation, in tandem with the deployment of electric power solutions. The use of gas and electricity in land-based transport, coupled with the use of lighter vehicle components, could potentially offer fuel cost savings and system efficiencies. The distribution of gas and power could be handled more efficiently by grid and pipeline, rather than the road-dependent tanker trucks used for liquid fuel deliveries. This change of transport fuel delivery mechanism would undoubtedly affect the vertically integrated oil businesses, and also to an extent the power supply companies, and so a compensatory mechanism should be developed. This should not be monetary, in the form of subsidies or tax breaks or similar transactions. The "subsidy candy" paradigm to enable change is perhaps broken, especially when everybody lobbies for it. Instead, the petrol (gasoline) and diesel fuel oil distribution and storage businesses should be permitted to have premium access to contracts for delivering the new integrated gas-and-power distribution networks. This could be justified on the basis of economic protection, as failing oil businesses would cripple a number of market activities and investment funds, so support for the low carbon evolution of these companies should be an inherent policy thread.

The efficiency of energy systems is a valuable goal in meeting the energy trilemma, and so policy should always be directed towards that aim. The sector with the highest capacity to deliver efficiency savings is in space heating – the heating of buildings. For oil companies to migrate to being gas-and-power integrated solutions delivery companies, it would be appropriate for them to be awarded the contracts for delivering District Heating (DH) networks, where the heat from the combustion of gas, or electrochemical conversion in fuel cells, is piped to the local neighbourhood in a configuration known as CHP or co-generation, with DH. The cost of land for new energy plant could be exorbitant, particularly in urban areas; however, diesel and petrol delivery businesses are already locally rooted in the form of filling stations, which would therefore be appropriate locations for new CHP/DH schemes as well as filling stations for renewable electricity, CNG (Natural Gas, compressed on site), fossil hydrogen, Renewable Hydrogen and Renewable Methane.

In the H2BER project, being developed by McPhy, ENERTRAG, Total and Linde, renewable electricity, water and both biogas and Natural Gas arrive at the filling station for the planned new Berlin airport; biogas is stored on-site; Renewable Hydrogen is produced from renewable electricity and also stored on-site; and vehicles can tank up with electricity, Natural Gas/biogas or Renewable Hydrogen; and any surplus biogas injected into the regional gas grid (ENERTRAG, n.d., 2013; McPhy, 2014). This is a large scheme, but could be replicated on the smaller scale, minus the biogas storage.

A local filling station in any urban setting could take the gas grid feed, make hydrogen using fuel cell technology (e.g. internally reforming Solid Oxide Fuel Cell, IR-SOFC), and provide for customers to fill up on methane, hydrogen or power as appropriate. It could additionally provide electricity for the grid or local area, and heat for the local DH network. It could store surplus heat under the forecourt in water, in the tanks that previously held petrol (gasoline) and diesel fuel. Alternatively, the tanks could hold cryogenically liquefied hydrogen (boiling point: –252.9°C) and methane (boiling point: –161.5°C), which could be used to deliver coolth in a District Cooling network, or heat exchanged and combusted to provide heat and power on demand. This local energy station would be efficient in responding to power and heat demand peaks in the area – for example, home car charging in the evening – if local area network power and heat metering is adopted. If the stored heat at the filling station is no longer required or has degraded in a water tank, if an SOFC is on-site, it could be switched to Solid Oxide Electrolysis Cell (SOEC) and convert stored water to additional hydrogen. An SOFC/SOEC unit could be relatively compact (Minh and Mogensen, 2013). In addition to this unit, a container-sized unit would be necessary for the storage of the manufactured Renewable Hydrogen (McPhy n.d.), and another could be needed for the production of Renewable Hydrogen from water (e.g. ITM Power, 2013b, PEM fuel cell). Local CHP/DH/filling stations such as this could respond to peak power demands in the morning and evening, that can correspond with heat demand peaks, and reduce inefficiency in the main power grid, as well as utilising the heat of combustion of the fuel, in contrast to centralised power generation plant. To supply Renewable Methane and Renewable Hydrogen to customers, the gas in the grid would need to have been decarbonised at source; that is, the gas pipeline network would need to be distributing Renewable Gas. Although this proposal would be suitable for urban settings, other arrangements would be needed for rural areas – based around locally produced biogas and locally generated renewable power.

It is likely with a transition of this scale that the current gas supply and distribution companies would be best-placed to take a large share of this work, at least in urban areas, and so there would not be market competition issues. A strategy to deploy localised CHP/DH in many places would not conflict with energy conservation programmes for buildings, as they would "meet in the middle" on demand and supply of energy, as they develop in parallel.

A very strong positive outcome of this system of vehicle fuelling would be that heavy goods vehicle (HGV) road traffic would decrease, as the energy could all be delivered by pipe and wire instead of tanker truck or lorry. Rural areas with sufficient renewable power and gas production potential could become partially off-grid, which would reduce the burden on centralised energy production systems.

At the scale of the individual building or household, because of their potential efficiency, gas-fuelled fuel cells used for co-generation, or tri-generation of heat, power and cooling, could theoretically shave carbon emissions from this sector, even before low carbon gas is provided in the grid.

A strategy for integrated gas-and-power appears to be optimal. It is not clear if a similar scope for increased energy system efficiencies can be offered by other resources or resource-and-technology combinations. For this reason, policy should prefer the deployment of gas-and-power solutions over other choices.

7.1.1.3 Convergent global strategies

The scope for global economic and political progress offered by the adoption of gas-and-power strategies is another reason why governments should not be technology-neutral in energy policy.

The ambitions of countries could align if a gas-and-power strategy is pursued, and this could enable improved international relations going forward. For example, the Russian Federation is expected to want to retain a favoured position in the global community based on its energy assets and its willingness to trade across the whole of Eurasia. International concern about its territorial ambitions could perhaps be negotiated down in exchange for preferential gas trading relationships. In particular, Russia could become a major producer of Renewable Gas owing to the biomass growth of its large boreal forests. Iran could be encouraged to increase its Natural Gas production to trade on the global markets, in exchange for inwards investment in integrated Renewable Gas and solar power systems – the kind of "technology transfer" enshrined in the United Nations Framework Convention on Climate Change (UNFCCC) Kyoto Protocol, Article 10 (UNFCCC, 1997). From a diplomatic point of view, for Iran to pursue integrated gas-and-power energy system designs, and abandon its civilian nuclear power ambition, could offer a significant advance, as the dialogue with the International Atomic Energy Agency (IAEA) has been mired in circling negotiations which seem to principally focus on the sourcing, supply chain and disposal of radioactive materials. Other Middle Eastern and North African (MENA) countries could also forward their aspirations by adopting gas-and-power energy systems, particularly given the volatility of petroleum oil prices and concerns about the arrival of Peak Sweet Crude – from both a supply and a demand viewpoint. Current and anticipated growth in demand for heavy and sour crude oils of all kinds would run counter to both carbon and sulphur emissions control measures, so an alternative track needs carving out. MENA countries could meet goals in economic development by trading gas and renewable electricity with Europe, as long as a timetable of decarbonisation of the gas flows using Renewable Gas technologies is designed in. This could help to maintain a balance in the relationships that Europe has with the MENA area, both at risk from Peak Oil – whether that stems from supply or demand factors. China and India could meet a considerable part of their clean air and other local environmental ambitions by adopting gas-and-power, and in addition make progress on their respective global warming commitments.

In having already implemented a gas-and-power system, albeit one that could be vastly improved, North America and the EU have led the way on energy system efficiency and integration. For other regions and countries to pursue the same system concepts, whether with or without universal grid networks, would enable technology trade without having to navigate matters of intellectual property – gas-and-power integration is not newfangled or a proprietary concept. In addition, a great deal of collective learning could be done, and advances made, if more countries were to follow the gas-and-power route and also head towards Renewable Gas.

There is expected to be strong and rising demand for Natural Gas, whether via pipeline or in the form of shipped Liquefied Natural Gas (LNG) in Asia. If the normal supply regions and routes become strained, the development of Renewable Gas would be a welcome in-country addition to supply volumes, even if new resources of Natural Gas can be brought to market.

7.1.2 Why we might need to develop gas as public infrastructure assets

Apart from road transport deliveries of bottled cryogenically liquefied energy gas – propane or Liquid Petroleum Gas (LPG) – the vast majority of gas in the British economy is supplied by pipeline. Because pipelines are laid out in a grid unchanged in the medium term, the gas distribution network is rightly considered part of the national infrastructure.

By contrast, since privatisation of the energy sector in the latter part of the 20th Century, gas-fired power generation has been considered private enterprise. However, since gas-to-power plants are responsive to demand, they have become prone to being left on standby, called on for responding to peaks in power demand, or losses of "baseload" generators such as nuclear power plants. This is one of a number of factors that can dent the economic balance of running the gas-to-power plants, if generating hours become too small to satisfy the financing arrangements. As a consequence, some gas-to-power plants have been mothballed or even shut down completely. Yet in the future, gas-to-power will be a vital component of the lower carbon electricity generation mix.

The Electricity Market Reform (EMR) in the UK has an element for a Capacity Mechanism to subsidise a bank of standby new and existing flexible generation – which would include gas-fired power plants (HMG, 2013; Phillips, 2014). This may or may not stimulate the construction of new gas-fired power plants, as the need for extra gas-to-power will mostly depend on the amount of variable generation absorbed into the grid, or the level of retention of unreliable units, such as nuclear power plants. However, the capacity market auctions should in theory keep some existing gas-fired power plants from closing.

There are several weaknesses with the Capacity Mechanism approach. For example, because of the "technology-neutral" or "technology agnostic" policy, coal-fired power generation is technically eligible to compete for Capacity Mechanism payments, but it is hoped that this eventually becomes untenable, considering other legislation. Additionally, there is no guarantee that capacity payments will be made to as-yet-unconstructed power plants, and the funds may be inefficiently used to maintain failing older plants. The Capacity Mechanism also runs the risk of supporting the inefficient conversion of energy, such as the use of gas in cheaper power plants, for instance, the Open Cycle Gas Turbine (OCGT) design, which are far less efficient than a modern Combined Cycle Gas Turbine (CCGT) installation. Crucially, the Capacity Mechanism is capped through auction, and the allocations may be clustered into expensive projects with long lead times, such as new nuclear power, meaning that there might come a point where desirable levels of flexible gas-fired generation capacity cannot be bought into the energy marketplace.

One way to answer a capacity crunch in the event of the Capacity Mechanism failing could be to create a special category of capacity subsidy, supporting "always-on" baseload gas-fired power plants that can increase capacity on demand by a certain amount. These plants would then also be eligible for another subsidy under the Energy Act – the "Contracts for Difference" or CfD, which are considered to be a replacement for the concept of Feed-in Tariff (FiT) subsidies, formerly used for solar power, and are directed at low carbon energy supply (DECC, 2013a; Fairley and Andrews, 2014; Low Carbon Contracts Company, 2014), and intended to support the financing of new generation. However, double subsidies would certainly garner much negative criticism, particularly if by extension of the principle they were also offered to coal-fired power plants. There might be a good case to make for offering double subsidies only to those gas-to-power generation technologies which are efficient but inflexible, such as some designs of fuel cells, for example, large-scale SOFC plants.

In the UK, the preliminary results of the first Capacity Auction (National Grid, 2014c) suggest that the Capacity Mechanism appears to have been interpreted as a means for "keeping the lights on" (Ofgem, 2013, 2014) in any way possible, without having to directly subsidise new generation plant or pay attention to the environmental emissions implications of energy technologies. In addition, Capacity Mechanism payments from this first round will be directed partly to "baseload", or always-on generators, such as already-existing coal-fired and nuclear power plants, undermining the rationale for the subsidy. Furthermore, the Contracts for Difference "strike prices" appear to have been set high enough to print money for costly new nuclear power projects, concentrating limited state subsidies on large, centralised projects, without considering the impact on the need to stimulate strong growth in decentralised renewable electricity such as wind and solar power. It is important that measures to guarantee power from flexible generators facing economic risks should be implemented in parallel with the deliberate deployment of variable renewable energy: no capacity payments should be made if renewable electricity projects are being halted, as this would be a direct windfall for existing coal-fired and nuclear power plants, which will gain economic advantage while renewable electricity deployment is scaled back.

If the Capacity Mechanism and the Contracts for Difference Feed in Tariffs do not stimulate new, flexible, electricity generation, another way to answer the problem of unmet low carbon generation capacity could be for the state to build and own gas-fired power plants. A case to justify this could easily be made – as this state-financed gas-fired generation would support the balancing of the supply of electricity, which is already a responsibility of the state, conducted through the offices of the National Grid. Gas-fired power plants would effectively be considered to be power system protection facilities, and as such could be built through tax revenue expenditure. An increased dependency upon Natural Gas for power generation is making the question of gas storage rise in importance, and here too, government financing could be easily justified.

A policy framework of this nature would place gas-fired power plant and gas storage facilities in the category of infrastructure, just as the gas and power grids are, and their construction could easily be accorded the value of assets of national interest.

Existing "baseload" power generators could survive the onslaught of renewable electricity, and capacity payments made to flexible gas power plants, if they were to re-purpose their coal-fired and nuclear power plants as hydrogen or methane production facilities – especially if they can build in carbon capture or carbon recycling.

It is not only the UK where the gas-to-power capacity problem is being felt – other countries in the EU are also facing a potential crisis of generation capacity (Caldecott and McDaniels, 2014), and similar solutions to the UK's Capacity Mechanism are being worked through. Some

European countries remain strongly dependent on coal-fired power generation, such as Germany, even despite its impressive advances in renewable electricity (Fraunhofer, 2013, 2014), and a relapse into coal-fired generation in countries like Germany and the UK needs to be prevented. Therefore, a policy focus with the ambition of increasing gas-fired generation will likely emerge. However, the investment landscape for new energy plant is likely to remain unstable, so there is therefore the potential need for CCGT power plants and gas storage facilities to be converted into a public asset class across the EU. Gas storage facilities – along with border interconnectors, and LNG shipping terminals – will attract high level civilian force protection as a consequence of energy security policy, and this may well become true of CCGT plant. It would therefore make a lot of sense for the security services to be guarding national assets.

Although heat and power services may become highly decentralised in future, it is hard to conceive of the manufacture of some forms of Renewable Gas in any setting apart from a centralised facility. Whilst biogas, biomethane and some biohydrogen can be produced "in the field", Renewable Gas processes that involve carbon recycling and high temperature reactors are only likely to be efficient at the industrial plant scale. It is therefore hard to envisage small players entering this particular market, and so competition is anticipated to be largely absent – along with the drivers for cost reductions and new Research and Development. This is an additional rationale for state-led investment and deployment of new low carbon gas manufacture and carbon recycling assets, especially as Renewable Gas systems with carbon recycling would need to have power generation co-located with gas manufacture, taking them into the nationally managed power grid system.

7.1.3 Why energy demand management is critical

The Natural Gas extracted from underneath the North Sea and the surrounding land has been a major resource for Europe, but the indicators are that this gas province is overall in significant decline. Although trading relationships for importing gas into Europe are currently favourable, in future they might become less so, for example, because of market competition from other parts of the world. It could take several decades to develop a significant Renewable Gas production industry, and this is likely to be essential, especially as it is not yet clear if shale gas extraction could provide a significant replacement for depleting Natural Gas – not even in the UK, where the prospect of shale gas is fêted. Energy security considerations therefore suggest that a crucial policy would be to implement strong energy efficiency and conservation measures, whilst also developing indigenous energy resources, including Renewable Gas. Efficiency with the use of energy is not only relevant for the end user – it is also important in energy conversions – the transformation from one kind of energy to another. The historical development path of energy systems has built-in inefficiencies that should ideally be addressed as soon as possible.

In the UK, for example, the "Dash for Gas", in the 1990s and beyond, saw Natural Gas used as a fuel for power generation rise from roughly 2.5 to 32.0 million tonnes of oil equivalent (mtoe) annually between 1992 and 2010 (DECC, 2013b). Since this was by and large in centralised power stations, with no local uses for the heat of combustion, this heat was all disposed of as waste – in 1995, this was something like 68.4 mtoe from all power stations. Although increasingly efficient plant has been deployed, which is possible with gas fuel, in 2013, 44 mtoe of energy is still being wasted, mostly as heat, in energy conversion at power stations. This is equivalent to the whole of end use energy demand in the residential or domestic built sector – 43.8 mtoe in 2013. Strikingly, this figure is also comparable to the amount of Natural Gas that is imported into the UK – 46 mtoe in 2013, (although 9.4 mtoe is then exported), making this issue central to energy security planning. The vast proportion of energy used in the home is for space and water heating, and so it emerges that centralised power plants are losing to the environment, unused, roughly as much heat energy as is needed in the entirety of the nation's homes.

The rational conclusion is that gas-fired power generation needs to become more localised to centres of population where the heat can be used instead of lost – a model of decentralisation, or distributed generation (DG). The model of local CHP with DH should therefore be adopted, and this will need incentivising in the British context, as it will be a reversion to an historical model of energy administration, and will require the co-operation of local town and city authorities. The work being done with urban solar power community-scale projects and independent energy co-operatives could be used as a foundation for local authorities to commence CHP/DH projects – using the same models of engagement. It is to be hoped that central government will see the national interest in removing gross energy system inefficiencies, and promote and financially support local authorities and communities in constructing renewable energy and CHP/DH energy efficiency schemes, despite austerity measures constraining the social budget, and despite the centralised energy sector continuing to lobby for advantages (Monbiot, 2015).

Despite some countries in Europe taking a strong lead on implementing decentralised energy projects, which will undoubtedly influence direction, it will still take several decades for the region as a whole to decentralise electricity generation sufficiently to remove most of the inefficiencies of heat losses from centralised power plants. Some centralised power plants will need to remain, for grid load balancing during peaks and troughs, but the heat could be used if the centralised electricity generators brought on the development of the manufacture of Renewable Gas – as the chemistry would be most efficient with heat input. For this to happen, there is likely to be a need for a Renewable Gas Obligation placed upon the centralised power generators – an obligation to manufacture renewably some of the gas that they combust to generate electricity. This would be a regulation mandate, not a subsidy, although early Renewable Gas plant would probably need state financial support, as "First Of A Kind" construction costs will put it at a disadvantage compared to incumbent technologies. In the UK, a Renewable Gas Obligation could be an extension to the existing Green Gas Certification Scheme operated by the Renewable Energy Association for Biomethane injected into the national gas grid (Green Gas, 2014).

A Renewable Gas Obligation is also likely to be necessary for the fossil fuel production companies – the obligation to substitute some of the Natural Gas that they are supplying into the economy, and some of the Natural Gas they are using in refinery, to be replaced by Renewable Gas. Where a fossil fuel production company is also an electricity generator, their obligations to both supply and consume Renewable Gas instead of Natural Gas could be met by the same investments. Their development would be an economic stimulus, as the Renewable Gas technology and plant would most likely be commissioned from third parties in engineering groups. A corollary to this would be that a Renewable Gas Obligation placed on companies like BP and Royal Dutch Shell could facilitate the necessary transition from the production and consumption of liquid transport fuels to gas phase transport fuels. As most Natural Gas production companies are also producers of crude petroleum oil, developing Renewable Gas production capability could compensate for the decline in demand for refined oil products as alternative vehicle fuelling increases. Again, the Renewable Gas Obligation would be a mandate, not a subsidy – fossil fuel production companies would be expected to invest in Renewable Gas production capability themselves. Car and HGV manufacturers, as usual, will need to be part of the contract of transition, and it is the history of these working relationships that holds within it the seeds of success.

If Renewable Gas power plants adopt carbon recycling, this could offer the prospect of lower input fuel and feedstock requirements. The same technology could be used for on-site power generation at petrorefineries, in fulfilment of the Renewable Gas Obligation, and thereby reduce overall energy demand.

While Renewable Gas is being developed over the next few decades, and while efficient decentralised power-and-gas-and-heat systems are being built, the energy security question for the residential/domestic sector will need to be answered by energy conservation at the end point of use. Subsidy-funded programmes for the insulation of buildings are probably the only way that the British, for example, can reduce energy demand appreciably in the home – where annual energy demand for all resources was 43.8 mtoe in 2013, and has averaged at around 43 mtoe since 1970, on a slow rising trend between 1970 and 2004, and a slow descent since then (DECC, 2014d).

Energy conservation can help supply meet demand even if Natural Gas supply is pinched before Renewable Gas resources are significantly developed. A useful example is the British case. In the 2014 edition of the UK's National Grid expert roundtable "Future Energy Scenarios", total gas demand for the UK in 2013 is marked as 834.9 terawatt hours per year (TWh/y) (National Grid, 2014b). Under the "Gone Green" scenario of "high affordability and high sustainability", in 2035, UK gas demand would be expected to be of the order of 705.1 TWh, a reduction of around 15.5% in consumption. During that period, however, UK Continental Shelf (UKCS) Natural Gas production is expected to fall from 32.0 billion cubic metres per year (bcm/y) to 10.0 bcm/y. Shale gas is expected to be developed to contribute 15.9 bcm/y. Biomethane – high-methane biogas from the anaerobic digestion of biomass – is expected to be contributing only 3.3 bcm/y by 2035 under the Gone Green scenario. Coalbed methane (CBM) is expected to produce at a volume of 0.9 bcm/y. It is not yet known if shale gas production can be brought on stream in quantity in line with this scenario, but if it were, it would contribute something of the order of 174.9 TWh, or nearly 25% of total gas supply in this projection; however, it could take until 2020 to begin to see significant progress in shale gas supply (EEF, 2014). The important point is that even if shale gas can be produced at the levels suggested in the scenario, indigenous production of gas would still only be 47% of consumption according to National Grid's figures. Progress in the development of shale gas, CBM and biomethane could all well experience setbacks, leading to pinch points in the supply of gas produced in the UK, and negative implications for energy security and energy prices.

The development of Renewable Gas may not be rapid enough to compensate for these eventualities, and so energy conservation is essential in parallel with work on alternatives to Natural Gas.

Energy conservation will grant alternative gas resources a higher utility in addressing energy security concerns, as they will contribute more, percentage-wise, to total supply. In National Grid's 2009 paper "Renewable Gas" (National Grid, 2009), the 2020 "stretch" scenario projects that biomethane, together with industrial Renewable Gas manufactured by the gasification of biomass, could contribute 18.432 bcm/y. If the Renewable Gas agenda is promoted, and strong energy conservation measures are also taken in the next decade, by 2025, UK national Natural Gas consumption could be of the order of 60 bcm/y, and Renewable Gas could fulfil 29.3% of that, and the combination of Renewable Gas and nationally produced Natural Gas from the UKCS could contribute just over 70% of the UK's gas demand. This would return the gas trade balance back to the situation in 2006/2007, when there was approximately 30% import dependence, compared to 2013, when there was roughly 65% import dependence (DECC, 2014a).

Shale gas, if it were not too costly to produce, and can be produced in significant volumes by that time, would then be useful in asserting gas energy independence, but there is no guarantee at this stage that it could offer comparable quantities to Renewable Gas. The successful production of shale gas at significant volumes in the United Kingdom has yet to be proven, whereas the manufacture of industrial-scale volumes of gas has not only been demonstrated, it was also the main gas resource for many decades. It could be considered incongruous to set great store by shale gas with all its concomitant uncertainties and risks, when Renewable Gas is already a viable prospect.

The parallel pathways of strong energy conservation measures and increasing the diversity of energy resources answer the energy trilemma; however, there is a need to keep a focus on the carbon dioxide and methane emissions implications for the whole energy system when considering new energy technologies. Any intervention that causes an adjustment to the whole energy system may tamper with measures to keep to the carbon budget, unless it is carefully introduced.

7.1.4 Carbon balance

Changes to energy systems should take particular care that they do not cause an increase in carbon dioxide and methane emissions. For example, it would be counter-productive if fugitive emissions of Natural Gas were to increase if a policy of decentralised CHP/DH systems were applied and new gas grid lines or gas power houses were not as gas-tight as they should be. Another example could be implementing a national building insulation project without compensating for changes in expectations of levels of comfort, that give a "rebound effect" in terms of increased net energy consumption. Programmes for building insulation must also consider the potential increase in emissions from the re-purposing of disposable income resulting from energy savings. National policy to decrease carbon dioxide emissions can even have an impact on international carbon balances. For example, the British policy of promoting the combustion of biomass, co-fired with coal in some instances, can have a significant impact on the cross-Atlantic trade in wood pellets, which some analysis suggests could lead to increased greenhouse gas emissions (Stephenson and MacKay, 2014). Changes to energy systems that reach wider into regional and global spheres should also take care not to break the "safe climate" carbon budget calculated by the science community (IPCC, 2014b). One example of this is in the promotion of hydrogen as an energy vector.

Although hydrogen is a zero carbon energy fuel at the point of use, it is in the development of hydrogen resources that the carbon balance of the whole energy system is perhaps placed most at risk. The International Energy Agency has recruited research and policy organisations and is working with private enterprise as part of the Hydrogen Implementing Agreement (HIA). This has focussed on bringing hydrogen into the transport sector. Hydrogen Filling Station (HFS) or Hydrogen Refuelling Station (HRS) networks are being considered in several regions (e.g. DOE, 2013; IGEM, 2014), yet the source of the hydrogen is not renewable yet. Roughly 80% of all hydrogen produced currently for all purposes is from hydrocarbon feedstocks (Spath and Dayton, 2003), the majority from Natural Gas. Despite the trend towards using Natural Gas as a feedstock for producing hydrogen, nearly 20% of hydrogen is still produced from coal (Romm, 2013). This means that to increase the use of hydrogen with no upwards pressure on carbon dioxide emissions would need a major, targeted and specific investment in Renewable Hydrogen production capability.

Most of the hydrogen that is produced already (see Table 7.1) has committed uses. Nearly 90% of all hydrogen produced is "captive" – for use by the producers, and not for sale as "merchant" hydrogen (PR Newswire, 2011). Of this non-marketed "captive" hydrogen, upwards of 75% is destined for the synthesis of ammonia to be used as agricultural fertiliser, and in petrorefinery (Buchel et al., 1989; Spath and Dayton, 2003; Wawrzinek and Keller, 2007). There is some flexibility however, as part of the hydrogen recovered as a by-product from industrial chemical processes, such as the ammonia synthesis plants, and from the production of chlorine, sodium chlorate, ethylene and styrene, could be re-purposed for the budding hydrogen transport sector (Ball et al., 2009; Schindler et al., 2006). In addition, petrorefineries could choose to increase their in-house production of hydrogen with the intention of selling it as "merchant" hydrogen for bulking up hydrogen transport sector supplies (Ball et al., 2009). However, the deployment of hydrogen-fuelled transport "over the fence" would not be the only driver of hydrogen production in refineries. Two important and parallel developments in the regulations for liquid fuels will not only increase hydrogen demand but also increase carbon-rich residues that will need to be disposed of.

7.1.4.1 Low sulphur fuels for land vehicles

Owing to changes in environmental and health and safety regulation, petrorefinery is making more use of hydrogen gas for processing. For example, the trend towards low sulphur and ultra-low sulphur road fuel standards in the US and Europe mean that deeper desulphurisation is required from petrorefinery. This means a greater use of hydrodesulphurisation (HDS) as well as other cleaning processing unit steps, all of which will create a greater demand for hydrogen (Abbess, 2014). The more hydrogen is used, the more the processing will cost, as more Natural Gas will need to be used to make it, through Steam Methane Reforming (SMR or SRM), the most common method. The more Natural Gas is used to produce hydrogen, the higher the plant carbon dioxide emissions will be. Furthermore, because of the higher cost of the low sulphur processing, refineries will undoubtedly prefer to use cheaper, heavier crude oil feedstocks. This will mean higher carbon dioxide emissions from the processing, because the hydrogen-to-carbon ratio of heavier crude oils is lower than for lighter crudes, so even more Natural Gas will need to be reformed for hydrogen. The use of heavier crude oils will also mean a higher "carbon rejection" rate – an increase in the amount of unused carbon-rich petroleum residues, because the processed crude oil fractions will not match the product slate required, even after thermal treatments such as coking or cracking.

7.1.4.2 Marine Pollution (MARPOL) treaty and shipping

With the next implementation stage of the MARPOL treaty, as of 1 January 2015, shipping in the EU ECA (SECA) area will need to have replaced the use of Heavy Fuel Oil (HFO) with Marine Gas Oil (MGO)

Table 7.1 Estimates of hydrogen production and consumption

Year	Geographical Region	Production/ Consumption	Production Method	Gas Mix	Hydrogen (mt/y)	Reference
2003	Global	Production	Gasification	Syngas	~7.03	Shoko et al. (2006)
2010	Global	Production	Gasification	Syngas	~6.69	NETL (2014)
2018	Global	Production	Gasification	Syngas	~11.09	PR Newswire (2013)
2008	France	Production	All	Hydrogen	0.09	Priem (2013)
1996	Europe	Production	All	Hydrogen	4.58	Buchel (1989)
2001	Europe	Production	All	Hydrogen	4.00	Wurster (2001a, 2001b)
2006	Europe	Production	All	Hydrogen	8.34	Priem (2013)
1999	US	Consumption	All	Hydrogen	7.64	Spath and Dayton (2003)
2007	US	Consumption	All	Hydrogen	5.81	Kerry (2007)
2009	Petrorefinery Global	Consumption	All	Hydrogen	35.99	Praxair (2011)
2011	Petrorefinery Global	Consumption	All	Hydrogen	22.47	Rostrup-Nielsen and Hansen (2011)
2014	Petrorefinery Global	Consumption	All	Hydrogen	23.51	SHFCA (2014)
2015	Petrorefinery Global	Consumption	All	Hydrogen	45.22	Praxair (2011)
2019	Petrorefinery Global	Consumption	All	Hydrogen	26.07	SHFCA (2014)
2025	Petrorefinery Global	Consumption	All	Hydrogen	51.68	Praxair (2011)
1996	Global	Production	All	Hydrogen	35.95	Buchel (1989)
1999	Global	Production	All	Hydrogen	38.20	Spath and Dayton (2003)
2001	Global	Production	All	Hydrogen	44.94	Wurster (2001a, 2001b)
2005	Global	Production	All	Hydrogen	53.93	Mueller-Langer et al. (2007)
2006	Global	Production	All	Hydrogen	44.94	Evers (2008a)
2006	Global	Production	All	Hydrogen	56.70	Priem (2013)
2007	Global	Production	All	Hydrogen	~50.00	Kerry (2007)
2007	Global	Production	All	Hydrogen	53.93	Wawrzinek and Keller (2007)
2009	Global	Production (Low)	All	Hydrogen	53.93	Ball et al. (2009)
2009	Global	Production (High)	All	Hydrogen	64.71	Ball et al. (2009)

2010	Global	Production	All	Hydrogen	53.93	Brisse et al. (2010)
2010	Global	Production	All	Hydrogen	42.64	Wokaun and Willhelm (2011)
2010	Global	Production	All	Hydrogen	53.00	PR Newswire (2011)
2011	Global	Consumption	All	Hydrogen	56.62	Rostrup-Nielsen and Hansen (2011)
2013	Global	Production	All	Hydrogen	44.94	Romm (2013)
2016	Global	Production	All	Hydrogen	69.60	PR Newswire (2011)
2006	Germany	Production	By-product	Hydrogen	0.04	Schindler et al. (2006)
2009	Germany	Production	By-product	Hydrogen	0.03	Baues (2009)
2010	Germany	Production	By-product (Avail)	Hydrogen	0.03	Damosso (2010)
2008	France	Production	By-product	Hydrogen	~0.06	Priem (2013)
2007	Europe	Production	By-product (Total)	Hydrogen	2.07	Ball et al. (2009)
2007	Europe	Production (Low)	By-product (Avail)	Hydrogen	0.18	Ball et al. (2009)
2007	Europe	Production (High)	By-product (Avail)	Hydrogen	0.45	Ball et al. (2009)
2007	Europe	Production	Excess Refinery	Hydrogen	0.45	Ball et al. (2009)
2015	Global	Production (Potl)	Biomass-to-H_2	Hydrogen	40.00	Czernik et al. (2004)
2015	Global	Production (Potl)	Waste-to-H_2	Hydrogen	6.00	Czernik et al. (2004)

mt/y = million tonnes per year.

or Marine Diesel Oil (MDO), unless they make use of on-board sulphur "scrubber" equipment (McGill et al., 2013; Rozmarynowska and Oldakowski, 2012; UKPIA, 2012) – or switch from oil to gas fuels. MGO is more costly than HFO (also known as Residual Fuel Oil – RFO), principally because it is a pure distillate, rather than a mixed residue. MDO is MGO blended with a small amount of residue (Viscopedia, n.d.), but it still has the same low sulphur profile as MGO. Since HFO is the "bottom of the barrel", these newly surplus-to-requirements heavy residual refinery products will need to find a new destination – for example in coking and cracking units, to recover shorter-chain hydrocarbon compounds that can be used for fuels. However, extra refinery processing inevitably means more energy consumption by the processing plant, and implies higher overall carbon dioxide emissions – and less feedstock-to-fuel conversion, or "yield".

The destination of the "rejected" carbon-rich residue is important to note. In refineries that use hydrocracking, for example, the processing will imply a greater demand for hydrogen – although some of this extra hydrogen demand would be offset because some hydrogen will be recuperated as a by-product of the cracking processes. A higher hydrogen demand would be met principally by higher Natural Gas use, which would imply higher carbon dioxide emissions – although some of the extra Natural Gas used could be offset by recycling light hydrocarbon by-products of refinery processing, which can substitute for Natural Gas in hydrogen production.

In petrorefineries where coking plant is already installed, or where new cokers are planned, the result of directing HFO residuum or "resid" to coking to convert it into more useful fuels will mean an increase in the by-production of petroleum coke or petcoke. This phenomenon is already being experienced in North America with the processing of heavy oil or oil sands (tar sands) (Andrews and Lattanzio, 2013). When refineries are operating in a fully balanced way, virtually all the carbon in the original feedstock eventually finds its way to the atmosphere in the form of carbon dioxide, either at the refinery, or when the final fuel products are burned, which is why using heavy crude oil has a worse global warming impact than using light crude oil, as it has a higher carbon-to-hydrogen ratio.

The accumulation of excess "rejected" carbon into petcoke provides an opportunity to lessen final carbon dioxide emissions by undertaking geological carbon sequestration – not of carbon dioxide gas, but of the petcoke, which is mostly solid carbon. Methods to permanently sequester petcoke could include land burial and deep ocean floor burial (de Bakker, 2014). However, in all likelihood, petcoke will become increasingly used to generate power via gasification – both inside and outside petrorefineries – releasing the carbon in the form of carbon dioxide emissions into the atmosphere. Alternatively, since incorporating hydrogen or water in the thermal processing of petcoke would lessen the carbon dioxide emissions impact per unit of final energy produced, the steam gasification (to produce hydrogen) or hydrogasification (to produce methane) of petcoke might be preferable to direct power generation via combustion or partial oxidation (Ray, 2014; Sutikino and Turini, 2012).

In petrorefineries where heavy and residual fuel oils are used as a feedstock for fluid catalytic cracking (FCC) units, hydrogen will be a useful gas phase by-product. However, hydrogen demand in all refineries will increase as hydrotreating replaces other processing, and outstrip the supply from hydrogen recovery, and this will be true even in refineries with high levels of by-product hydrogen.

From this cursory analysis, it can be seen that one of the counterproductive outcomes of improving the control of sulphur content in fuels for both shipping and land vehicles is likely to be a tendency towards petrorefineries sourcing more heavy crude oils, which are usually much less costly than lighter crude. Besides the higher need for desulphurisation that results from the legislation, the use of heavier crude oils with a higher carbon-to-hydrogen ratio will create imbalances in the flow of carbon in refineries, and because of parasitic energy use and yield inefficiencies arising from coking and cracking, set back progress on carbon budgets – that is, carbon dioxide emissions from petrorefineries will increase.

Petrorefinery processes are expected to create increased demand for hydrogen (EIA, 2006; Gonzalez, 2009; Karatzos et al., 2014, Section 1.10), whether because of environmental reasons, or to manage changing crude oil profiles, or product demand profiles. Therefore, a logical pathway would be for petrorefineries to enhance their hydrogen production capacity. Given that the HIA is beginning to affect mobility (transport) policy in industrialised countries, the petrorefineries will be in a good position to overproduce hydrogen, and sell "merchant" hydrogen "over the fence" into the new HRS/HFS network (Ball et al., 2009). However, if this hydrogen is produced from fossil fuels, especially coals, cokes, petcoke, petroleum residues and the heavier crude oils, then this will not represent progress.

The environmental legislation on fuels, the HIA and modernised refinery processing are all drivers for hydrogen demand which, without increased focus on and support for Renewable Hydrogen, would lock the energy system into an extended reliance on fossil fuels. The issue of "carbon slip" in cleaning up marine and motoring fuels may be just a small step change upwards in greenhouse gas emissions and resource consumption (ECN, 2007b), but it is indicative of what could happen on a much larger scale if significantly more hydrogen is demanded for an energy supply sector still wedded to coal, oil and Natural Gas and still building megarefineries, especially in Asia, for processing heavy and sour crude oil.

There is therefore the need to ask whether the HIA is trying to leap before it can crawl. In order to ensure that rolling out a hydrogen distribution network and a fleet of hydrogen vehicles does not negatively impact carbon dioxide emissions controls, some basic regulatory measures will be needed to guard against "carbon leakage" in the entire energy system. It is not sufficient to set an emissions performance standard for the use of finished refinery fuels – the whole production life cycle must be considered. There should be no hiding place for carbon.

In this regard, one important policy indicator has to be the life cycle carbon intensity of each of the range of products from a petroleum refinery; calculation of which is always possible based on the data from the computer software used to programme and monitor refinery workflows, from raw crude oil to finished fuels, fuel additives and other chemicals. In the EU, the amended Fuel Quality Directive (FQD) (EC, 2009) sets the ambition for a Low Carbon Fuel Standard, to reduce carbon dioxide emissions from the full life cycle of fuels by 6% by 2020 (EC, 2014b). The use of "drop-in" (advanced) biofuels or suitably stabilised bio-oils would give refiners cleaner credentials, but could then allow them to lean towards including products from lower quality fossil fuels into blends, perhaps even claiming emissions offsets in their mining and transportation (Carbon Market Watch, 2014; Cheadle, 2014). Policy work should seek to ensure that regulation adapts to demand "light sourcing", ensuring that all components of a fuel are under a certain carbon-intensive bar. In order to ensure this, the carbon intensity of all refinery operations, transportation and mining operations should be made fully visible. A cap on carbon intensity of each individual component of a fuel would influence the choice of crude oils used as feedstocks, and also the processing plan. It would not be anti-competitive in energy markets as it would need to be applied across the board. Applied at the global level, it would disproportionately affect some oil- and gas-producing nations which have heavy and sour fossil fuel resources, so diplomatic work would be required to build converging narratives. It would also impact those countries and companies that have made the decision to focus their infrastructure and installations on the refining of heavy crude oils, and so they would need to be encouraged to pursue low carbon opportunities to compensate.

As a point of principle, "light sourcing" conditions could be an extra justification for eliminating very heavy oil products from the supply chain, such as the synthetic crude oil made from tar sands and other highly bituminous resources, as these have significant local environmental impacts that potentially fall foul of a raft of international environmental protection treaties. Such a carbon intensity indicator could incentivise the burial of excess carbon such as petcoke and very heavy oil residues – and could even prevent their source fuels mining in the first place.

Another policy strand could be to pursue a "Methane Implementing Agreement" (MIA) as a low carbon transitional arrangement before "zero carbon" hydrogen can be supplied at volume into the economy. For example, in the marine sector in Europe, a growth in fuelling with either CNG or LNG – and one could add compressed or liquefied Renewable Methane – is anticipated for shipping (McGill et al., 2013, Section 7.4 and Table 10). With forethought it should be possible to re-fit the ships with gas-fired propulsion systems that are flexible enough to use a range of gaseous fuels, perhaps fuel cell stacks, or modified traditional diesel engines. Another area in which an MIA would be helpful would be in meeting short-term goals to address urban air quality, for example, in Paris, which has made the decision to ban diesel-fuel vehicles, but where diesel fleets would be costly to replace, and China, where cities are working urgently to lower particulate emissions.

7.1.5 Smooth transition

Although it may never be adequately reflected in GDP, a healthy proportion of the industrialised economies is essentially dependent on the energy sector – without energy, they would not function. Not only does energy make the world go round for transport, trade, comfort, refrigeration, lighting, power and data, many pension and private funds are invested in energy stocks. Despite the recent "credit crunch", the energy sector needs new investment, in grids, in power plants and in storage, but this needs to be arranged in such a way that it does not deflate large pension funds and disappoint high yield financial products. As awareness of climate change grows, so does understanding of the risks of "unburnable carbon". It could be speculated that continued exploitation of emissions-intensive fossil fuels such as tar sands, heavy crude oils and coal, in a framework of tightening carbon controls, could culminate in the puncturing of a "carbon bubble" – destroying the value of shares in energy businesses, and thereby destabilising economies. Developments in the energy sector should therefore not be considered in isolation from the rest of the economy. The energy sector needs to be encouraged to turn away from the most risky behaviour of exploiting heavier hydrocarbon resources, and the energy system as a whole needs to have high efficiencies built in. A focus on building capacity to supply and consume Natural Gas in the most efficient ways could provide economic protection – as this fossil fuel will remain viable for longer, even in an environment of strong carbon control or competition for supplies. Transition is possible, as there are ways to provide low carbon gas to substitute for Natural Gas, so investment in plant and appliances that use gas will not need replacing. The level of ambition for policy should not be in granting subsidies for individual energy installations, research into exciting but unproven technologies or sub-sector measures – it should be more comprehensive, not rely on government subsidy stimuli, and it must point the way clearly to an entirely new configuration of energy services. Mass action by individual and institutional investors to disinvest or divest from fossil fuel shareholdings risks the dissolution of oil and gas sector players, so it is important to have a strategy to enable fossil fuel production and supply companies to transition to providing low carbon gas. States could perhaps choose to issue Renewable Gas Bonds instead of applying Quantitative Easing to capital markets in general, as energy bonds of any kind will create economic impetus across markets and sectors. Policymakers could also consider demanding low carbon transition plans from oil and gas companies that operate within their territory.

7.1.6 The economics of Renewable Gas

Although only a few Renewable Methane production facilities have been built, the prospective Levelised Cost of Energy (LCOE) has been modelled by a number of researchers and governments (see Table 7.2 for a selection), both on the basis of generated electricity and on gas produced. The first thing to note is that as far as supplied gas is concerned, the anticipated cost of Natural Gas (valued on a gas basis) is expected to remain far less than that for Renewable Gas – both biogas and bioSNG (or synthetic methane) – even in 2030. However, learning-by-doing may reduce Renewable Gas life cycle costs considerably. The second thing to note is that the low and some central estimates for electricity generation by Renewable Gas, valued on an electricity basis, are projected to be

Table 7.2 Analysis of the Levelised Cost of Energy (LCOE) for Renewable Gas and Natural Gas

Commissioning Year	Discount Rate	Region	Low	Mid	High	Basis	Note
A. Levelised Cost of Electricity Estimates							
1. For Natural Gas-fired power plants, utilising Combined Cycle and Gas Turbines (CCGT, NGCC)							
IEA (2010c)							
2010	5%	US	–	103	–	Electricity	Natural Gas to Power
2010	10%	US	–	111	–	Electricity	Natural Gas to Power
2010	5%	Switzerland	–	126	–	Electricity	Natural Gas to Power
2010	10%	Switzerland	–	141	–	Electricity	Natural Gas to Power
DECC (2013d)							
2013	10%	UK	69	95	121	Electricity	Natural Gas to Power, High fuel costs estimate
2019	10%	UK	75	101	130	Electricity	Natural Gas to Power, High fuel costs estimate
2030	10%	UK	102	105	110	Electricity	Natural Gas to Power, High fuel costs not assumed
WEC (2013)							
2013	–	UK	151	–	187	Electricity	Natural Gas to Power
2013	–	US	81	–	91	Electricity	Natural Gas to Power
2013	–	Australia	122	–	143	Electricity	Natural Gas to Power
2013	–	Global	–	91	–	Electricity	Natural Gas to Power
2013	–	Japan	–	196	–	Electricity	Natural Gas to Power
EIA (2014c)							
2019	–	–	–	92	–	Electricity	Natural Gas to Power
2040	–	–	–	112	–	Electricity	Natural Gas to Power

(continued)

Table 7.2 (continued)

Commissioning Year	Discount Rate	Region	Low	Mid	High	Basis	Note
2. For a hybrid Combined Cycle and Gas Turbine, co-firing Natural Gas and Biomass (Hybrid CCGT, Hybrid NGCC) Pihl et al. (2010)							
2010	5%	–	~60	–	~72	Electricity	Biomass/NG to Power, biomass 10 EUR/MWh
2010	5%	–	~105	–	~125	Electricity	Biomass/NG to Power, biomass 30 EUR/MWh
2010	10%	–	~90	–	~108	Electricity	Biomass/NG to Power, biomass 10 EUR/MWh
2010	10%	–	~135	–	~160	Electricity	Biomass/NG to Power, biomass 30 EUR/MWh
3. Advanced Conversion Technologies (without Combined Heat and Power (CHP))							
TINA Bioenergy (2012)							
2012	–	UK	–	111	–	Electricity	Biopower, Gasification
2050	–	UK	–	46	–	Electricity	Biopower, Gasification
DECC (2013d)							
2013	10%	UK	68	171	289	Electricity	"ACT Standard"
2019	10%	UK	65	165	282	Electricity	"ACT Standard"
2030	10%	UK	76	158	249	Electricity	"ACT Standard", fuel costs not assumed
2013	10%	UK	138	188	224	Electricity	"ACT Advanced"
2019	10%	UK	133	182	218	Electricity	"ACT Advanced"
2030	10%	UK	140	175	190	Electricity	"ACT Advanced", fuel costs not assumed
WEC (2013)							
2013	–	US	66	–	186	Electricity	Biomass and Waste to Power, Gasification
2013	–	W. Europe	66	–	186	Electricity	Biomass and Waste to Power, Gasification
2013	–	Global	~106	~172	~278	Electricity	Biomass to Power, Gasification
EIA (2014c)							
2019	–	–	160	–	204	Electricity	Integrated Gasification Combined Cycle (IGCC)
2040	–	–	138	–	167	Electricity	Integrated Gasification Combined Cycle (IGCC)

4. Biogas							
IEA (2010c)							
2010	5%	France	–	107	–	Electricity	Biogas to Power
2010	10%	France	–	128	–	Electricity	Biogas to Power
2010	5%	Switzerland	–	338	–	Electricity	Biogas to Power
2010	10%	Switzerland	–	439	–	Electricity	Biogas to Power
2010	5%	US	–	64	–	Electricity	Biogas to Power
2010	10%	US	–	85	–	Electricity	Biogas to Power
B. Levelised Cost of Gas Estimates							
Zwart et al. (2004)							
2004	–	–	–	31	–	Gas	BioSNG, O_2-blown gasification
2004	–	–	–	28	–	Gas	BioSNG, indirect gasification
2004	–	–	–	20	–	Gas	BioSNG, hydrogasification
Zwart et al. (2006)							
2006	–	–	–	22	–	Gas	Natural Gas
2006	–	–	–	111	–	Gas	BioSNG, 10 MW (thermal), atmospheric pressure
2006	–	–	–	55	–	Gas	BioSNG, 100 MW (thermal), atmospheric pressure
2006	–	–	–	53	–	Gas	BioSNG, 100 MW (thermal), 7 bar pressure
2006	–	–	–	33	–	Gas	BioSNG, 1,000 MW (thermal), 7 bar pressure
Chandel and Williams (2009)							
2009	–	–	–	54	–	Gas	BioSNG, biomass less than 10 EUR/MWh
Gassner and Marechal (2009)							
2009	–	–	76	–	107	Gas	BioSNG, 20 MW (thermal), Biomass gasification
2009	–	–	59	–	97	Gas	BioSNG, 150 MW (thermal), Biomass gasification

(continued)

Table 7.2 (continued)

Commissioning Year	Discount Rate	Region	Low	Mid	High	Basis	Note
Seiffert et al. (2009)							
2009	–	–	75	–	~103	Gas	BioSNG, largest plant size
CNG Services (2010)							
2010	–	UK	16	–	24	Gas	Natural Gas
2010	–	UK	77	–	118	Gas	BioSNG, Small-scale facility
2010	–	UK	37	–	84	Gas	BioSNG, Large-scale facility
TINA Bioenergy (2012)							
2012	–	UK	–	63	–	Gas	Biogas, Anaerobic Digestion
2012	–	UK	–	94	–	Gas	BioSNG
2050	–	UK	–	32	–	Gas	Biogas, Anaerobic Digestion
2050	–	UK	–	22	–	Gas	BioSNG
Heyne et al. (2013)							
2013	–	–	103	–	112	Gas	BioSNG
2030	–	–	32	–	36	Gas	Natural Gas (Reference: IEA)
Scheepers (2013)							
2013	–	–	36	–	65	Gas	BioSNG
Aranda et al. (2014)							
2014	–	–	66	80	113	Gas	BioSNG
C. Subsidies							
DECC (2013a)							
2014	–	UK	–	191	–	Electricity	Advanced Conversion Technologies (ACT)

Schulz et al. (2006)							
2006	–	–	17	–	44	Gas	BioSNG, range of subsidy scenarios proposed
CNG Services (2010)							
2010	–	UK	49	–	75	Gas	BioSNG, tariff-adjusted cost, small-scale facility
2010	–	UK	9	–	38	Gas	BioSNG, tariff-adjusted cost, large-scale facility
DECC (2014c)							
2014	–	UK	43	–	95	Gas	BioSNG, Renewable Heat Incentive (RHI), grid injection
NL EA (2014)							
2014	–	Netherlands	2	–	6	Gas	Renewable Gas, Biomass gasification, SDE+ 2014

Notes:

1. Currency exchange rates have been taken from the closest date to the report from weekly historic data given by OandA.com.
2. Costs are given on either an electricity or gas basis. These are not equivalent, as using gas to generate electricity is a further energy transformation and will involve losses.
3. Where figures have been read from charts rather than tables, the "~" character is used to show an approximation.
4. Unit conversions have been approximated to: 1 GJ = 0.27778 MWh; 1 Nm^3 = 7.1447 MWh; 1 MMBtu = 0.2933 MWh.
5. All amounts are converted to EUR/MWh (euro per megawatt-hour) for comparison purposes.

comparable to power generation by Natural Gas, although the high estimates are way beyond the most expensive Natural Gas projections. This suggests that power generation offers good possibilities for efficiencies from system integration, and is probably highly sensitive to the price of biomass and waste feedstocks. It is also useful to note the contrasting subsidies being deployed, remembering to take into account the differences between costs made on an electricity or gas basis, for example, Electricity Market Reform Contracts for Difference (CfD) for Advanced Conversion Technologies in the UK, compared to the Renewable Heat Incentive (RHI) for gas grid injection, and the parallel SDE+ subsidies in The Netherlands. Considering the early state of technological development, the SDE+ subsidy for biomass gasification appears too low to stimulate the financing of entirely new processing plant for bioSNG production. Nonetheless, in some possible future scenarios, subsidies to support Renewable Gas could be inappropriate, for example, if some technologies for the production of Renewable Hydrogen emerge as very low cost. If this occurs, then Renewable Methane manufactured for power generation could also have a lower cost profile, where repeated carbon dioxide recycling is practised in thermochemical treatments, as input biomass fuel costs would be dramatically reduced, and the production of the gas would cost not much more than the hydrogen (e.g. Aranda et al., 2014, "Power-to-Gas"). This scenario, if it unfolds, would make industrially manufactured Renewable Methane far less costly to produce than biogas from anaerobic digestion. Should subsidies continue to be applied to the production of bioSNG (e.g. DECC, 2013a, 2014c; NL EA, 2014), there should be a degression formula that compensates for cheaper inputs, such as Renewable Hydrogen produced from excess, zero-cost renewable electricity, or low-net-cost waste biomass.

Appropriate levels of subsidy are at this stage of development bound to be differential between countries. For example, in Switzerland, a high cost of bioSNG production could easily be sustained, with a fairly generous subsidy, as the country has no coasts, and so cannot import LNG by sea, and currently has no indigenous Natural Gas production. Sweden, also with no domestic Natural Gas production, is well forested, and so bioSNG production at a medium price range could be offered an uncontested subsidy, especially if the evolving political flavour of the government moves to steer an exit from nuclear power. The UK, always with an eye on Chinese markets, may want to develop bioSNG technologies as soon as possible with a view to exporting them, which could underpin the rationale for the generous "First Of A Kind" subsidies to be bid for. The CfD tariffs are being offered through sealed bid auction, which may result in only high prestige projects being beneficed in the early rounds – for example, National Grid's plasma project, in demonstration at Swindon (National Grid, n.d.).

The systems analysis of industrially manufactured Renewable Methane also points towards economies of scale (see Table 7.2), and so it is likely that large engineering firms such as Siemens, GE and Alstom would be building the most cost-efficient Renewable Gas installations – at least in the first decade. Besides petrorefinery and chemical engineering, the most suitable large-scale area of industry where Renewable Gas production should be considered is in iron and steel manufacture – considered effectively part of the energy sector in the UK – where coke is still produced from coal in large quantities (DUKES, 2014, "Manufacture of solid fuels" and "Manufactured fuels"). The production of biogas, Renewable Hydrogen, bioSNG (Renewable Methane, synthetic methane), biocoal and biocoke could provide a significant carbon dioxide emissions reduction opportunity – both internally to the industry, and also as "merchant" sales to wider industry and energy networks. Considering the simplicity of some of the Renewable Gas engineering, it is possible that the lifetimes of most of the plant could be considerable. There are many parallels with other similar manufacturing plant, which have proven capacity over several decades – or more. Safe lifetime extensions could drag down the overall LCOE to be comparable to that of coke oven gas, or other manufacturing or refining offgases. Several countries have plans for constructing carbon dioxide "gathering" pipelines, to collect waste carbon dioxide for burial at sea or onshore in CCS schemes (Brownsort, 2013; DECC, 2012). Positioning Renewable Gas plant where it can have access to this resource could bring the cost of Renewable Gas down sharply.

8
Reflections and Conclusions

8.1 The hydrogen-carbon question

Significant change in the hydrocarbon energy industry appears inevitable, and it can be simply summarised as a progressively worsening ratio of hydrogen to carbon molecules in the processes that provide liquid and gaseous fuels. The more complicated the fossil fuel resources brought to refinery and processing plant, the more "churn" is required to raise the hydrogen content and thus the energy value of the final products, which implies rejecting excess carbon, with inefficiency and therefore cost ramifications. In addition, without careful consideration of the path that the rejected carbon molecules take, this could have serious greenhouse gas emission implications. The current way to address the problem of the worsening hydrogen-to-carbon ratio is to contribute the hydrogen from thermochemically treated Natural Gas to the processes at oil refinery, although this also contributes a certain amount of fossil carbon to the system, which will be disposed of as increased carbon dioxide emissions. An improvement would be to introduce emissions-neutral carbon from biomass into the processes being used to form hydrogen. Although this does not correct the overall balance of hydrogen and carbon, it does mitigate against net greenhouse gas emissions. A further transitional step would be to introduce hydrogen from carbon-free sources, such as water. Hydrogen is already being contributed at some refineries through the thermochemical reaction of unwanted heavy hydrocarbon residues and water in the form of steam. Although this provides renewable hydrogen from the steam, this partial oxidation, or gasification, results in excess fossil carbon for disposal. Alternatively, adding renewable hydrogen production to petrorefinery could be done through the electrolytic or electrochemical liberation of hydrogen from water, which could theoretically become entirely emissions-free. The requirement for extra hydrogen in crude petroleum oil refining needs to be answered in the most low carbon way possible, or the energy sector could face failure in the context of imposed emissions control. Renewable hydrogen could thus save the oil and gas industry.

8.2 The Natural Gas and ammonia question

Natural Gas of all flavours is a valuable tool with which to feed, light, power, warm and cool human society, through either being used as a fuel, or for fertiliser production via ammonia synthesis, for which we need to consider the relationship between hydrogen and nitrogen. Human population is still expanding, which, coupled with climate change, will sharpen demand for both water and fertilisers. Although we shall never run short of nitrogen, global demand for Natural Gas for all purposes is set to rise, accentuating competition for the fuel, which may disadvantage fertiliser manufacturers, and therefore affect food prices. It would be carbon-inefficient to instead get the necessary hydrogen from the more complex hydrocarbons or carbon-rich residues being rejected from petrorefinery. Here, too, a viable molecular-level strategy would be to aim for carbon-neutral or carbon-free hydrogen production, either from biomass or water.

8.3 Sustainable business

The oil and gas industry look set to rely increasingly on the hydrogen from Natural Gas, refinery off-gas and severe thermochemical treatments of very heavy hydrocarbons to continue to produce vehicle fuels (Dodds and McDowall, 2012; MHLnews, 2014; Sutikino and Turini, 2012). This increase in the relative amounts of lighter chemistry and increased supplies of Natural Gas Liquids (NGLs) will help meet the increased petrorefinery demand for hydrogen. Increased hydroprocessing and other treatments imposed by the changing palette of crude oils and regulations on fuel standards mean there is almost no part of raw crude petroleum oil that persists into the final products of oil refinery – "straight run" distillation fractions are a thing of the past. This means that virtually all petroleum products are essentially manufactured and not "natural". This suggests a level of chemical intervention by process engineering that demonstrates the capacity of the fossil fuel sector to initiate a transition to low carbon manufactured oil and gas. Although there may be concerns over the economic impact on the "bottom line" of implementing

Renewable Gas systems as an adjunct to oil and gas refinery, most of the costs will be initial, infrastructural and at the individual plant level. Besides offering a measure of climate protection, Renewable Gas may also become the sustainer of oil and gas businesses, as a gap between demand and supply for hydrogen opens up. Besides offering to bridge the hydrogen availability gap, Renewable Gas can also offer a way to recoup otherwise disposable carbon and recycle it into useful new fuel, essentially offering a form of refinery gain.

8.4 Renewable Gas is imminent

Now could be considered an opportune moment to make decisions to invest in Renewable Gas – both Renewable Hydrogen and Renewable Methane, as decisions made now will put solutions in place before the situation becomes critical. Consideration of countervailing examples is relevant. For example, in December 2013, BP announced that they had completed most of the work on a heavy fuel oil delayer coker facility at their Whiting plant in the US. Although their decision to use cheaper, sour, heavy oil feedstocks rather than more costly, light, sweet crude oil seems rational, it locks them into a future where hydrogen and carbon are imbalanced in their processing, where they will be creating carbon-rich wastes, and where the costs of refining may turn out to be higher, as more feedstock processing creates inefficiencies. They would also be responsible, and ultimately could be held accountable, for higher carbon dioxide emissions, if carbon-rich by-product petroleum coke, or "petcoke", or residuum hydrocarbons are used for power generation. However, despite this negative trend towards using lower grade fossil fuels, there is a positive side effect in that large refineries, such as those operated by BP, Royal Dutch Shell and ExxonMobil, are working on increasing their hydrogen production capacity, many of them choosing the Natural Gas-to-hydrogen route, or even going one step better, by recycling refinery gas. It would not take too much of a stretch of the political imagination to envisage strict carbon emissions legislation causing regulatory control of the levels of carbon rejection by refineries, leading to a decision to ramp up Renewable Gas capacity. It would take focus for this option to grow, however.

8.5 Renewable Gas can already displace some fossil fuels

Although to many Renewable Hydrogen and Renewable Methane may seem like a fable or far-off into the future, to the contrary, the well-established chemistry and process engineering of Renewable Gas can already displace the use of some fossil fuels, however, perhaps more in the liquid transport fuels sector than the gas-to-power sector at the present time.

8.6 Renewable Gas and uranium

There are many ageing fission nuclear power reactors worldwide, and it has been suggested that their increasing unreliability should be met by requirements that they disconnect from the power grids, as they require very high levels of backup in case of unplanned (and scheduled) outage. What to do with these still steaming reactors before they should be decommissioned? The answer may be to divert them into producing Renewable Gas.

8.7 Development pathways

Given that the centres of oil and gas refining are turning to heavy waste hydrocarbons or even coal to supply their hydrogen in some cases, they must be unsure of being able to secure sufficient suitably priced Natural Gas in future, and they are in effect locking themselves into a rising carbon emissions profile. It would seem sensible to offer a low emissions alternative in the form of Transitional Hydrogen. Economic studies into the use of grid electricity to make hydrogen, making use of green electricity where this is available, show that under some circumstances, Transitional Hydrogen can be produced on a market competitive basis without the need for subsidies (ITM Power, 2013a).

8.8 The hydrogen numbers game

In order to move on Renewable Gas, it is essential to increase non-fossil hydrogen production. The Thüga Mainova hydrogen demonstration plant in Germany produces 60 Nm^3 per hour, and therefore around 500,000 Nm^3 per year (44 tonnes). Estimates vary (see Table 8.1), but this amount is only a millionth of approximate global annual hydrogen production – approximately 500 billion cubic metres (42 megatonnes). However, aiming to provide significant quantities of hydrogen into the petrorefinery market using this technology is more feasible than attempting to displace any appreciable quantity of Natural Gas. The world produced 3,369.9 billion cubic metres of Natural Gas in 2013 (BP, 2014c), which if considered to be entirely methane represents 2,261.2 megatonnes (Linde, n.d.). Because of the

Table 8.1 Estimates of annual global hydrogen production (all markets – captive and merchant)

	2007	2009	2013	2015	2018	Reference
(a) Million tonnes of H_2						
CAN-Europe	–	45	–	–	–	CAN-Europe (2009)
Freedonia	–	~21	–	–	–	MHLnews (2014)
Praxair	–	~25	–	~31	–	Praxair (2011)
IEA	65	–	–	–	–	IEA (2007)
(b) Billion Normal cubic metres (Nm^3) of H_2						
CAN-Europe	–	535	–	–	–	CAN-Europe (2009)
Freedonia	–	254.5	–	–	–	MHLnews (2014)
Praxair	–	294	–	~372.5	–	Praxair (2011)
IEA	~773	–	–	–	–	IEA (2007)

Unit conversion (Barchard n.d.; Linde n.d.).

relative molecular weights, roughly a quarter will be hydrogen atoms, representing 6,721.8 billion cubic metres of hydrogen bound in the Natural Gas – 565 megatonnes – an order of magnitude greater than global hydrogen production.

At the present time, less than 5% of global hydrogen is produced via electrolysis (IEA, 2007). However, Proton Exchange Membrane Fuel Cells (PEMFC) electrolysers can easily be scaled up. For example, Siemens plan a 200 tonnes per year plant at the Energiepark Mainz (dena, 2014a; Hotellier, 2014), where the power rating is 6.3 MW, with a power consumption of 55 GWh per year if they are always on. Germany is currently exporting something of the order of 30 TWh of electricity per year (Fraunhofer, 2013, Slide 8), the vast majority due to excess renewable electricity generation, so if instead that was used to manufacture hydrogen, that could amount to over 100,000 tonnes of hydrogen (0.1 megatonnes, ~1.3 million Nm^3), but it would take a deployment of around 545 plants of Energiepark Mainz size. If the kind of acceleration in renewable electricity seen in Germany were replicated in ten European Member States, and they followed through with double the numbers of hydrogen electrolyser plants, with each deployment scaled up by 200%, because of increasing amounts of surplus renewable electricity generation available, indigenous European Renewable Hydrogen production would be of the order of 4.4 megatonnes (52 million Nm^3), a tenth of present Western European hydrogen demand (MHLnews, 2014). Since this could easily be absorbed by demand from petrorefinery, or the early hydrogen vehicle market, not much of this is likely to be available for manufacturing Renewable Methane for gas grids. If the demand pressure from petrorefinery wins out, this Renewable Hydrogen would go to creating "greener" gasoline (and diesel) or, rather, to ensuring that petrorefinery does not expand its carbon emissions profile because of the increasing carbon-to-hydrogen ratio resulting from future processing.

A level conclusion that can be drawn is that whilst "Power-to-Gas" via the production of electrolytic hydrogen is going to play a vital role, it will not be a major player in Renewable Gas as a whole, unless there is a continuing acceleration in renewable electricity capacity that leads to huge excess generation. The scale of Renewable Hydrogen required to bring about a large volume displacement of Natural Gas for all energy applications in the short to medium term therefore necessitates reliance on biomass routes to hydrogen, and carbon-abated fossil fuel routes to hydrogen (e.g. Soetaert and Vandamme, 2009, Chapter 6).

8.9 Good "black swans"

Disruptive technologies with benefits for both carbon control and fossil fuel replacement could float into view in a short time frame. "Black swans" would ideally be in the production of hydrogen in the first instance.

8.9.1 Hydrogen catalysis

Research announcements are regularly made in the area of improving the hydrogen evolution reaction from a range of feedstocks, and using a range of catalysts, on a range of substrates. Although an emergent catalytic hydrogen option would need targeted support to grow its potential, there is already a market for hydrogen, so if production can be ramped up, it can be used, even if the wider Renewable Gas machinery is not yet in place.

The ideal place to add catalytic hydrogen facilities would be where there are high carbon emissions, ensuring market acceptance, such as at heavy oil or tar sands processing plants.

It would also be suitable to place catalytic hydrogen facilities at fertiliser production plants, thus dropping the carbon emissions of the plants drastically.

8.9.2 Electromagnetic radiation-assisted hydrogen and methane production

Advancements in the plasmolysis of water vapour (steam) suggest that hydrogen production with this method might become less costly than electrolysis (Rehman et al., 2013). The equipment needed for plasmolysis could probably be added almost anywhere in a power grid, and its deployment could initially be precipitated by organisations wishing to store hydrogen to ensure uninterruptible power supplies.

Microwave-assisted gas reactions could lead to improvements in the synthesis of methane (Ferrari et al., 2014; Hunt et al., 2013). Microwave radiation could also aid in forming suitable methanation catalysts (e.g. Burbey and Joseph, 2004).

8.9.3 A breakthrough in biomass gasification

If a reliable mechanism could be developed for the prevention of tar formation in the gasification of a range of biomass feedstocks that did not require post-processing of the product gas, this would pose a huge advantage. Potentially, useful tar moderants could be clays or muds, some salts or even biochar (Abu El-Rub, 2008).

8.9.4 A breakthrough in the methanation of sour/acid Natural Gas

In order to make extensive use of sour/acid Natural Gas with high levels of carbon dioxide in it, it would be best to convert this carbon dioxide into fuel – rather than separating it and venting it, or emitting it after combusting or steam reforming the Natural Gas.

To add the carbon dioxide value to Natural Gas will optimally require methanation processing that can operate with the carbon dioxide still in situ in the gas.

8.10 The policy ambit

There are two policy objectives that are crucially important: maintaining flexible gas-to-power in industrialised countries, and enabling a transition to Renewable Gas.

8.10.1 Retaining gas-to-power capacity

Natural Gas-fired power generation is flexible and produces roughly half the carbon dioxide emissions of using coal as fuel. It is vital that Natural Gas-fired capacity and availability is maintained, despite worsening economics, partly owing to its flexibility. Natural Gas-fired power plants have been put "on call" for grid load balancing, and so operate for fewer hours a year, which drives down returns on investment. Rather than create Capacity Mechanism subsidies, industrialised countries should perhaps consider gas-fired power generation as a form of infrastructure, and make tax-funded acquisitions of plant to ensure its availability.

8.10.2 Enabling the transition to Renewable Gas

The use of Natural Gas in power generation in industrialised countries will remain acceptable in carbon terms for several decades, but it is vital to consider mechanisms for enabling a transition to low carbon gas. An obligation on gas supply companies to provide "green" gas may go some way to creating a traded market in Renewable Gas, but will not secure investment in plant. Various organisations in Germany have spelled out the energy and climate security reasons for a transition to home-grown Renewable Gas, and are supporting key opening projects, and setting targets.

There will still be a need to involve large engineering companies and engage finance. This could perhaps emerge as oil and gas companies consider the diminishing returns on investment in exploration and production of fossil fuels and seek to place their capital elsewhere. They could perhaps be encouraged to do so by strict carbon accounting of their businesses, and a requirement that they do not undo progress made in the abatement of carbon dioxide emissions from their businesses in the last thirty years. A regulatory framework on the transition to Renewable Gas would enable oil and gas companies to retain market and shareholder confidence.

Whichever route is chosen for re-directing capital to the task of gas transition, there will need to be a stronger collaboration between chemical engineering, process engineering and petrorefinery companies.

8.11 Recommendations

Renewable Gas systems with carbon recycling can theoretically meet the objective of very low carbon emissions power generation, whilst also reducing fuel input requirements. Pursuing a fleet of centralised power plants around designs of this nature, even if they use fossil fuels as input initially, could leverage further Renewable Gas development.

If Renewable Gas production facilities can be encouraged to adopt Renewable Carbon, this could lead to the opportunity to decarbonise not only power generation, but the other main energy sectors as well – transport and space heating/cooling. This should thus become core energy policy, even if it can only be effected by state intervention on the basis of securing indigenous energy supplies.

Hydrogen research should continue to be supported strongly, particularly Renewable Hydrogen, as this is a key tool in growing Renewable Gas production volumes.

Hydrogen should be considered in the context of its requirements by the broader energy system (e.g. ADEME, 2014; Agora Energiewende, 2014;

Dodds et al., 2015; E-Cube, 2013; Grond and de Joode, 2014; IEA, 2014c; SBC, 2014) and not just for Fuel Cell Vehicles (e.g. BMVI, 2014; IPCC, 2014b, Chapter 8; Mohseni, 2012).

Power-to-Gas is not a final destination. It is only a beginning for hydrogen. Work to increase hydrogen production should be prioritised, taking a lead from the large global production of syngas, and substituting biomass fuels as feedstocks.

Besides playing an essential role in decarbonising energy, and providing a stimulus to the global economy by providing capital, energy and carbon "paybacks" on investment within a short period, renewable electricity deployment should be accelerated to maximise the potential of Power-to-Gas.

Old and new nuclear fission reactors should be re-purposed for constant hydrogen and methane generation, as nuclear power plants are a risk to power grid security. The best strategy for grid security would be to drop nuclear power plants out in favour of gridded renewable electricity capacity with gas-fired power plant backup. Nuclear reactor heat would instead be used to make gas to replenish storage. Small modular nuclear reactors may at some point be available for localised electricity grid support, and therefore potentially available for producing Renewable Gas; however, these will take some time to develop, so there is no rationale to build a plan around them as yet.

Current coal-fired generation plant should be removed from the electricity grid. Coal should not be used as an energy fuel unless most of its carbon can be sequestered or recycled.

Gas-fired plants are vital for current and future energy security, so perhaps they should be owned by governments or local authorities, as they are equivalent to infrastructure assets.

Renewable Gas plant should be designed with a view to possibly needing to switch from fuels rich in Renewable Methane to fuels rich in Renewable Hydrogen after some decades, and should show input fuel flexibility over all time frames.

Biomass resources need developing to provide Renewable Hydrogen (and Renewable Carbon) for Renewable Gas production at a scale that can easily outstrip electrolytic hydrogen production.

Biocoals and biocoke need developing to replace the use of coal in steel manufacture and other industries.

Bioammonia needs developing at the same pace and scale as Renewable Gas, in order to protect food security, and can make use of the same input resources.

The price of Natural Gas is likely to be more strongly affected by conditions of its production than by market trading. It is expected that these circumstances will keep Natural Gas prices low for some time to come. Therefore, any Capacity Mechanism policy to maintain gas-fired power generation capacity should be crafted as subsidies dependent on an obligation to take Renewable Gas fuel – to stimulate its introduction.

The deployment of renewable electricity and energy conservation should be promoted by all means available.

In the end, energy policy should not be a battle between technologies, but a pathway towards a cohesive and efficient whole energy system. An energy system built around renewable electricity and Renewable Gas could not only be efficient, it could also integrate all power technologies and fuels, until the fossil fuels need to be phased out. It could thus garner co-operation from all the existing energy sector actors, even as it prepares us for the low carbon future. All energy sector players can produce and consume renewable electricity and Renewable Gas, offering up the possibility of a smooth transition.

Vehicle manufacturers should be encouraged to build gas-drive options as well as electric vehicles.

Fossil fuel oil and gas production corporations should be encouraged to develop a vision of a transition to becoming Renewable Gas manufacture organisations.

References

Aali and Rahmani (2011). "Evidences for secondary cracking of oil in South Pars field, Persian Gulf, Iran", by Jafar Aali and Omeid Rahmani, in Journal of Petroleum Science and Engineering, Volume 76, Issues 3–4, March 2011, Pages 85–92, http://www.dx.doi.org/10.1016/j.petrol.2011.01.016

Abbess (2014). "Clean Burn: The Transition to Renewable Gas", Supplementary Material, Technical Essay, "The Chemistry of Gas", 15 October 2014. Accessed online.

Abdel-Aal et al. (2010). "Hydrogen Production Using Sea Water Electrolysis", by H. K. Abdel-Aal, K. M. Zohdy and M. Abdel Kareem, in The Open Fuel Cells Journal, ISSN: 1875-9327, Volume 3, 2010, Pages 1–7.

ABO (2014). "The first GTL (gas-to-liquids) plant in Iran by 2018", by Editorial Staff, About Oil, 12 August 2014. Accessed online.

Abu Bakar et al. (2012). "Catalytic methanation reaction over supported nickel-ruthenium oxide base for purification of simulated natural gas", by Wan Azelee Wan Abu Bakar, Rusmidah Ali and S. Toemen, in Scientia Iranica C, Volume 19, Issue 3, June 2012, Pages 525–534, http://www.dx.doi.org/10.1016/j.scient.2012.02.004

Abu El-Rub (2008). "Biomass Char as an In-Situ Catalyst for Tar Removal in Gasification Systems", by Ziad Yousef Kamel Abu El-Rub, PhD Thesis, University of Twente, 7 March 2008.

ADEME (2014). "Study on Hydrogen and methanation as means to give value to electricity surpluses", by E&E Consultant, Hespul and Solagro, for ADEME, GRT-Gaz and GRdF, Executive Summary, September 2014.

AEMO (2013). "100 Per Cent Renewables Study – Modelling Outcomes", published by the Australian Energy Market Operator Ltd (AEMO), July 2013.

AEMO (2014). "Fuel and Technology Cost Review", by ACIL Allen Consulting, report for Australian Energy Market Operator (AEMO), 10 June 2014.

Agora Energiewende (2014). "Electricity Storage in the German Energy Transition", by Agora Energiewende, Report: 055/11-S-2014/En, December 2014.

Ahmed (2013). "Former BP geologist: peak oil is here and it will 'break economies'", by Nafeez Ahmed, "Earth Insight", The Guardian, 23 December 2013.

Ahmed (2014). "The inevitable demise of the fossil fuel empire", by Nafeez Ahmed, The Guardian, 10 June 2014.

Aleklett et al. (2010). "The Peak of the Oil Age – Analyzing the world oil production Reference Scenario in World Energy Outlook 2008", "Uppsala critique", by Kjell Aleklett, Mikael Höök, Kristofer Jakobsson, Michael Lardelli, Simon Snowden and Bengt Söderbergh, in Energy Policy, Volume 38, 2010, Pages 1398–1414, http://www.dx.doi.org/doi:10.1016/j.enpol.2009.11.021

Al-Hafedh et al. (2014). "Experiments on an integrated aquaculture system (seaweeds and marine fish) on the Red Sea coast of Saudi Arabia: efficiency comparison of two local seaweed species for nutrient biofiltration and production", by Yousef S. Al-Hafedh, Aftab Alam, Alejandro H. Buschmann and

Kevin M. Fitzsimmons, in Reviews in Aquaculture, Volume 4, Issue 1, March 2012, Pages 21–31, http://www.dx.doi.org/10.1111/j.1753-5131.2012.01057.x

Alicat (n.d.). "Gas Viscosities, Densities and Compressibilities at 25 degrees C", prepared by Alicat Scientific Inc., Data: from National Institute of Standards and Technology (NIST) Software REFPROP version 7. Data: at 25 degrees Celsius and 14.696 Pounds per square inch absolute (PSIA or psia), Methane = 0.6569 grammes per litre (g/l), Hydrogen = 0.08235 g/l.

Allam (2013). "NET Power's CO2 cycle: the breakthrough that CCS needs", by Rodney J. Allam, in Modern Power Systems, 10 July 2013. Accessed online.

Aminzadeh et al. (2001). "Determining Migration Path from Seismically Derived Gas Chimney", by Fred Aminzadeh, Paul de Groot, Timothy B. Berge, Tanja Oldenziel and Herald Ligtenberg; presentation to American Association of Petroleum Geologists (AAPG), Hedberg Conference, "Near-Surface Hydrocarbon Migration: Mechanisms and Seepage Rates", 2001, Page 2, "Background".

Amiri and Eslamian (2010). "Investigation of Climate Change in Iran", by M. J. Amiri and S. S. Eslamian, in Journal of Environmental Science and Technology, Volume 3, 29 May 2010, Pages 208–216, http://www.dx.doi.org/10.3923/jest.2010.208.216

Ancheyta and Rana (n.d.). "Future Technology in Heavy Oil Processing", by Jorge Ancheyta and Mohan S. Rana, in Encyclopedia of Life Support Systems (EOLSS), Developed under the Auspices of the UNESCO, Eolss Publishers, Paris, France.

Andrews (2013). "The Carboniferous Bowland Shale gas study: geology and resource estimation", by I. J. Andrews, produced by the British Geological Survey for HMG DECC, 2013.

Andrews and Lattanzio (2013). "Petroleum Coke: Industry and Environmental Issues", by Anthony Andrews and Richard K. Lattanzio, Congressional Report Service, Report: 7–5700, R43263, 29 October 2013.

AOGD (2006). "Iran", Arab Oil and Gas Directory, 2006. Accessed online.

Aranda et al. (2014). "The Economy of Large Scale Biomass to Substitute Natural Gas (bioSNG) plants", by G. Aranda, A. van der Drift and R. Smit, Energy Research Centre of The Netherlands (ECN), January 2014, Report: ECN-E--14-008, EDGaR Project on SNG impact, Project Number: 51244.

Arapostathis et al. (2013). "Governing transitions: Cases and insights from two periods in the history of the UK gas industry", by Stathis Arapostathis, Anna Carlsson-Hyslop, Peter J. G. Pearson, Judith Thornton, Maria Gradillas, Scott Laczay and Suzanne Wallis, in Energy Policy, Volume 52, January 2013, Pages 25–44, Section 3.2, http://www.dx.doi.org/10.1016/j.enpol.2012.08.016

Archer and Jacobson (2013). "Geographical and seasonal variability of the global "practical" wind resources", by Cristina L. Archer and Mark Z. Jacobson, in Applied Geography, Volume 45, Pages 119–130, 2013, http://www.dx.doi.org/10.1016/j.apgeog.2013.07.006

ARI (2013). "EIA/ARI World Shale Gas and Shale Oil Resource Assessment: Technically Recoverable Shale Oil and Shale Gas Resources: An Assessment of 137 Shale Formations in 41 Countries Outside the United States", prepared by Advanced Resources International (ARI), published by Government of United States of America (USA), Energy Information Administration (EIA), 10 June 2013.

Arvizu et al. (2011). "Renewable Energy Sources and Climate Change Mitigation", by Dan Arvizu, Palani Balaya, Luisa F. Cabeza, K. G. Terry Hollands, Arnulf Jäger-Waldau, Michio Kondo, Charles Konseibo, Valentin Meleshko, Wesley Stein,

Yutaka Tamaura, Honghua Xu and Roberto Zilles, edited by Ottmar Edenhofer, Patrick Eickemeier, Gerrit Hansen, Susanne Kadner, Ramón Pichs-Madruga, Patrick Matschoss, Steffen Schlömer, Kristin Seyboth, Youba Sokona, Christoph von Stechow and Timm Zwickel, Special Report of the Intergovernmental Panel on Climate Change, Chapter 3, "Direct Solar Energy", Pages 333–400, published by: Cambridge University Press, Cambridge, UK and New York, NY, USA, December 2011, Online ISBN: 9781139151153, http://www.dx.doi.org/10.1017/CBO9781139151153.007

Asre KimyaGari (n.d.). "Natural gas reserves in Iran: Kish Gas Field Stands On 2nd Position", by Asre Kimia Gari, News. Accessed online.

Bahmannia (2006). "Life Cycle Assessment of a Natural Gas Plant: Case Study – Sarkhoon Gas Treating Plant in Iran", by Gh. Bahmannia, presented to 23rd Gas Conference, Amsterdam, 2006, Table 3.

Ball et al. (2009). "Hydrogen production", by Michael Ball, Werner Weindorf and Ulrich Bünger, Chapter 10 in "The Hydrogen Economy: Opportunities and Challenges", by Michael Ball and Martin Wietschel, published by Cambridge University Press, 24 September 2009, ISBN: 978-0521882163, Section 10.1, Section 10.9.

Barchard (n.d.). "Notes on Volumes at Pressure, Temperature & Heat Values", Barchard Engineering. Accessed online.

Barkan (1999). "Walther Nernst and the Transition to Modern Physical Science", by Diana Kormos Barkan, published by Cambridge University Press, second edition, 3 March 2011, first edition, 13 January 1999, ISBN: 978-0-521-44456-9.

Barrett et al. (2013). "Ensuring the Resilience of the U. S. Electrical Grid", by J. Michael Barrett, Jeff Harner and John Thorne, published by the Lexington Institute, January 2013.

Basu (2007). "Materials for Solid Oxide Fuel Cells", by Rajendra N. Basu, Chapter 12 in "Recent Trends in Fuel Cell Science and Technology", edited by Suddhasatwa Basu, published by Springer, 24 January 2007, ISBN: 978-0387355375, Section 7.3.3.1.

Basu (2013). "Biomass Gasification, Pyrolysis and Torrefaction: Practical Design and Theory", by Prabir Basu, 2nd edition, published by Academic Press, Elsevier Inc., 12 September 2013, ISBN: 978-0123964885.

Baues (2009). "NRW Hydrogen Highway – Activities in North Rhine-Westphalia (NRW), by Dr Heinz Baues, presentation to International Partnership for Hydrogen and Fuel Cells in the Economy (IPHE) Implementation-Liaison Committee, Oslo, 10 March 2009.

BCC (2015). "A City Blueprint for Low Carbon Fuel Refuelling Infrastructure", by Celine Cluzel and Sophie Lyons for Birmingham City Council, February 2015. Accessed online.

Behling (2012). "Fuel Cells: Current Technology Challenges and Future Research Needs", by Noriko Hikosaka Behling, published by Elsevier, 29 October 2012, Section 6.2.3.3.

Bellona (n.d.). "Post combustion", published online by Bellona Foundation. Accessed online.

Bensmann et al. (2014). "Biological methanation of hydrogen within biogas plants: A model-based feasibility study", by A. Bensmann, R. Hanke-Rauschenbach, R. Heyer, F. Kohrs, D. Benndorf, U. Reichl and K. Sundmacher, in Journal of Applied Energy, Volume 134, 1 December 2014, Pages 413–425, http://www.dx.doi.org/10.1016/j.apenergy.2014.08.047

Berge (2011). "Recognition and Meaning of Hydrocarbon Seeps", by Timothy B. Berge, adapted from presentation to AAPG Annual Convention 2011. Accessed online.

Berge (2013). "Hydrocarbon Seeps: Recognition and Meaning", by Timothy B. Berge, Chapter 1 in "Hydrocarbon Seepage: From Source to Surface", edited by Fred Aminzadeh, Timothy B. Berge and David L. Connolly; Geophysical Developments Series No. 16, published by Society of Exploration Geophysicists (SEG) and AAPG, 2013, ISBN: 9781560803102.

Bermudez et al. (2013). "An overview of novel technologies to valorise coke oven gas surplus", by José M. Bermúdez, Ana Arenillas, Rafael Luque and J. Angel Menéndez, in Fuel Processing Technology, Volume 110, June 2013, Pages 150–159, http://www.dx.doi.org/10.1016/j.fuproc.2012.12.007

BGS (2010). "The Unconventional Hydrocarbon Resources of Britain's Onshore Basins – Shale Gas", produced by the British Geological Survey for HMG DECC, 2012.

Biello (2014). "Google's Cars Sniff Out Natural Gas Leaks to Deliver Cleaner Air", by David Biello, Scientific American blog, 16 July 2014. Accessed online.

Biogas Renewable Energy (2009). "Biogas composition", 2009. Accessed online.

Bloom Energy (2014). "Solid Oxide Fuel Cells", by Bloom Energy. Accessed online.

BMVI (2014). "Power-to-Gas (PtG) in transport: Status quo and perspectives for development", by Dr. U. Bünger, H. Landinger, E. Pschorr-Schoberer, P. Schmidt, W. Weindorf, J. Jöhrens, U. Lambrecht, K. Naumann and A. Lischke, published by German Government, Federal Ministry of Transport and Digital Infrastructure (BMVI), Report: AZ Z14/SeV/288.3/1179/UI40, Munich, Heidelberg, Leipzig, Berlin, 11 June 2014.

Borrill and Easterby (1985). "Exploitation of Heavy Oil: The FBH Process", by P. A. Borrill and R. J. Easterby, in "New technologies for the exploration and exploitation of oil and gas resources", Proceedings of the 2nd E. C. Symposium held in Luxembourg, 5–7 December 1984, Volume 2, EUR 10 168, published by Graham & Trotman, for the Commission of the European Communities, 1985, Section (03.67/68), Pages 725–730.

Boucher and Ready (2008). "Climate trade-off between black carbon and carbon dioxide emissions", by Dr Olivier Boucher and M. Shekar Reddy, in Energy Policy, Volume 36, Issue 1, January 2008, Pages 193–200, http://www.dx.doi.org/10.1016/j.enpol.2007.08.039

Boukis et al. (2007). "Biomass Gasification in Supercritical Water. Experimental Progress Achived with the Verena Pilot Plant", by N. Boukis, U. Galla, H. Müller and E. Dinjus, paper presented to 15th European Biomass Conference and Exhibition, 7–11 May 2007, Berlin, Germany.

BP (1972). "Gas Making and Natural Gas", BP Trading Ltd, 1972. Published book.

BP (2003). "BP Statistical Review of World Energy: June 2003", by BP plc. Accessed online.

BP (2004). "BP Statistical Review of World Energy June 2004", by BP plc. Accessed online.

BP (2010). "BP Statistical Review of World Energy: June 2010", by BP plc. Accessed online.

BP (2011b). "BP Statistical Review of World Energy June 2011", BP plc. Accessed online.

BP (2011c). "BP Statistical Review of World Energy June 2011", BP plc, Workbook. Accessed online.

BP (2012b). "BP Statistical Review of World Energy June 2012", BP plc. Accessed online.

BP (2013a). "Energy Outlook 2030", BP, January 2013. Accessed online.

BP (2013c). "BP Statistical Review of World Energy June 2013", BP plc, Workbook. Accessed online.

BP (2014a). "Energy Outlook 2035", BP plc, January 2014. Accessed online.

BP (2014c). "BP Statistical Review of World Energy June 2014", BP plc, Workbook, Revised version, 18 June 2014. Accessed online.

Breault (2010). "Gasification Processes Old and New: A Basic Review of the Major Technologies", by Ronald W. Breault, in Energies, Volume 3, 2010, Pages 216–240, Section 4.1, http://www.dx.doi.org/10.3390/en3020216

Brimont et al. (2012). "Waste Gas: a Crucial Component of the Energy Poverty Dilemma", by Renaud Brimont, Antoine Aris and Peter von Campe, in SBC Energy Perspectives, published by Schlumberger Business Consulting, Winter 2012.

Brisse et al. (2010). "Electrolysis Using Fuel Cell Technology", by A. Brisse, J. Schefold, C. Stoots and J. O'Brien, in Chapter 9 in "Innovations in Fuel Cell Technologies", edited by Robert Steinberger-Wilckens and Werner Lehnert, published by Royal Society of Chemistry, 18 October 2010, ISBN: 978-1-84973-033-4.

Brouwer et al. (2014). "Impacts of large-scale Intermittent Renewable Energy Sources on electricity systems, and how these can be modeled", by Anne Sjoerd Brouwer, Machteld van den Broek, Ad Seebregts and André Faaij, in Renewable and Sustainable Energy Reviews, Volume 33, Pages 443–466, 2014, http://www.dx.doi.org/10.1016/j.rser.2014.01.076

Brown (2014a). "Coal still counts (1): keeping options open in the Energy Act 2013", by Adam Brown, Target UK Energy, 10 January 2014. Accessed online.

Brown (2014b). "Coal still counts (2): decision time for generators (or it will be soon)", by Adam Brown, Target UK Energy, 17 February 2014. Accessed online.

Brownsort (2013). "Briefing: CCS for Industrial Sources of CO2 in Europe", by Dr Peter Brownsort, Scottish Carbon Capture & Storage (SCCS), 30 August 2013.

Buchel (1989). "Industrial Inorganic Chemistry", by Karl Heinz Büchel, Hans-Heinrich Moretto and Peter Woditsch, published by Wiley-VCH, 5 February 1989, Chapter 1.2 "Hydrogen".

Bullis (2013). "An Inexpensive Fuel-Cell Generator", by Kevin Bullis, for Technology Review, 22 August 2013. Accessed online.

Bungay (2009). "Operational Experience of Advanced Anaerobic Digestion", by Steve Bungay, Principal Process Engineer, Monsal Ltd, paper presented to 14th European Biosolids and Organic Resources Conference and Exhibition, 2009, Technical Paper No. 45.

Bungay and Abdelwahab (2008). "Monsal Enzymic Hydrolysis – New Developments and Lessons Learnt", by Steve Bungay and Mohamed Abdelwahab, paper presented to 13th European Biosolids and Organic Resources Conference, Workshop and Exhibition, 10–11 November 2008, Manchester, UK, Technical Paper 40.

Burbey and Joseph (2004). "Microwave-Assisted Fluidized Bed Synthesis", by Jason Burbey and Naja Joseph, REU-NSF 2004 Summer Research Program, University of Illinois at Chicago, Final Report.

Burgers et al. (2011). "Worldwide development potential for sour gas", by W. F. J. Burgers, P. S. Northrop, H. S. Kheshgi and J. A. Valencia, in Energy Procedia, Volume 4, 2011, Pages 2178–2184, http://www.dx.doi.org/10.1016/j.egypro.2011.02.104

Butterman and Castaldi (2009). "CO2 as a carbon neutral fuel source via enhanced biomass gasification", by Heidi C. Butterman and Marco J. Castaldi, in Environmental Science and Technology, Volume 43, Issue 23, 5 November 2009, Pages 9030–9037, http://www.dx.doi.org/10.1021/es901509n

Caldecott and McDaniels (2014). "Stranded generation assets: Implications for European capacity mechanisms, energy markets and climate policy", Smith School of Enterprise and the Environment, University of Oxford, Stranded Assets Programme, Working Paper, Final Version, January 2014.

CAN-Europe (2009). "Climate technology sheet #9: Hydrogen production", 29 November 2009. Accessed online.

Cao and Caldeira (2010). "Atmospheric carbon dioxide removal: Long-term consequences and commitment", by Long Cao and Ken Caldeira, in Environmental Research Letters, Volume 5, Issue 2, 024011, 6 pages, http://www.dx.doi.org/10.1088/1748-9326/5/2/024011

Carbon Market Watch (2014). "Press Release: New report shows carbon offsets from Alberta's tar sands jeopardize EU fuel quality rules", by Femke de Jong, 17 December 2014. Accessed online.

Carbon Tracker (2014a). "Unburnable Carbon – Are the world's financial markets carrying a carbon bubble?", published by Carbon Tracker Initiative, 8 May 2014.

Carbon Tracker (2014b). "Key Terms", by Carbon Tracker Initiative. Accessed online.

Carter (2013). "Fuel Cells and Hydrogen in Scotland", by Dan Carter, for Fuel Cells Today, 23 October 2013. Accessed online.

Cassidy et al. (2010). "Going Beyond Hydrogen: Non-hydrogen Fuels, Re-oxidation and Impurity Effects on Solid Oxide Fuel Cell Anodes", by Mark Cassidy, Jan Pieter Ouweltjes and Nico Dekker, Chapter 5 in "Innovations in Fuel Cell Technologies", edited by Robert Steinberger-Wilckens and Werner Lehnert, published by Royal Society of Chemistry, 18 October 2010, ISBN: 978-1-84973-033-4.

CAT (2013). "Zero Carbon Britain: Rethinking the Future", by Paul Allen, Laura Blake, Peter Harper, Alice Hooker-Stroud, Philip James and Tobi Kellner, Vijay Bhopal, Guppi Bola, Isabel Bottoms, Julie Bromilow, Louisa Casson, Richard Hebditch, Ling Li, Nuria Mera-Chouza, Lucy Neal, Darcy Pimblett, Sophie Wynne-Jones, Liz Zeidler and Mike Zeidler, Centre for Alternative Technology (CAT), 2013.

Cathles et al. (2012). "A commentary on "The greenhouse-gas footprint of natural gas in shale formations", by R. W. Howarth, R. Santoro, and Anthony Ingraffea", by Lawrence M. Cathles III, Larry Brown, Milton Taam and Andrew Hunter, in Climatic Change, Commentary, Volume 113, Issue 2, Pages 525–535, July 2012, http://www.dx.doi.org/10.1007/s10584-011-0333-0

CCC (2011). "The Renewable Energy Review", by the Committee on Climate Change, May 2011, Box 3.2, "Scope for increasing the contribution from biogas".

CCC (2013). "Fourth Carbon Budget Review", by the Committee on Climate Change, Part 1 "Assessment of climate risk and the international response", Part 2 "The cost-effective path to the 2050 target".

CCCC (2009). "Heading for 100% Renewable Urban Transport", The Expert Commission on Cities and Climate Change, published by the World Future Council and HafenCity University Hamburg, Hamburg, Germany, September 2009.

CCCC (2010). "100% Renewable Energy – and Beyond – for Cities", The Commission on Cities and Climate Change, published by HafenCity University Hamburg and World Future Council Foundation, Hamburg, Germany, March 2010.

Cenovus (2014a). "Annual report", by Cenovus Energy, "Investors", "Annual documents".

Chandel and Williams (2009). "Synthetic Natural Gas (SNG): Technology, Environmental Implications, and Economics", by Munish Chandel and Eric Williams, Climate Change Policy Partnership, Duke University, January 2009.

Che (2013). "Nobel Prize in chemisty 1912 to Sabatier: Organic chemistry or catalysis?", by Michael Che, Catalysis Today, Volumes 218–219, December 2013, Pages 162–171, http://www.dx.doi.org/10.1016/j.cattod.2013.07.006

Cheadle (2014). "Oilsands exporters narrowly survive 'dirty' oil vote in Europe", by Bruce Cheadle, in Cleantech Canada, for The Canadian Press, 17 December 2014.

Cheng et al. (2006). "A Solid Oxide Fuel Cell Running on H2S/CH4 Fuel Mixtures", by Zhe Cheng, Shaowu Zha, Luis Aguilar, Dean Wang, Jack Winnick and Meilin Liu, in Electrochemical and Solid State Letters, Volume 9, Issue 1, Pages A31–A33, 2006, http://www.dx.doi.org/10.1149/1.2137467

Cherry et al. (2014). "Environmental Impacts of Shale Gas Extraction in Canada", by John Cherry, Michael Ben-Eli, Lalita Bharadwaj, Richard Chalaturnyk, Maurice B. Dusseault, Bernard Goldstein, Jean-Paul Lacoursière, Ralph Matthews, Bernhard Mayer, John Molson, Kelly Munkittrick, Naomi Oreskes, Beth Parker and Paul Young, of the Expert Panel on Harnessing Science and Technology to Understand the Environmental Impacts of Shale Gas Extraction, published by Council of Canadian Academies, 2014, ISBN: 978-1-926558-77-6.

Clark (2014). "Biomass report adds to debate on power station subsidies", by Pilita Clark, in Financial Times, 23 July 2014.

Client Earth (2012). "Black Carbon Summit", report from the Black Carbon Campaign summit, convened at the Royal Society, London, 22 March 2012.

CNG Services (2010). "Bio-SNG: Feasibility Study. Establishment of a Regional Project", by Progressive Energy and CNG Services, a report for NEPIC, National Grid and Centrica, 10 November 2010.

ComAp (2009). "Bi-Fuel (Dual Fuel) Conversion of Diesel & HFO Generating Sets", by ComAp Systems. Accessed online.

Cormos (2014). "Economic implications of pre- and post-combustion calcium looping configurations applied to gasification power plants", by Calin-Cristian Cormos, in International Journal of Hydrogen Energy, Volume 39, Issue 20, 3 July 2014, Pages 10507–10516, http://www.dx.doi.org/10.1016/j.ijhydene.2014.05.033

CRD (2015). "LPG/CNG Conversion – Diesel Vehicles", by Common Rail Diesel. Accessed online.

Creutzig et al. (2014). "Catching two European birds with one renewable stone: Mitigating climate change and Eurozone crisis by an energy transition", by Felix Creutzig, Jan Christoph Goldschmidt, Paul Lehmann, Eva Schmid, Felix von Blücher, Christian Breyer, Blanca Fernandez, Michael Jakob, Brigitte Knopf, Steffen Lohrey, Tiziana Susca and Konstantin Wiegandt, in Renewable and

Sustainable Energy Reviews, Volume 38, Pages 1015–1028, 2014, http://www.dx.doi.org/10.1016/j.rser.2014.07.028

Crowther (2011). "Hydrogen for grid-2-gas load balancing in future energy systems", presentation given by Mark Crowther of Kiwa GASTEC at CRE, 15 September 2011, Slide 4, "What next?".

CSC (2013). "From Vision to Action: A Workshop Report on 100% Renewable Energies in European Regions", published by Climate Service Center, Renewables 100 Policy Center and World Future Council, March 2013.

Curran et al. (1967). "Carbon Dioxide Acceptor Gasification Process", or "CO2 Acceptor Gasification Process", by George P. Curran, Carl E. Fink and Everett Gorin, Chapter 10, in "Fuel Gasification", edited by Frank C. Schora Jr., Volume 69, Pages 141–165, ISBN 13: 9780841200708, published by Advances in Chemistry, 1 June 1967, http://www.dx.doi.org/10.1021/ba-1967-0069.ch010

Czernik et al. (2004). "Hydrogen from Biomass: Process Research", by S. Czernik, R. French, K. Magrini and R. Evans, presentation for National Renewable Energy Laboratory (NREL), USA DOE Hydrogen, Fuel Cells and Infrastructure Technologies Program Review, 24–27 May 2004.

Czisch (2011a). "Scenarios for a Future Electricity Supply: Cost-Optimized Variations on Supplying Europe and Its Neighbours with Electricity from Renewable Energies", by Dr.-Ing. Dipl.-Phys. Gregor Czisch, published by The Institution of Engineering and Technology (IET), 2011, ISBN: 978-1-84919-156-2.

Czisch (2011b). "The Supergrid: 100% Renewable Electricity Supply for Europe and Its Neighborhood", by Dr.-Ing. Dipl.-Phys. Gregor Czisch, presentation to Grid Integration of Renewables Workshop, held at Huang Engineering Center, TomKat Center for Sustainable Energy, Stanford University, 13 January 2011, Session 1, "Large-scale Integration of Renewables".

Czisch and Schmid (2006). "Low Cost but Totally Renewable Electricity Supply for a Huge Supply Area – a European/Trans-European Example", by Dr.-Ing. Dipl.-Phys. Gregor Czisch and Jürgen Schmid, Deutsche Physikalische Gesellschaft (DPG), Arbeitskreis Energie (AKE), "Klimaschutz und Energie Versorgung", 2006, "10 Vorträge und 1 Abstract der Münchner Tagung", Pages 133–160.

Damosso (2010). "First steps in regional hydrogen infrastructure buildup in Europe", by Davide Damosso, HyRaMP Annual Hydrogen Infrastructure Report, presentation to Fuel Cells and Hydrogen Joint Undertaking (FCH JU), 3rd FCH JU Stakeholders General Assembly, 9 November 2010 (HyRaMP is now known as HyER).

Davidson (2007). "Post Combustion Carbon Capture from Coal Fired Plants – Solvent Scrubbing", by Robert M. Davidson, for IEA Clean Coal Centre, International Energy Agency (IEA) Greenhouse Gas R&D Programme, Report: 2007/15, July 2007, ISBN: 92-9029-444-2, Introduction.

Davies et al. (2014). "Oil and gas wells and their integrity: Implications for shale and unconventional resource exploitation", by Richard J. Davies, Sam Almond, Robert S. Ward, Robert B. Jackson, Charlotte Adams, Fred Worrall, Liam G. Herringshaw, Jon G. Gluyas and Mark A. Whitehead, in Journal of Marine and Petroleum Geology, Volume 56, September 2014, Pages 239–254, http://www.dx.doi.org/10.1016/j.marpetgeo.2014.03.001

Day et al. (2013). "Low carbon gas from mixed waste, biomass and coal: a low cost route to CCS", by Anthony Day, Dr Andrew Williams and Ronald Christopher Hodrien, in Carbon Capture Journal, 29 November 2013. Accessed online.

de Bakker (2014). Informal unrecorded conversation with Adriaan de Bakker of Gasunie, 24 September 2014.

de Castro et al. (2009). "The role of non conventional oil in the attenuation of peak oil", by Carlos de Castro, Luis Javier Miguel and Margarita Mediavilla, in Energy Policy, Volume 37, Issue 5, May 2009, Pages 1825–1833, http://www.dx.doi.org/10.1016/j.enpol.2009.01.022

DECC (2012). "CCS Roadmap", published by HMG DECC, 3 April 2012. Accessed online.

DECC (2013a). "Investing in renewable technologies – CfD contract terms and strike prices", by HMG DECC, December 2013.

DECC (2013b). "Energy Flow Chart 2013", HMG DECC. Accessed online.

DECC (2013c). "Electricity Market Reform: Capacity Market – Detailed Design Proposals", by HMG DECC, Cmd 8637, June 2013.

DECC (2013d). "Electricity Generation Costs", HMG DECC, 19 December 2013, Document: URN 14D/005.

DECC (2014a). "Energy Trends", HMG DECC, 31 July 2014, Section 4, Section 5.

DECC (2014b). "Delivering UK Energy Investment", published by HMG DECC, July 2014, URN: 14D/180.

DECC (2014c). "RHI Biomethane Injection to Grid Tariff Review", by HMG DECC, 9 December 2014.

DECC (2014d). "Energy Consumption in the UK (ECUK): Domestic Data Tables: 2014 update", by HMG DECC, part of National Statistics, Table 3.03, "Domestic final energy consumption by end use 1970 to 2013".

DECC (2014e). "Delivering UK Energy Investment 2014", by HMG DECC, January 2015.

DECC (2015). "Delivering UK Energy Investment: Networks 2015", by HMG DECC, January 2015.

Defra (2011). "The Economics of Waste and Waste Policy", by HMG Defra, June 2011.

Defra (2012). "UK Bioenergy Strategy", by HMG Defra, DfT and DECC, April 2012.

Defra (2014). "Energy from waste: A guide to the debate", by HMG Defra and DECC, February 2014 (revised edition), Document: PB14130, Crown Copyright 2014.

De Klerk (2011). "Fischer-Tropsch Refining", by Arno de Klerk, published by Wiley-VCH, 20 July 2011, ISBN: 978-3527326051.

Deloitte (2013). "Potential Bowland Basin shale gas development: Economic and fiscal impacts", Deloitte LLP, 2013. Accessed online.

dena (2014a). "Pilot- und Demonstrationsprojekte im Power-to-Gas-Konzept". Accessed online.

dena (n.d.) "Power to Gas Strategy Platform: Power to Gas – an innovative system solution for the energy turnaround", Federal Republic of Germany, Deutsche Energie-Agentur GmbH (dena).

Denmark (2011). "Our Energy Future", The Government of Denmark, November 2011. Accessed online.

Denmark (2012). "Accelerating Green Energy Towards 2020", The Danish Energy Agreement of March 2012, The Government of Denmark, March 2012.

Desclaux et al. (2010). "Direct Carbon Fuel Cells", by P. Desclaux, S. Nürnberger and U. Stimming, Chapter 6 in "Innovations in Fuel Cell Technologies", edited by Robert Steinberger-Wilckens and Werner Lehnert, published by Royal Society of Chemistry, 18 October 2010, ISBN: 978-1-84973-033-4, Section 6.2.1.

de Souza et al. (2008). "Molybdenum electrodes for hydrogen production by water electrolysis using ionic liquid electrolytes", by Roberto F. de Souza, Gabriel Loget, Janine C. Padilha, Emilse M. A. Martini, Michele O. de Souza, in Electrochemistry Communications, Volume 10, Issue 11, November 2008, Pages 1673–1675, http://www.dx.doi.org/10.1016/j.elecom.2008.08.029

DG ENER (2013). "Capacity Mechanisms In Individual Markets Within the IEM", by European Commission, Directorate General Energy (DG ENER), Directorate B, Project: ENER/B2/175/2012, Document: TE-2013-06, 28 May 2013.

Dittus and Johnson (2001). "The Hidden Value of Lignite Coal", by Myles Dittus and Dale Johnson, Dakota Gasification Company, Great Plains Synfuels Plant, paper presented at 2001 Gasification Technologies Conference.

Dlugokencky et al. (2011). "Global atmospheric methane: budget, changes and dangers", by Edward J. Dlugokencky, Euan G. Nisbet, Rebecca Fisher and David Lowry, in Philosophical Transaction of the Royal Society A, Volume 369, Issue 1943, 18 April 2011, 28 May 2011, Pages 2058–2072, http://www.dx.doi.org/10.1098/rsta.2010.0341

DNV (2012). "Fuel cells for ships", by Det Norske Veritas (DNV), now DNV GL, Research and Innovation, Position Paper 13, 2012. Accessed online.

Dodds and McDowall (2012). "A review of hydrogen production technologies for energy system models", by Paul E. Dodds and Will McDowall, UK Sustainable Hydrogen Energy Consortium (UKSHEC), Working Paper Number 6, 2012.

Dodds et al. (2015). "Hydrogen and fuel cell technologies for heating: A review", by Paul E. Dodds, Iain Staffell, Adam D. Hawkes, Francis Li, Philipp Grünewald, Will McDowall and Paul Ekins, in International Journal of Hydrogen Energy, Volume 40, Issue 5, 9 February 2015, Pages 2065–2083, DOI: http://www.dx.doi.org/10.1016/j.ijhydene.2014.11.059

DOE (2006). "Practical Experience Gained During the First Twenty Years of Operation of the Great Plains Gasification Plant and Implications for Future Projects", by USA DOE, Office of Fossil Energy, April 2006.

DOE (2013). "Report of the Hydrogen Production Expert Panel: A Subcommittee of the Hydrogen & Fuel Cell Technical Advisory Committee", by US Department of Energy, May 2013.

DUKES (2014). "Digest of United Kingdom Energy Statistics 2014", Government of the United Kingdom, Department of Energy and Climate Change, A National Statistics Publication, Printed version, ISBN: 9780115155307.

Dumas (1859) (2006). "Le Caucase: impressions de voyage", by Alexandre Dumas, 1859, in print from Editions Le Joyeux Roger, Montreal, 2006, ISBN: 978-2-923523-10-1, ISBN-10: 2-923523-10-5. Chapter XXII, "Bakou".

Dupont (2011). "hydrogen production from biomass feedstock", by Valerie Dupont, presentation to UK Hydrogen Researchers' Workshop, United Kingdom Energy Research Centre (UKERC), The Meeting Place, Oxford, 28 February 2011.

Dyman and Cook (1998). "Summary of Deep Oil and Gas Wells in the United States Through 1998", by Thaddeus S. Dyman and Troy A. Cook, Chapter B in "Geologic Studies of Deep Natural Gas Resources", edited by Thaddeus S. Dyman and Vello A. Kuuskraa, USGS Digital Data Series 67, ISBN: 0-607-95407-8, Version 1, 2001. Accessed online.

E-Cube (2013). "Vers un système gazier 100% «décarboné»: Etude technico-économique des conditions de développement d'un réseau gaz décarboné en France", by E-Cube Strategy Consultants, July 2013.

E4tech (2010). "The potential for bioSNG production in the UK", E4Tech consulting with National Non-Food Crops Centre (NNFCC), the UK's National Centre for Biorenewable Energy, NNFCC project 10/008, April 2010.
EBTP (2014). "Biofuels Policy and Legislation", by European Biofuels Technology Platform (EBTP), "Recent EC legislation, policies and communciations relating to biofuels". Accessed online.
EC (2009). "Directive 2009/30/EC of the European Parliament and the Council", 23 April 2009.
EC (2013a). "The Clean Air Policy Package", by European Commission, 18 December 2013, including "Industrial emissions".
EC (2013b). "Single market for gas & electricity: Studies – Gas", by European Commission, Reports to 2013. Accessed online.
EC (2014a). "Single market for gas & electricity: Gas quality harmonisation", by European Commission, Workshops, 1 July 2014. Accessed online.
EC (2014b). "Fuel Quality", by European Commission, 2014. Accessed online.
ECN (2007a). "Gasification technology", 2007. Accessed online.
ECN (2007b). "Quick Scan of the Economic Consequences of Prohibiting Residual Fuels in Shipping", by H. P. J. de Wilde, P. Kroon, M. Mozaffarian and T. Sterker, for ECN, Report ECN-E--07-051, July 2007, "Conclusions".
EcoDual (2012). "Dual Fuel (Natural Gas/Diesel) Conversion Systems for Heavy Duty Trucks", by EcoDuel, presentation, June 2012. Accessed online.
ECUK (2014). "Energy Consumption in the UK", Government of the United Kingdom, Department of Energy and Climate Change, Document: URN 14D/218, July 2014, All data tables (Terawatt hours), 2014 update.
EEF (2014). "UK Shale Gas: The Manufacturers' View", by Engineering Employers' Federation (EEF), November 2014, Chart 4 "Source of UK natural gas supply: Actual and projected".
EIA (2004). "The Basics of Underground Natural Gas Storage", by USA EIA. Accessed online.
EIA (2006). "Changing Trends in the Refining Industry", USA EIA, Annual Energy Outlook (AEO), 2006. Accessed online.
EIA (2012). "Annual Energy Outlook 2012: With Projections to 2035", USA DOE EIA, Annual Energy Outlook (AEO) 2012, Report: DOE/EIA-0383(2012), June 2012.
EIA (2013). "Technically Recoverable Shale Oil and Shale Gas Resources: An Assessment of 137 Shale Formations in 41 Countries Outside the United States", by USA EIA, 10 June 2013.
EIA (2014a). "Annual Energy Outlook 2014: with projections to 2040", USA, DOE, EIA, AEO 2014, Report: DOE/EIA-0383(2014), April 2014.
EIA (2014b). "World Oil Transit Chokepoints", by USA EIA, 10 November 2014. Accessed online.
EIA (2014c). "Levelized Cost and Levelized Avoided Cost of New Generation Resources in the Annual Energy Outlook 2014", USA DOE EIA, Annual Energy Outlook (AEO), April 2014.
Electricity Council (1982). "Electricity Supply in the United Kingdom: A Chronology", by The Electricity Council, 3rd edition, 1982, ISBN: 978-0851880877.
Electrochaea (2014). "Technology", by Electrochaea. Accessed online.
ENERTRAG (2013). "Working with Energy: wind2hydrogen; the energy storage solution", by ENERTRAG, presentation to All Energy Conference 2013. Accessed online.

ENERTRAG (n.d.). "The ENERTRAG-Hybrid Power Plant". Accessed online.

Enviros (2008). "Barriers to Renewable Heat: Analysis of Biogas Options", Enviros Consulting Limited, report for the UK Government Department of Business Enterprise and Regulatory Reform (BERR), CAN: DE1510001, Part 2 Final Biogas Report v4_0, September 2008.

EPA (2014). "Inventory of U. S. Greenhouse Gas Emissions and Sinks: 1990–2012", by USA EPA, Report: EPA 430-R-14-003, 15 April 2014.

Ernst & Young (2014). "Spotlight on oil and gas megaprojects", by Ernst & Young Global Limited (EYGM). Accessed online.

Er-rbib and Bouallou (2013). "Modelling and Simulation of Methanation Catalytic Reactor for Renewable Electricity Storage", by Hanaâ Er-rbib and Chakib Bouallou, in Chemical Engineering Transactions, Volume 35, 2013, ISBN: 978-88-95608-26-6, ISSN: 1974-9791, http://www.dx.doi.org/10.3303/CET1335090

Etiope (2010). "Geological Methane", by Giuseppe Etiope, in "Methane and Climate Change", edited by Dave Reay, Pete Smith and Andre van Amstel, published by Routledge, 26 May 2010, Chapter 4.

Etiope et al. (2013a). "Natural seepage of shale gas and the origin of "eternal flames" in the Northern Appalachian Basin, USA", by Giuseppe Etiope, Agnieszka Drobniak and Arndt Schimmelmann, Journal of Marine and Petroleum Geology, Volume 43, May 2013, Pages 178–186, http://www.dx.doi.org/10.1016/j.marpetgeo.2013.02.009

Etiope et al. (2013b). "Geochemical technology for petroleum exploration and monitoring gas seepage and groundwater contamination during shale gas development", by Giuseppe Etiope, Martin Schoell and Calin Baciu; in a presentation given by GasConsult International, Berkeley, USA.

EURLEX (1999). "Council Directive 1999/31/EC of 26 April 1999 on the landfill of waste", The Council of the European Union, 26 April 1999.

EURLEX (2001). "Directive 2001/80/EC of the European Parliament and of the Council of 23 October 2001 on the limitation of emissions of certain pollutants into the air from large combustion plants", by European Parliament and Council of the European Union, 23 October 2001.

EURLEX (2008). "Directive 2008/98/EC of the European Parliament and of the Council of 19 November 2008 on waste and repealing certain Directives (Text with EEA relevance)", by European Parliament and of the Council of the European Union, 19 November 2008.

EURLEX (2010). "Directive 2010/75/EU of the European Parliament and of the Council of 24 November 2010 on industrial emissions (integrated pollution prevention and control) Text with EEA relevance", by European Parliament and Council of the European Union.

Evans-Pritchard (2014). "Fossil industry is the subprime danger of this cycle", by Ambrose Evans-Pritchard, published by The Daily Telegraph newspaper, 9 July 2014. Accessed online.

Evers (2008a). "Actual Worldwide Hydrogen Production from . . .", by Arno A. Evers, December 2008. Data for 2006 = 45 million tonnes, 500 million cubic metres, "(Source: Chemical Economics Handbook, SRI – July 2001 and Industrial Gases by the Chemical Economics Handbook, SRI – October 2007)".

EWEA (2005). "Large Scale Integration of Wind Energy in the European Power Supply: Analysis, Issues and Recommendations", A report by The European Wind Energy Association (EWEA), December 2005.

ExxonMobil (2011). "Flexicoking: Coking and Integrated Steam/Air Gasification", presentation by Dr Girish Chitnis and Dr Paul Kamienski of ExxonMobil Research & Engineering Company, and Mr Boyd Rumball, of Syncrude Canada Ltd, to the 8th Annual Canadian Oil Sands Summit, Calgary, Alberta, Canada, 2–3 February 2011.

Fairley and Andrews (2014). "Electricity market reform: an update on contracts for difference", by Ross Fairley and Emma Andrews, 5 March 2014. Accessed online.

Fantazzini et al. (2011). "Global oil risks in the early 21st century", by Dean Fantazzini, Mikael Höök and André Angelantoni, in Energy Policy, Volume 39, Issue 12, December 2011, Pages 7865–7873, http://www.dx.doi.org/10.1016/j.enpol.2011.09.035

FAO (1986). "Wood gas as engine fuel", by United Nations (UN), Food and Agriculture Organization (FAO), Forestry Paper 72, M-38, ISBN: 92-5-102436-7, 1986. Accessed online.

FAO (2001). "Sea Farming and Sea Ranching in China", by Jiansan Jia and Jiaxin Chen, United Nations (UN), Food and Agriculture Organization (FAO), Fisheries Technical Paper 418, ISSN: 0429-9345, Section 2.1. Accessed online.

Farzaneh-Gord et al. (2010). "Computing Thermal Properties of Natural Gas by Utilizing AGA8 Equation of State", by Mahmood Farzaneh-Gord, Azad Khamforoush, Shahram Hashemi and Hossein Pourkhadem Namin, in International Journal of Chemical Engineering and Applications, Volume 1, Issue 1, June 2010, ISSN: 2010-0221, Table 1.

Farzaneh-Gord et al. (2013). "Effects of Natural Gas Compositions on CNG Fast Filling Process for Buffer Storage System", by Mahmood Farzaneh-Gord, H. Reza Rahbari and M. Deymi-Dashtebayaz, in Oil & Gas Science and Technology, Revue IFP Energies Nouvelles, Volume 69, Issue 2, March–April 2014, Pages 319–330, http://www.dx.doi.org/10.2516/ogst/2012010

FCO (2014). "Opening statement of welcome to Enterprise Forum in Kazakhstan", by Foreign and Commonwealth Office, 20 May 2014. Accessed online.

Ferrari et al. (2014). "Microwave-Specific Effects on the Equilibrium Constants and Thermodynamics of the Steam-Carbon and Related Reactions", by Anthony Ferrari, Jacob Hunt, Adrian Lita, Bridgett Ashley and Professor Albert E. Stiegman, in Journal of Physical Chemistry C, Volume 118, Issue 18, 8 April 2014, Pages 9346–9356, http://www.dx.doi.org/10.1021/jp501206n

Finkenrath et al. (2012). "CCS Retrofit: Analysis of the Globally Installed Coal-Fired Power Plant Fleet", by Matthias Finkenrath, Julian Smith and Dennis Volk, International Energy Agency (IEA), 2012.

Fitzgibbon and Rogers (2014). "Implications of Light Tight Oil Growth for Refiners in North America and Worldwide", by Tim Fitzgibbon and Matt Rogers, of McKinsey & Company, January 2014. Accessed online.

Forbes (1939). "Petroleum and Bitumen in Antiquity", by Robert James Forbes, summary of a lecture to the Society for the Study of Alchemy and Early Chemistry, 19 October 1938 at University College London, published in Journal of the Institute of Petroleum, Volume 25, Issue 183, January 1939.

Forbes (1958). "Studies in Early Petroleum History", by Robert Jacobus Forbes, published by E. J. Brill, Leiden, Netherlands. Chapter XII, "Ex Oriente Bitumen", Pages 152–153.

Foster Wheeler (2008). "New Challenges & Solutions in Designing Large Sour Gas Projects", presentation by John Carroll and John Foster, of Gas Liquids Engineering Ltd and Foster Wheeler Ltd, 18 February 2008.

Fraunhofer (2013). "Electricity production from solar and wind in Germany in 2013", compiled by Professor Dr Bruno Burger, Fraunhofer Institut für Solare Energiesysteme (ISE), Renewable Energy Data, 9 January 2014.

Fraunhofer (2014). "Electricity production from solar and wind in Germany", compiled by Professor Dr Bruno Burger, Fraunhofer Institut für Solare Energiesysteme (ISE), Renewable Energy Data, Updates, "Electricity production from solar and wind in Germany in 2014", Slide 8, "Export and import balance since 2001".

Fthenakis (2012). "How Long Does it Take for Photovoltaics To Produce the Energy Used?", by Vasilis Fthenakis, in PE Magazine, National Society of Professional Engineers, January/February, 2012.

Fthenakis et al. (2011). "Methodology Guidelines on Life Cycle Assessment of Photovoltaic Electricity", by Vasilis Fthenakis, Rolf Frischknecht, Marco Raugei, Hyung Chul Kim, Erik Alsema, Michael Held and Mariska de Wild-Scholten, 2nd edition, IEA PVPS Task 12, Subtask 20, Life Cycle Analysis/Assessment (LCA), Report IEA-PVPS T12-03:2011, International Energy Agency (IEA), Photovoltaic Power Systems Programme (PVPS), Task 12, November 2011.

Galetti (2005). "Kirkuk: The Pivot of Balance in Iraq Past and Present", by Mirella Galetti PhD, Journal of Assyrian Academic Studies, Volume 19, Number 2, 2005, Pages 21–52. Accessed online.

Galimov and Rabbani (2001). "Geochemical Characteristics and Origin of Natural Gas in Southern Iran", by E. M. Galimov and A. R. Rabbani, in Geochemistry International, Volume 39, Issue 8, 2001, Pages 780–792, translated from Russian, from Geokhymiya, Issue 8, 2001, Pages 862–874, English translation, 2007.

Gandia et al. (2013). "Renewable Hydrogen Energy: An Overview", by Luis M. Gandía; Gurutze Arzamendi and Pedro M. Diéguez, Chapter 1 in "Renewable Hydrogen Technologies", edited by Luis M. Gandía; Gurutze Arzamendi and Pedro M. Diéguez, ISBN: 978-0080444291, Section 1.3.1.2, Page 14.

Garcier (2009). "The nuclear 'renaissance' and the geography of the uranium fuel cycle", by Romain Garcier, in Geography, Volume 94, Issue 3, Autumn 2009, Pages 198–206. Accessed online.

Gardner and Leff (2014). "U. S. opening of oil export tap widens battle for global market", by Timothy Gardner and Jonathan Leff, for Reuters, 31 December 2014. Accessed online.

Gassner and Marechal (2009). "Thermo-economic process model for thermochemical production of Synthetic Natural Gas (SNG) from lignocellulosic biomass", by Martin Gassner and François Maréchal, in Biomass and Bioenergy, Volume 33, Issue 11, November 2009, Pages 1587–1604, http://www.dx.doi.org/10.1016/j.biombioe.2009.08.004

Gas Turbine World (2011). "Nuon postpones implementation of IGCC technology at Magnum plant", by Harry Jaeger, in Gas Turbine World, 8 April 2011. Accessed online.

GE (2014a). "The New Power Generation: This Fuel Cell Startup Could Spark a Revolution", by General Electric (GE), 22 July 2014. Accessed online.

GE (2014b). "Solutions for the Biowaste Industry: Monsal Anaerobic Digestion Technology", by GE Power and Water, Water and Process Technologies, December 2014. Accessed online.

GEA (2012). "Energy Resources and Potentials", by Hans-Holger Rogner, Roberto F. Aguilera, Cristina L. Archer, Ruggero Bertani, S. C. Bhattacharya, Maurice B. Dusseault, Luc Gagnon, Helmut Haberl, Monique Hoogwijk, Arthur Johnson, Mathis L. Rogner, Horst Wagner, Vladimir Yakushev, Doug J. Arent, Ian Bryden,

Fridolin Krausmann, Peter Odell, Christoph Schillings and Ali Shafiei, edited by Ji Zou, Chapter 7, in "Global Energy Assessment: Towards a Sustainable Future", edited by Thomas B. Johansson, Nebojsa Nakicenovic, Anand Patwardhan and Luis Gomez-Echeverri, published by the International Institute for Applied Systems Analysis (IIASA), Laxenburg, Austria and Cambridge University Press, Cambridge, UK and New York, NY, USA. Released 19 June 2012 at the Rio + 20 United Nations Conference on Sustainable Development, Rio, Brazil.

German Renewable Energy Agency (2013). "Kombikraftwerk 2 demonstrates: Reliable Electricity Grid Operation Possible in Future with 100% Renewable Energy", published by Agentur für Erneuerbare Energien, 30 October 2013.

GGFR (2013). "Global Gas Flaring Reduction Partnership (GGFR)", presentation by Bjorn Hamso, GGFR Program Manager, to the Internernational Petroleum Industry Environmental Conservation Association (IPIECA) Workshop, Rome, 9 October 2013.

Ghassemi et al. (1979). "Environmental Assessment Report: High-Btu Gasification Technology", by M. Ghassemi, K. Crawford, S. Quinlivan and D. Strehler, paper presented at Symposium Proceedings: Environmental Aspects of Fuel Conversion Technology, IV, April 1979, Hollywood, Florida, an Interagency Energy/ Environment R&D Program Report, Office of Research and Development, USA EPA, Pages 203–250.

Giering et al. (2014). "Reconciliation of the carbon budget in the ocean's twilight zone", by Sarah L. C. Giering, Richard Sanders, Richard S. Lampitt, Thomas R. Anderson, Christian Tamburini, Mehdi Boutrif, Mikhail V. Zubkov, Chris M. Marsay, Stephanie A. Henson, Kevin Saw, Kathryn Cook and Daniel J. Mayor, in Nature, Volume 507, 27 March 2014, Pages 480–483, Letters, http://www.dx.doi.org/10.1038/nature13123

Gismatullin (2014). "Shell Sells U. S. Natural Gas Fields for $2.1 Billion", by Eduard Gismatullin, in Bloomberg News, 14 August 2014. Accessed online.

Glancy (2013). "Quantifying Fugitive Emission Factors from Unconventional Natural Gas Production Using IPCC Methodologies", by Ryan P. Glancy, Institute for Global Environmental Studies (IGES), Hayama, Japan, 2013.

Global CCS Institute (2012). "CO2 Capture Technologies: Technology Options for CO2 Capture", published by Global CCS Institute, January 2012.

Global CCS Institute (n.d.). "How CCS works – capture", published online by Global CCS Institute. Accessed online.

Gonzalez (2009). "Hydrocracking and hydrotreating developments", by René Gonzalez, in Digital Refining, PTQ Catalysis, 2009. Accessed online.

Google Maps (2015a). Jam, Iran. Accessed online.

GPOBA (2007). "World Bank-administered Trust Fund GPOBA will Fund 37,000 Biogas Plants in Rural Nepal", 4 October 2007. Accessed online.

Graboski and Donath (1973). "Combined Shift and Methanation Reaction Process for the Gasification of Carbonaeceous Materials", by Michael S. Graboski and Ernest E. Donath, US Patent 3,904,386.

Gray (n.d.). "Tutorial on Upgrading of Oilsands Bitumen", by Murray R. Gray, presentation at University of Alberta. Accessed online.

Green Gas (2014). "The Green Gas Certification Scheme", Renewable Energy Association. Accessed online.

Greenpeace (2010). "Energy [R]evolution: Towards a Fully Renewable Energy Supply in the EU 27", published by the European Renewable Energy Council (EREC) and Greenpeace, June 2010. Accessed online.

Grond and de Joode (2014). "The role of power-to-gas in the future Dutch energy system", by Lukas Grond and Jeroen de Joode, Report: ECN-O-14-030, Summary, July 2014. Accessed online.

Gru (2012). "Professor uses bacteria to catalyze carbon-neutral methane", by Jacqueline Gru, in The Stanford Daily, Stanford University, 2 August 2014. Accessed online.

Guell (2012). "Hydrothermal gasification of seaweed: a promising technology to biofuels production", by Berta Matas Güell, of SINTEF Energy Research, presentation on 25 September 2012.

Guezuraga et al. (2012). "Life cycle assessment of two different 2 MW class wind turbines", by Begoña Guezuraga, Rudolf Zauner and Werner Pölz, in Renewable Energy, Volume 37, Issue 1, January 2012, Pages 37–44, http://www.dx.doi.org/10.1016/j.renene.2011.05.008

Habazaki et al. (1998). "Co-methanation of carbon monoxide and carbon dioxide on supported nickel and cobalt catalysts prepared from amorphous alloys", by Hiroki Habazaki, Michiaki Yamasaki, Bo-Ping Zhang, Asahi Kawashima, Shunpei Kohno, Takuro Takai and Koji Hashimoto, in Applied Catalysis A: General, Volume 172, Issue 1, 24 August 1998, Pages 131–140, http://www.dx.doi.org/10.1016/S0926-860X(98)00121-5

Hagoort (1988). "Fundamentals of Gas Reservoir Engineering", by Jacques Hagoort, published by Elsevier, 1 June 1988, Page 29, ISBN: 978-0444554116. Accessed online.

Haines et al. (2003). "Reduction of CO2 emissions by addition of hydrogen to natural gas", by M. R. Haines, Erik A. Polman and J. C. de Laat (internal document reference "E. A. Polman, J. C. de Laat, M. Crowther et al."), 16 September 2003. Also: "Greenhouse Gas Control Technologies", Proceedings of the 7th International Conference on Greenhouse Gas Control Technologies 5–9 September 2004, Vancouver, Canada, 2005, Volume I: "Peer Reviewd Papers and Overviews", Pages 337–345: http://www.dx.doi.org/10.1016/B978-008044704-9/50035-5

Haldor Topsoe (2009). "From solid fuels to substitute natural gas (SNG) using TREMP (TM)", "Topsoe Recycle Energy-efficient Methanation Process", by Haldor Topsoe, "What is TREMP?", March 2009. Accessed online.

Hall et al. (2014). "EROI of different fuels and the implications for society", by Charles A. S. Hall, Jessica G. Lambert and Stephen B. Balogh, in Energy Policy, Volume 64, 2014, Pages 141–152, http://www.dx.doi.org/10.1016/j.enpol.2013.05.049

Hammond (2013). "Economic Analysis of Shale Gas Wells in the United States", by Christopher D. Hammond, MIT, June 2013. Accessed online.

Hamzehei et al. (2013). "Red Tide Monitoring in the Persian Gulf and Gulf of Oman Using MODIS Sensor Data", by Samad Hamzehei, A. A. Bidokhti, Mohammad Seddiq Mortazavi and Abolhasan Gheiby, in Technical Journal of Engineering and Applied Sciences (TJEAS), ISSN: 2051-0853, Volume 3, Issue 12, Pages 1100–1107.

Harrabin (2013). "Biomass fuel subsidies to be capped says energy secretary", by Roger Harrabin, BBC News Online, 16 July 2013. Accessed online.

Harvey (2013). "Solar companies to sue UK government for £140m over feed-in tariff cuts", by Fiona Harvey, The Guardian, 23 January 2013. Accessed online.

Heaven (1996). "Technology Gasification converts a variety of problem feedstocks and wastes", by David L. Heaven, Oil and Gas Journal, Volume 94, Issue 22, 27 May 1996. Accessed online.

He et al. (2013). "Hydrogen from Biomass: Advances in Thermochemical Processes", by Li He, Jia Yang and De Chen, Chapter 6 in "Renewable Hydrogen Technologies", edited by Luis M. Gandía; Gurutze Arzamendi and Pedro M. Diéguez, ISBN: 978-0080444291, Section 6.5.7, Page 126.

Heffels et al. (2014). "An ecological and economic assessment of absorption-enhanced-reforming (AER) biomass gasification", by Tobias Heffels, Russell McKenna and Wolf Fichtner, in Energy Conversion and Management, Volume 77, January 2014, Pages 535–544, http://www.dx.doi.org/10.1016/j.enconman.2013.09.007

Heide et al. (2011). "Reduced storage and balancing needs in a fully renewable European power system with excess wind and solar power generation", by Dominik Heide, Martin Greiner, Lüder von Bremen and Clemens Hoffmann, in Renewable Energy, Volume 36, Pages 2515–2523, 2011, http://www.dx.doi.org/10.1016/j.renene.2011.02.009

Heidenreich and Foscolo (2015). "New concepts in biomass gasification", by Steffen Heidenreich and Pier Ugo Foscolo, in Progress in Energy and Combustion Science, Volume 46, February 2015, Pages 72–95, http://www.dx.doi.org/10.1016/j.pecs.2014.06.002

Hemmes (2010). "Hydrogen Production by Internally Reforming Fuel Cells", by Kas Hemmes, Chapter 10, in "Innovations in Fuel Cell Technologies", edited by Robert Steinberger-Wilckens and Werner Lehnert, published by Royal Society of Chemistry, 18 October 2010, ISBN: 978-1-84973-033-4, Section 10.3, Section 10.4.

Herzog et al. (2009). "Advanced Post-Combustion CO2 Capture", by Howard Herzog, Jerry Meldon and Alan Hatton, report for Clean Air Task Force, April 2009. Accessed online.

Heyne (2010). "Process Integration Opportunities for Synthetic Natural Gas (SNG) Production by Thermal Gasification of Biomass", by Stefan Heyne, Thesis for Licentiate of Engineering, Chalmers University of Technology, Division of Heat and Power Technology, Department of Energy and Environment, Göteborg, Sweden 2010, publication 2010:2, ISSN: 1404-7098.

Heyne et al. (2013). "Bio-SNG Production via Gasification – Process Integration Aspects for Improving Process Performance", by Stefan Heyne, Henrik Thunman, Martin Seemann and Simon Harvey, Proceedings of the 21st European Biomass Conference and Exhibition, 2013, ISSN: 2282-5819.

Highwood and Kinnersley (2006). "When smoke gets in our eyes: The multiple impacts of atmospheric black carbon on climate, air quality and health", Environment International, Volume 32, Issue 4, May 2006, Pages 560–566, http://www.dx.doi.org/10.1016/j.envint.2005.12.003

Higman and Eppinger (1994). "The Zero-residue refinery", by Christopher Higman and Manfred Eppinger, paper presented at ACHEMA, Frankfurt, Germany, 8 June 1994. Accessed online.

Higman and van der Burgt (2003). "Gasification", by Christopher Higman and Maarten van der Burgt, published by Gulf Professional Publishing, Elsevier Inc., 2nd edition, 26 February 2008, ISBN: 978-0750685283.

HKCG (2013). "Natural Gas as Feedstock for Town Gas Production", by Ir. Chris W. C. Man, presentation to HKIE/IGEM Joint Gas Seminar, 24 April 2013. Accessed online.

Hlina et al. (2014). "Production of high quality syngas from argon/water plasma gasification of biomass and waste", by M. Hlina, M. Hrabovsky, T. Kavka and M. Konrad, in Waste Management, Volume 34, 2014, Pages 63–66, http://www.dx.doi.org/10.1016/j.wasman.2013.09.018

HMG (2013). "Energy Act 2013", Her Majesty's Government. Accessed online.

Hong et al. (2009). "Analysis of oxy-fuel combustion power cycle utilizing a pressurized coal combustor", by Jongsup Hong, Gunaranjan Chaudhry, J. G. Brisson, Randall Field, Marco Gazzino and Ahmed F. Ghoniem, in Energy, Volume 34, Issue 9, September 2009, Pages 1332–1340, Introduction, http://www.dx.doi.org/10.1016/j.energy.2009.05.015

Honjo (1997). "The Rain of Ocean Particles and Earth's Carbon Cycle", by Susumu Honjo, Woods Hole Oceanographic Institution (WHOI), "Oceanus" series, December 1997. Accessed online.

Honjo and Weller (1997). "Monsoon Winds and Carbon Cycles in the Arabian Sea", by Susumu Honjo and Robert A. Weller, Woods Hole Oceanographic Institution (WHOI), "Oceanus" series, December 1997. Accessed online.

Honore (2014). "The Outlook for Natural Gas Demand in Europe", by Anouk Honoré, Oxford Institute for Energy Studies, University of Oxford, Paper: NG 87, ISBN: 978-1-78467-003-0. Accessed online.

Hook (2014). "Depletion rate analysis of fields and regions: A methodological foundation", by Mikael Höök, in Fuel, Volume 121, 1 April 2014, Pages 95–108, http://www.dx.doi.org/10.1016/j.fuel.2013.12.024

Hook et al. (2010). "Global coal production outlooks based on a logistic model", by Mikael Höök, Werner Zittel, Jörg Schindler and Kjell Aleklett, in Fuel Volume 89, 2010, Pages 3546–3558, http://www.dx.doi.org/10.1016/j.fuel.2010.06.013

Hosgormez (2007). "Origin of the natural gas seep of Cirali (Chimera), Turkey: Site of the first Olympic fire", by Hakan Hosgormez, Journal of Asian Earth Sciences, Volume 30, Issue 1, 1 April 2007, Pages 131–141, http://www.dx.doi.org/10.1016/j.jseaes.2006.08.002

Hosgormez et al. (2008). "New evidence for a mixed inorganic and organic origin of the Olympic Chimaera fire (Turkey): a large onshore seepage of abiogenic gas", by H. Hosgormez, G. Etiope and M. N. Yalcin, in Geofluids, Special Issue: Special Thematic Edition on Gas Geochemistry, Volume 8, Issue 4, Pages 263–273, November 2008, http://www.dx.doi.org/10.1111/j.1468-8123.2008.00226.x

Hotellier (2014). "PEM-Electrolysis – a technology bridge for a more flexible energy system", presentation by Gaëlle Hotellier, Head of Hydrogen Solutions, Siemens AG, 2014, Slide 9. Accessed online.

Howarth (1864a). "Making Illuminating Gas", by John Howarth, US Patent, Number 42,771, 17 May 1864. Accessed online.

Howarth (1864b). "Improvement in Apparatus for Distilling Off Gases and Vapors", by John Howarth, US Patent, Number 42,772, 17 May 1864. Accessed online.

Howarth et al. (2011). "Methane and the greenhouse-gas footprint of natural gas from shale formations", by Robert W. Howarth, Renee Santoro and Anthony Ingraffea, in Climatic Change, Letters, Volume 106, Issue 4, Pages 679–690, June 2011, http://www.dx.doi.org/10.1007/s10584-011-0061-5

Howell (2014). "British fracking policy – a change of direction needed", by Lord David Howell, Journal of Energy Security, Note, 12 May 2014, to Spring 2014 Issue. Accessed online.

HSE (2011). "Hazards arising from the conveyance and use of gas from Non-Conventional Sources (NCS)", by Diane Broomhall, Martin Brown, Tim Illson, Yoke Ling Lee, Martin Maple, Glyn Morgan, Luisa Shelenko, Angie Siddle and Julie Truong, of GL Noble Denton, for the Health and Safety Executive, 2011, Research Report: RR882.

Hsu et al. (2012). "Life Cycle Greenhouse Gas Emissions of Crystalline Silicon Photovoltaic Electricity Generation", by David D. Hsu, Patrick O'Donoughue, Vasilis Fthenakis, Garvin A. Heath, Hyung Chul Kim, Pamala Sawyer, Jun-Ki Choi and Damon E. Turney, from Yale University, a Systematic Review and Harmonization, in Journal of Industrial Ecology, Volume 16, Issue Supplement s1, Pages S122–S135, Special Issue: "Meta-Analysis of Life Cycle Assessments", April 2012, http://www.dx.doi.org/10.1111/j.1530-9290.2011.00439.x

Huang et al. (2014). "Biomass Char Direct Chemical Looping Gasification Using NiO-Modified Iron Ore as an Oxygen Carrier", by Zhen Huang, Fang He, Yipeng Feng, Kun Zhao, Anqing Zheng, Sheng Chang, Guoqiang Wei, Zengli Zhao and Haibin Li, in Energy Fuels, Volume 28, Issue 1, 2014, Pages 183–191, http://www.dx.doi.org/10.1021/ef401528k

Hughes (2013). "Drill, Baby, Drill: Can Unconventional Fuels Usher in a New Era of Energy Abundance?", by J. David Hughes, Post Carbon Institute, February 2013. Accessed online.

Hunt et al. (2013). "Microwave-Specific Enhancement of the Carbon-Carbon Dioxide (Boudouard) Reaction", by Jacob Hunt, Anthony Ferrari, Adrian Lita, Mark Crosswhite, Bridgett Ashley and Albert E. Stiegman, in Journal of Physical Chemistry C, Volume 117, Issue 51, 30 October 2013, American Chemical Society, Pages 26871–26880, http://www.dx.doi.org/10.1021/jp4076965

Hurst and Gartner (2013). "Electric Vehicle Market Forecasts Global Forecasts for Light Duty Hybrid, Plug-in Hybrid, and Battery Electric Vehicles: 2013–2020", by Dave Hurst and John Gartner, Executive Summary, published by Navigant Research, 2Q 2013.

Hussen et al. (2012). "Reservoir stimulation for enhanced gas recovery: An economic evaluation", by Chawarwan Hussen, Robert Amin, Gary Madden and Brian Evans, in Journal of Natural Gas Science and Engineering, Volume 5, March 2012, Pages 42–50, http://www.dx.doi.org/10.1016/j.jngse.2012.01.010

Hutchison (1987). "High Speed Gas: An Autobiography", by Sir Kenneth Hutchison, published by Duckworth, 12 November 1987, ISBN: 978-0715622001. Published book.

Hydrocarbons Technology (n.d.). "Shah Sour Gas Field, United Arab Emirates", published by Hydrocarbons Technology, a product of Kable, Kable Intelligence Limited. Accessed online.

IAEA (2014). "Nuclear Power Reactors in the World", by International Atomic Energy Agency (IAEA), Power Reactor Information System (PRIS), Reference Data Series No. 2, 2014 Edition.

ICAE (2012). "Methanol Production from Steel-work-off-gases and Biomass Based Synthesis Gas", by T. Ekbom, C. Hulteberg, C-E. Grip, M. Larsson, J. Lundgren, L. Nilsson and P. Tuna, International Conference on Applied Energy (ICAE), July 5–8 2012, Paper ID ICAE2012-A10512, Table 1.

ICAO (2012). "Annual Passenger Total Approaches 3 Billion According to ICAO 2012 Air Transport Results", International Civil Aviation Organization (ICAO), COM 30/12, 18 December 2012. Accessed online.

IEA (2003a). "Energy to 2050: Scenarios for a Sustainable Future", International Energy Agency (IEA), Organisation for Economic Co-operation and Development (OECD), Paris, 2003, ISBN: 92-64-01904-9.

IEA (2003b). "Reduction of CO2 Emissions by Adding Hydrogen to Natural Gas", International Energy Agency (IEA), Report Number PH4/24, October 2003, Page iii, Section "CO2 abatement potential".

IEA (2006). "Energy Technology Perspectives: Scenarios and Strategies to 2050", International Energy Agency (IEA), Organisation for Economic Co-operation and Development (OECD), Paris, 2006, ISBN: 92-64-10982-X.

IEA (2007). "Hydrogen Production & Distribution", International Energy Agency (IEA), IEA Technology Essentials, ETE 05, April 2007, Table 1. Accessed online.

IEA (2008). "Energy Technology Perspectives: Scenarios and Strategies to 2050", International Energy Agency (IEA), Organisation for Economic Co-operation and Development (OECD), Paris, 2008, ISBN: 92-64-04142-4.

IEA (2010a). "Energy Technology Perspectives: Scenarios and Strategies to 2050", by International Energy Agency (IEA), Organisation for Economic Co-operation and Development (OECD), Paris, 2010, ISBN: 978-92-64-08597-8.

IEA (2010b). "Natural Gas Liquids: Supply Outlook 2008–2015", International Energy Agency (IEA), Oil Industry and Markets Division, April 2010. Accessed online.

IEA (2010c). "Projected Costs of Generating Electricity", by International Energy Agency (IEA) and Nuclear Energy Agency (NEA), 2010 Edition, ISBN: 978-92-64-08430-8.

IEA (2011). "Are We Entering a Golden Age of Gas?", World Energy Outlook (WEO), Special Report, International Energy Agency (IEA), Organisation for Economic Co-operation and Development (OECD), World Paris, 2011.

IEA (2012a). "Energy Technology Perspectives 2012: Pathways to a Clean Energy System", International Energy Agency (IEA), Organisation for Economic Co-operation and Development (OECD), Paris, 2012.

IEA (2012b). "Golden Rules for a Golden Age of Gas", International Energy Agency (IEA), Organisation for Economic Co-operation and Development (OECD), World Energy Outlook Special Report on Unconventional Gas, Paris, 2012.

IEA (2014a). "World Energy Investment Outlook", by International Energy Agency (IEA), Special Report, 3 June 2014.

IEA (2014b). "Energy Technology Perspectives", by International Energy Agency (IEA), 12 May 2014.

IEA (2014c). "Technology Roadmap: Energy Storage", by International Energy Agency, 19 March 2014.

IEA (2014d). "Technology Roadmaps", by International Energy Agency. Accessed online.

IEA (2015). "Technology Roadmap: Nuclear Energy", by International Energy Agency (IEA), Nuclear Energy Agency (NEA), in Energy Technology Perspectives series, 2015 edition, Nuclear reactors under construction: Section "Key Findings": 72; Section "OECD Europe": 70; Table 1: Total: 63.

IEA (n.d.-a). "IEA – About us", by International Energy Agency (IEA). Accessed online.

IEA (n.d.-b). "Annual Reports", published by International Energy Agency (IEA), Hydrogen Implementing Agreement (HIA). Accessed online.

IEA HIA (n.d.-a). "IEA HIA Annexes (Research Tasks)", by International Energy Agency (IEA), Hydrogen Implementing Agreement (HIA), Research Tasks. Accessed online.

IEA HIA (n.d.-b). "IEA/HIA Task 25: High Temperature Hydrogen Production Process", by International Energy Agency (IEA), Hydrogen Implementing Agreement (HIA), Process Sheet Synopsis. Accessed online.

IGEM (2014). "Hydrogen: Untapped Energy?", by Institution of Gas Engineers and Managers, report author Olu Ajayi-Oyakhire, 2012. Accessed online.

IGU (2012). "Renewable Gas: The Sustainable Energy Solution", International Gas Union (IGU), paper presented to the 2012 World Gas Conference, the 25th World Gas Conference, Kuala Lumpur, held 4–8 June 2012, "2009–2012 Triennium Work Report", June 2012, produced by Program Committee A: Sustainability.

IMO (2014). "International Convention for the Prevention of Pollution from Ships (MARPOL)", by International Maritime Organisation (IMO). Accessed online.

Ingersoll (1996). "Natural Gas Vehicles", by John G. Ingersoll, Fairmont Press, Lilburn Georgia, 1996, ISBN: 978-0881732184. Chapter 3, "Domestic Natural Gas Availability". Published book.

Inman (2013). "The True Cost of Fossil Fuels", by Mason Inman, in Scientific American, Volume 308, Issue 4, April 2013. Accessed online.

Innovate UK (n.d.). "Natural gas vehicles", by Innovate UK. Accessed online.

IPCC (2005). "Carbon Dioxide Capture and Storage", Special Report by Intergovernmental Panel on Climate Change (IPCC), edited by Bert Metz, Ogunlade Davidson, Heleen de Coninck, Manuela Loos and Leo Meyer, published by Cambridge University Press, ISBN: 978-0-521-86643-9.

IPCC (2013). "Long-term Climate Change: Projections, Commitments and Irreversibility", by Matthew Collins, Reto Knutti, Julie Arblaster, Jean-Louis Dufresne, Thierry Fichefet, Pierre Friedlingstein, Xuejie Gao, William J. Gutowski Jr., Tim Johns, Gerhard Krinner, Mxolisi Shongwe, Claudia Tebaldi, Andrew J. Weaver and Michael Wehner, Chapter 12 in International Panel on Climate Change (IPCC), "Climate Change 2013", Working Group I Contribution to the Fifth Assessment Report of the Intergovernmental Panel on Climate Change, edited by Thomas F. Stocker, Dahe Qin, Gian-Kasper Plattner, Melinda M. B. Tignor, Simon K. Allen, Judith Boschung, Alexander Nauels, Yu Xia, Vincent Bex and Pauline M. Midgley, published by Cambridge University Press, Cambridge, UK and New York, NY, USA, ISBN: 978-1-107-66182-0, 2013, Section 12.5.

IPCC (2014a). "Energy Systems", Intergovernmental Panel on Climate Change (IPCC), Working Group III (WG3) – Mitigation of Climate Change, "Climate Change 2014", Chapter 7 "Energy Systems", final draft, 17 December 2013.

IPCC (2014b). "Fifth Assessment Report", by Intergovernmental Panel on Climate Change (IPCC). Accessed online.

IPHE (2011). "Utsira Wind Power and Hydrogen Plant", by International Partnership for Hydrogen and Fuel Cells in the Economy, March 2011. Accessed online.

IRENA (2014). "REmap 2030: A Renewable Energy Roadmap", International Renewable Energy Agency (IRENA), June 2014. Accessed online.

IRNA (2014). "Iran to expand sour gas production capacity", published online by Islamic Republic News Agency (IRNA), 25 May 2014. Accessed online.

Iskov and Rasmussen (2013). "Global screening of projects and technologies for Power-to-Gas and Bio-SNG", by Henrik Iskov and Niels Bjarne Rasmussen, published by the Danish Gas Technology Centre, 8 November 2013, ISBN: 978-87-7795-373-6, Section 5.6, Section 7.1, Section 7.12.

ITM Power (2013a). "Methanation: CO2 as a Feedstock for Synthetic Natural Gas and Energy Storage", by ITM Power plc, Final Report, Public Version (redacted), signed off by Dr Kris Hyde, Technology Development Manager, ITM Power plc, supplied to HMG DECC competition, 30 November 2013. Supplied by DECC in response to a Freedom of Information Request, Reference: FOI2014/11761, by letter, 3 July 2014.

ITM Power (2013b). "Injection of Hydrogen into the German Gas Distribution Grid", 4 December 2013. Accessed online.

ITM Power (2014a). "Power-to-Gas & Methanation – Pathways to a Hydrogen Economy", presentation by Dr John Newton to the Advanced Power Generation Technology Forum (APGTF), 12–13 March 2014, Slides 21–24.

ITM Power (2014b). "Thüga Group's Power-to-Gas plant officially commissioned and operational", 8 May 2014. Accessed online.

Jackson (1911). "From Constantinople to the Home of Omar Khayyam: Travels in Transcaucasia and Northern Persia for Historic and Literary Research", by A. V. Williams Jackson, published by The Macmillan Company, New York, 1911. Chapter IV, "The Oil-Fields and the Fire-Temple of Baku", Pages 39–57.

Jacobson and Delucchi (2009). "A Path to Sustainable Energy by 2030", by Mark Z. Jacobson and Mark A. Delucchi, Scientific American, Volume 301, Number 5, November 2009, Headline: "A Plan for a Sustainable Future".

Janajreh et al. (2013). "Plasma gasification process: Modeling, simulation and comparison with conventional air gasification", by Isam Janajreh, Syed Shabbar Raza and Arnar Snaer Valmundsson, in Energy Conversion and Management, Volume 65, January 2013, Global Conference on Renewable Energy and Energy Efficiency for Desert Regions, Pages 801–809, DOI: http://www.dx.doi.org/10.1016/j.enconman.2012.03.010, http://www.sciencedirect.com/science/article/pii/S0196890412001549

Jechura (2014). "Hydrogen from natural gas via SMR", presentation by John Jechura, for the Colorado School of Mines course, "CBEN472 Introduction to Energy Technologies", Slide 3, "Energy Efficiency", Data from GPSA Data Book: Hydrogen Lower Heating Value (LHV) = 273.8 British Thermal Unit/standard cubic foot (Btu/scf), Methane LHV = 909.4 (Btu/scf), Data from Aspen Plus, version 2006.5: Hydrogen LHV = 273.8 Btu/scf; Methane LHV = 909.1 Btu/scf.

Johansson et al. (2012). "Hydrogen production from biomass gasification in the oil refining industry – A system analysis", by Daniella Johansson, Per-Åke Franck and Thore Berntsson, in Energy, Volume 38, Issue 1, February 2012, http://www.dx.doi.org/10.1016/j.energy.2011.12.011

Jones et al. (2011). "Addressing Gas Turbine Fuel Flexibility", by Robert Jones, Jeffrey Goldmeer and Bruno Monetti, of General Electric Company (GE), White Paper, Document: GER4601 (05/11), 2011.

Jones and Worthington (2014). "Europe's failure to tackle coal", by Dave Jones and Bryony Worthington, of Sandbag Climate Campaign. Accessed online.

JRC (2014). "Nitrous oxide (N2O)", by Joint Research Centre of the European Commission. Accessed online.

Kaewpanha et al. (2014). "Steam co-gasification of brown seaweed and land-based biomass", by Malinee Kaewpanha, Guoqing Guan, Xiaogang Hao, Zhongde Wang, Yutaka Kasai, Katsuki Kusakabe and Abuliti Abudula, in Fuel Processing Technology, Volume 120, April 2014, Pages 106–112, http://www.dx.doi.org/10.1016/j.fuproc.2013.12.013

Kamali and Rezaee (2012). "Identification and Evaluation of Unconventional Hydrocarbon Reserves: Examples from Zagros and Central Iran Basins", Mohammad R. Kamali and M. Reza Rezaee, in Journal of Petroleum Science and Technology, Research Institute of Petroleum Industry (RIPI), Volume 2, Issue 1, Spring 2012, Article 3, Pages 27–36, Section: "Oil Shale deposits in Zagros Basin Iran".

Kang (2014). "Natural gas supply-demand situation and prospect in China", by Zhang Kang, in Natural Gas Industry B, Volume 1, Issue 1, October 2014, Pages 103–112, http://www.dx.doi.org/10.1016/j.ngib.2014.10.014

Kang et al. (2014). "Direct measurements of methane emissions from abandoned oil and gas wells in Pennsylvania", by Mary Kang, Cynthia M. Kanno, Matthew C. Reid, Xin Zhang, Denise L. Mauzerall, Michael A. Celia, Yuheng Chen and Tullis C. Onstott, in Proceedings of the National Academy of Sciences of the United States of America (PNAS), http://www.dx.doi.org/10.1073/pnas.1408315111

Karatzos et al. (2014). "The Potential and Challenges of Drop-in Biofuels", by Sergios Karatzos, James D. McMillan and Jack N. Saddler, International Energy Agency (IEA), IEA Bioenergy Task 39, Report: T39-T1, July 2014, Section 1.10, Figure 1–18, ISBN: 978-1-910154-07-6.

Karunadasa et al. (2010). "A molecular molybdenum-oxo catalyst for generating hydrogen from water", by Hemamala Karunadasa, Christopher J. Chang and Jeffrey R. Long, in Nature, Volume 464, 29 April 2010, Letters, http://www.dx.doi.org/10.1038/nature08969

Keairns and Newby (2010). "Integrated Gasification Fuel Cycle (IGFC) Systems", by Dale L. Keairns and Richard A. Newby, presentation to 11th Annual SECA Workshop (Solid State Energy Conversion Alliance), Pittsburgh, July 2010.

Keay-Bright (2013). "Capacity Mechanisms for Power System Reliability", by Sarah Keay-Bright, published by Regulatory Assistance Project (RAP), October 2013. Accessed online.

Kee (2015). "Can nuclear succeed in liberalized power markets?", by Edward Kee, in World Nuclear News, 4 February 2015, Nuclear reactors under construction: 69.

Ker Porter (1822). "Travels in Georgia, Persia, Armenia, Ancient Babylonia, &c. &c. During the Years 1817, 1818, 1819, and 1820", by Sir Robert Ker Porter, printed by Longman, Hurst, Rees, Orme, and Brown, Paternoster Row, 1822, Volume II, Pages 438–445, "Kirkook", "Naphtha Springs", "Burning Hill".

Kerry (2007). "Industrial Gas Handbook: Gas Separation and Purification", by Frank G. Kerry, published by CRC Press, 22 February 2007, ISBN: 978-0849390050, Section 8.1.6, Section 8.1.7.6.

Khodapanah and Tabatabaei-Nejad (2014). "A consistent method for simultaneous calculation of upper and lower dew point pressures of gas condensate fluids", by Elnaz Khodapanah and Seyyed Alireza Tabatabaei-Nejad, in Journal of Natural Gas Science and Engineering, Volume 21, 2014, Pages 556–567, http://www.dx.doi.org/10.1016/j.jngse.2014.09.023

Kim et al. (2010). "Chemical-Looping Steam Methane Reforming for Hydrogen Production in a Circulating Fluidized Bed Reactor", by Sang Done Kim, Kang

Seok Go, Young Wook Jeon and Chu Sik Park, Refereed Proceedings, The 13th International Conference on Fluidization – New Paradigm in Fluidization Engineering", 2010.

Klass (1998). "Biomass for Renewable Energy, Fuels, and Chemicals", by Donald L. Klass, published by Academic Press, San Diego, CA, and Academic Press Ltd, London, 1998, ISBN: 978-0-12-410950-6. Chapter 9, Section IV, Page 289, Section V, Page 308, Section: "Pyrolysis and Partial Oxidation with Air of MSW in a Rotating Kiln".

Klotz et al. (2014). "Electrochemical model for SOFC and SOEC mode predicting performance and efficiency", by D. Klotz, A. Leonide, A. Weber and E. Ivers-Tiffée, in International Journal of Hydrogen Energy, Volume 39, Issue 35, 3 December 2014, http://www.dx.doi.org/10.1016/j.ijhydene.2014.08.139

Komilis et al. (2012). "Revisiting the elemental composition and the calorific value of the organic fraction of municipal solid wastes", by Dimitrios Komilis, Alexandros Evangelou, Georgios Giannakis and Constantinos Lymperis, in Waste Management, Volume 32, Issue 3, March 2012, Pages 372–381, http://www.dx.doi.org/10.1016/j.wasman.2011.10.034

Koper (2014). "Poland cuts estimate for shale gas exploration wells", by Anna Koper, in Reuters online, UK edition, "Regulatory News", 31 July 2014. Accessed online.

Kopyscinski (2010). "Production of synthetic natural gas in a fluidized bed reactor", by Jan Kopyscinski, PhD dissertation, Dissertation Number 18800, Paul Scherrer Institut (PSI), Eidgenössische Technische Hochschule Zürich (ETH), Swiss Federal Institute of Technology Zurich, 2010.

Koukouzas et al. (2008). "Co-gasification of solid waste and lignite – A case study for Western Macedonia", by N. Koukouzas, A. Katsiadakis, E. Karlopoulos and E. Kakaras, in Waste Management, Volume 28, Issue 7, 2008, Pages 1263–1275, http://www.dx.doi.org/10.1016/j.wasman.2007.04.011

Kumar and Narvekar (2005). "Seasonal variability of the mixed layer in the central Arabian Sea and its implication on nutrients and primary productivity", by S. Prasanna Kumar and Jayu Narvekar, in Deep Sea Research Part II: Topical Studies in Oceanography, Volume 52, Issues 14–15, July 2005, Pages 1848–1861, http://www.dx.doi.org/10.1016/j.dsr2.2005.06.002

Kuramochi et al. (2009). "Techno-economic prospects for CO2 capture from a Solid Oxide Fuel Cell–Combined Heat and Power plant. Preliminary results", by Takeshi Kuramochi, Hao Wu, Andrea Ramírez, André Faaij and Wim Turkenburg, in Energy Procedia, Volume 1, Issue 1, February 2009, Pages 3843–3850, http://www.dx.doi.org/10.1016/j.egypro.2009.02.186

Kuzemko (2013). "The Energy Security-Climate Nexus: Institutional Change in the UK and Beyond", by Caroline Kuzemko, Pages 80, 87, published by Palgrave Macmillan, 7 May 2013, ISBN: 978-1137307828.

Laherrere (2013a). "Oil & Gas Production Forecasts 1900–2100", by Jean Laherrère, 4 February 2013. Accessed online.

Laherrere (2013b). "Shortened World Oil & Gas Production Forecasts 1900–2100", by Jean Laherrère, 30 May 2013. Accessed online.

Lallemand et al. (2012). "Solutions for the treatment of highly sour gases", by François Lallemand, Gauthier Perdu, Cristina Maretto, Claire Weiss, Julia Magne-Drisch and Anne-Claire Lucquin, in Gas Treatment/Processing, Digital Refining, Gas 2012, April 2012.

Larminie and Dicks (2003). "Fuel Cell Systems Explained", by James Larminie and Andrew Dicks, 2nd edition, published by Wiley-Blackwell, 18 February 2003, ISBN: 978-0470848579.

Lebarbier et al. (2014). "Sorption-enhanced synthetic natural gas (SNG) production from syngas: A novel process combining CO methanation, water-gas shift and CO2 capture", by Vanessa M. Lebarbier, Robert A. Dagle, Libor Kovarik, Karl O. Albrecht, Xiaohong Li, Liyu Li, Charles E. Taylor, Xinhe Bao and Yong Wang, in Applied Catalysis B: Environmental, Volume 144, January 2014, Pages 223–232, http://www.dx.doi.org/10.1016/j.apcatb.2013.06.034

Lecarpentier (2012). "Short term trends in the gas industry", by Armelle Lecarpentier, IFP Energies Nouvelles, "Panorama 2013". Accessed online.

Le Fevre (2014). "The Prospects for Natural Gas as a Transport Fuel in Europe", by Chris Le Fevre, published by The Oxford Institute for Energy Studies, March 2014. Accessed online.

Leggett (2013). "The Energy of Nations: Risk Blindness and the Road to Renaissance", by Jeremy Leggett, published by Routledge, Earthscan, 23 September 2013, ISBN: 978-0415857826, Chapter 1: "Lies, scaremongering, and affordable oil". Published book.

Le Strange (1905) (2011). "The Lands of the Eastern Caliphate", by Guy Le Strange, Cambridge Geographical Series, Cambridge University Press, 1905, republished 2011, ISBN: 978-1-107-60014-0, Paperback.

Lewis (2014a). "'Peak demand' oil theory fails scrutiny test", by Mark C. Lewis, Financial Times, "Markets Insight", 20 January 2014. Accessed online.

Lewis (2014b). "Book of Extremes: Why the 21st Century Isn't Like the 20th Century", by Ted G. Lewis, published by Copernicus, 20 May 2014, ISBN: 978-3319069258, Section 9.6.

Lewis and Gilliland (1954). "Production of Pure Carbon Dioxide", by Warren K. Lewis and Edwin R. Gilliland, United States Patent: 2,665,972, Patented 12 January 1954.

Liebetrau et al. (2013). "Analysis of greenhouse gas emissions from 10 biogas plants within the agricultural sector", by J. Liebetrau, T. Reinelt, J. Clemens, C. Hafermann, J. Friehe and P. Weiland, in Water and Science Technology, Volume 67, Issue 6, Pages 1370–1379, IWA Publishing, 2013, http://www.dx.doi.org/10.2166/wst.2013.005

Linde (2012). "Valuable liquids from refinery gases", by Zaheer Malik and John Slack, of Linde Process Plants, in Digital Refining, PTQ Q4 2012. Accessed online.

Linde (n.d.). "Gas units converter", "Gas calculator". Accessed online.

Lockhart (1939). "Iranian Petroleum in Ancient and Medieval Times", by Laurence Lockhart PhD, in Journal of the Institute of Petroleum, Volume 25, Issue 183, January 1939. Accessed online.

Lokhorst et al. (1997). "NW European Gas Atlas", by British Geological Survey (BGS), Bundesanstalt für Geowissenschaften und Rohstoffe (BGR), Danmarks og Grønlands Geologiske Undersøgelse (GEUS), Panstwowy Instytut Geologiczny (PGI), Nederlands Instituut voor Toegepaste Geowetenschappen TNO (NITG-TNO), edited by Ad Lokhorst, published by NITG-TNO, Haarlem, Nederland, 1997.

Low Carbon Contracts Company (2014). "Electricity Market Reform Contracts for Difference: GB Implementation Plan", by Low Carbon Contracts Company, 15 December 2014.

Luo and Angelidaki (2012). "Integrated Biogas Upgrading and Hydrogen Utilization in an Anaerobic Reactor Containing Enriched Hydrogenotrophic Methanogenic Culture", by Gang Luo and Irini Angelidaki, in Biotechnology and Bioengineering, Volume 109, Issue 11, Pages 2729–2736, November 2012, http://www.dx.doi.org/10.1002/bit.24557

Lyngfelt (2007). "Chemical-Looping Reforming (a) and (s): Two processes for producing hydrogen from natural gas with separation of carbon dioxide", by Anders Lyngfelt, presentation made in Lyon, from the Department of Energy and Environment, Chalmers University of Technology, Göteborg, Sweden, 23 January 2007.

Ma et al. (2010). "Performance and emission characteristics of a turbocharged spark-ignition hydrogen-enriched compressed natural gas engine under wide open throttle operating conditions", by Fanhua Ma, Mingyue Wang, Long Jiang, Jiao Deng, Renzhe Chen, Nashay Naeve and Shuli Zhao, in International Journal of Hydrogen Energy, Volume 35, Issue 22, November 2010, Pages 12502–12509, http://www.dx.doi.org/10.1016/j.ijhydene.2010.08.053

MacKay and Stone (2013). "Potential Greenhouse Gas Emissions Associated with Shale Gas Extraction and Use", by Professor David J. C. MacKay and Dr Timothy J. Stone, published by HMG DECC, 9 September 2013.

Marcel (2012). "Finding the Right Partners for National Oil Companies", by Valérie Marcel, Chapter 6 in "What Next for the Oil and Gas Industry?", by John Mitchell, Valérie Marcel and Beth Mitchell, published by The Royal Institute of International Affairs (Chatham House), October 2012.

Marra and Barber (2005). "Primary productivity in the Arabian Sea: A Synthesis of JGOFS data", by John Marra and Richard T. Barber, Progress in Oceanography, Volume 65, 2005, Pages 159–175, http://www.dx.doi.org/10.1016/j.pocean.2005.03.004

Marston (2013). "Tracking down fugitive methane leaks", by James D. Marson, Environmental Defense Fund blog, 17 January 2013. Accessed online.

Martinez-Merino et al. (2013). "Biological Hydrogen Production", by Víctor Martínez-Merino, María José Gil and Alfonso Cornejo, Chapter 8, in "Renewable Hydrogen Technologies", edited by Luis M. Gandía; Gurutze Arzamendi and Pedro M. Diéguez, ISBN: 978-0080444291.

Marvin (1884) (1891). "The Region of the Eternal Fire: An Account of a Journey to the Petroleum Region of the Caspian in 1883", by Charles Thomas Marvin, published by W. H. Allen & Co. Limited, 13, Waterloo Place, S.W., 1884.

Materazzi et al. (2013). "Thermodynamic modelling and evaluation of a two-stage thermal process for waste gasification", by Massimiliano Materazzi, Paola Lettieri, Luca Mazzei, Richard Taylor and Chris Chapman, in Fuel, Volume 108, June 2013, Pages 356–369, http://www.dx.doi.org/10.1016/j.fuel.2013.02.037

Maurstad et al. (2006). "SOFC and Gas Turbine Power Systems – Evaluation of Configurations for CO2 Capture", by Ola Maurstad, Rune Bredesen, Olav Bolland, Hanne M. Kvamsdal and Morten Schell, of SINTEF Energy Research, 2006.

McCreanor et al. (2007). "Respiratory Effects of Exposure to Diesel Traffic in Persons with Asthma", by James McCreanor, Paul Cullinan, Mark J. Nieuwenhuijsen, James Stewart-Evans, Eleni Malliarou, Lars Jarup, Robert Harrington, Magnus Svartengren, In-Kyu Han, Pamela Ohman-Strickland, Kian Fan Chung and Junfeng Zhang, in The New England Journal of Medicine, Volume 357, Pages 2348–2358, 6 December 2007, http://www.dx.doi.org/10.1056/NEJMoa071535

McGill et al. (2013). "Alternative Fuels for Marine Applications", by Ralph McGill, William Remley and Kim Winther, "A Report from the IEA Advanced Motor Fuels Implementing Agreement", Annex 41, May 2013.

McGlade (2013). "Uncertainties in the Outlook for Oil and Gas", by Chrisophe E. McGlade, PhD Thesis, University College, London, 6 November 2013. Accessed online.

McGlade (2014). "Comment on the 'Uppsala critique'", by Christophe E. McGlade, in Energy Policy, Volume 67, April 2014, Pages 154–108, http://www.dx.doi.org/10.1016/j.enpol.2013.11.036

McGlade et al. (2012a). "A review of regional and global estimates of unconventional gas resources", by Christophe McGlade, Jamie Speirs and Steve Sorrell, Report to the Energy Security Unit of the Joint Research Centre (JRC) of the European Commission (EC), September 2012.

McGlade et al. (2012b). "Unconventional gas – A review of estimates", by Christophe McGlade, Jamie Speirs and Steve Sorrell, Imperial College Centre for Energy Policy and Technology (ICEPT), Working Paper: ICEPT/WP/2012/05, September 2012.

McGlade et al. (2013a). "Unconventional gas – A review of regional and global resource estimates", by Christophe McGlade, Jamie Speirs and Steve Sorrell, in Energy, Volume 55, 15 June 2013, Pages 571–584, http://www.dx.doi.org/10.1016/j.energy.2013.01.048

McGlade et al. (2013b). "Methods of estimating shale gas resources: Comparison, evaluation and implications", by Christophe McGlade, Jamie Speirs and Steve Sorrell, in Energy, Volume 59, 15 September 2013, Pages 116–125, http://www.dx.doi.org/10.1016/j.energy.2013.05.031

McPhy (2014). "The H2BER Project", by McPhy Energy. Accessed online.

McPhy (n.d.) "Solid hydrogen storage", by McPhy Energy. Accessed online.

Melaina et al. (2010). "Blending Hydrogen into Natural Gas Pipeline Networks: A Review of Key Issues", by M. W. Melaina, O. Antonia and M. Penev, National Renewable Energy Laboratory (NREL), Task Number: HT12.2010, Technical Report: NREL/TP-5600-51995, March 2013.

Meng and Bentley (2008). "Global oil peaking: Responding to the case for 'abundant supplies of oil'", by Q. Y. Meng and R. W. Bentley, in Energy, Volume 33, Issue 8, August 2008, Pages 1179–1184, http://www.dx.org/10.1016/j.energy.2008.04.001

MHLnews (2014). "World Hydrogen Market Set for 3.5 Percent Annual Growth through 2018", by Tom Andel, in Material Handling & Logistics magazine, 3 July 2014. Accessed online.

Microsoft (2014). "Microsoft and Its Partners Were Awarded $5 million by ARPA-E To Develop Fuel Cell Technologies", by Pradeep, reporter for Microsoft News. Accessed online.

Miller and Sorrell (2014). "The future of oil supply", by Richard G. Miller and Steven R. Sorrell, in Philosophical Transaction of The Royal Society A, Volume 372, Issue 2006 20130179, 2 December 2013, 13 January 2014, http://www.dx.doi.org/10.1098/rsta.2013.0179

Miller et al. (2013). "Anthropogenic emissions of methane in the United States", by Scot M. Miller, Steven C. Wofsy, Anna M. Michalak, Eric A. Kort, Arlyn E. Andrews, Sebastien C. Biraud, Edward J. Dlugokencky, Janusz Eluszkiewicz, Marc L. Fischer, Greet Janssens-Maenhout, Ben R. Miller, John B. Miller, Stephen A. Montzka, Thomas Nehrkorn and Colm Sweeney, in Proceedings of the National Academy of Sciences of the United States of America, Volume 110, Issue 50, 10 December 2013, Pages 20018–20022, http://www.dx.doi.org/10.1073/pnas.1314392110

Minh and Mogensen (2013). "Reversible Solid Oxide Fuel Cell Technology for Green Fuel and Power Production", by Nguyen Q. Minh and Mogens B. Mogensen, in Interface, The Electrochemical Society's magazine publication, Winter 2013, Pages 55–62. Accessed online.

Ministry of Power (1952). "Report of the Committee on National Policy for the Use of Fuel and Power Resources", published by H. M. Stationery Office (HMSO), for Ministry of Fuel and Power, Document: Cmd 8647, 1952. Committee chair: Viscount Ridley. Published papers.

Ministry of Power (1967). "Fuel Policy. Presented to Parliament by the Minister of Power Nov 1967", Government of the United Kingdom, Ministry of Power, Document: Cmd 3438, 1967. Published papers.

Minjares et al. (2013). "Alignment of policies to maximize the climate benefits of diesel vehicles through control of particulate matter and black carbon emissions", by Ray Minjares, Kate Blumberg and Francisco Posada Sanchez, in Energy Policy, Volume 54, March 2013, Special Issue, "Decades of Diesel", Pages 54–61, http://www.dx.doi.org/10.1016/j.enpol.2012.09.053

Mirmoshtaghi et al. (2013). "Bio-Methane Production Through Different Biomass Gasifiers", by Guilnaz Mirmoshtaghi, Hailong Li, Erik Dahlquist and Eva Thorin, paper presented to International Conference on Applied Energy (ICAE), 1–4 July 2013, Pretoria, South Africa, Paper ID: ICAE2013-454.

MIT (2013). "Nuon Magnum Fact Sheet: Carbon Dioxide Capture and Storage Project", published by Massachusetts Institute of Technology (MIT), 16 December 2013. Accessed online.

Mitchell (2012). "Financing the Future", by Beth Mitchell (Elizabeth Mitchell), Chapter 7 in "What Next for the Oil and Gas Industry?", by John Mitchell, Valérie Marcel and Beth Mitchell, published by The Royal Institute of International Affairs (Chatham House), October 2012.

Mitchell et al. (2012). "What Next for the Oil and Gas Industry?", by John Mitchell, Valérie Marcel and Beth Mitchell, published by The Royal Institute of International Affairs (Chatham House), October 2012.

Mohr and Evans (2009). "Forecasting coal production until 2100", by Steve H. Mohr and Geoffrey M. Evans, in Fuel, Volume 88, Issue 11, November 2009, Pages 2059–2067, http://www.dx.doi.org/10.1016/j.fuel.2009.01.032

Mohr and Evans (2011). "Long term forecasting of natural gas production", by S. H. Mohr and G. M. Evans, in Energy Policy, Volume 39, Issue 9, September 2011, Pages 5550–5560, http://www.dx.doi.org/10.1016/j.enpol.2011.04.066

Mohseni (2012). "Power to Gas: Bridging renewable electricity to the transport sector", by Farzad Mohseni, Licentiate thesis, 2012, KTH Royal Institute of Technology, ISBN: 978-91-7501-597-2, 1 November 2011.

Monbiot (2015). "With this attack on community energy the big six win out over 'big society'", by George Monbiot, in The Guardian newspaper, 23 January 2015. Accessed online.

Monsal (2009). "Anaerobic Digestion of regional food waste – A scenario development", by Aidan Cumiskey, Managing Director of Monsal Ltd, Biomethane workshop, 22 June 2009. Accessed online.

Moser et al. (2014). "100 RE Regions in Germany, Europe and the World", "100ee-Regionen in Deutschland, Europa und der Welt", by Pia Buschmann, Dr Peter Moser, Stefanie Roth and Katharina Schenk, published by Institut dezentrale Energietechnologien (IdE) and Kompetenznetzwerk dezentrale Energietechnologien e.V. (deENet), January 2014.

Moulijn et al. (2013). "Chemical Process Technology", by Jacob A. Moulijn, Michiel Makkee and Annelies E. van Diepen, 2nd edition, published by John Wiley & Sons Ltd, 2013, ISBN: 9781444320244, ISBN: 9781444320251.

Mueller-Langer et al. (2007). "Techno-economic assessment of hydrogen production processes for the hydrogen economy for the short and medium term", by Franziska Mueller-Langer, Evangelos Tzimas, Martin Kaltschmitt and S. D. Peteves, in International Journal of Hydrogen Energy, Volume 32, Issue 16, November 2007, Pages 3797–3810, http://www.dx.doi.org/10.1016/j.ijhydene.2007.05.027

Muhich et al. (2013). "Efficient Generation of H2 by Splitting Water with an Isothermal Redox Cycle", by Christopher L. Muhich, Brian W. Evanko, Kayla C. Weston, Paul Lichty, Xinhua Liang, Janna Martinek, Charles B. Musgrave and Alan W. Weimer, in Science, Volume 341, Issue 6145, 2 August 2013, Pages 540–542, http://www.dx.doi.org/10.1126/science.1239454

Muscatello and Santiago-Maldonado (2012). "Mars In-Situ Resource Utilization Technology Evaluation", by Anthony C. Muscatello and Edgardo Santiago-Maldonado, National Aeronautics and Space Administration (NASA), 9 January 2012.

Myllyntaus (2010). "Switching to a Biofuel at the Pinch: Wood Gas in Finnish Motoring during World War II", by Timo Myllyntaus, in Journal of the International Committee for the History of Technology (ICON), Volume 16, 2010, Pages 101–122.

Najafi et al. (2011). "Algae as a sustainable energy source for biofuel production in Iran: A case study", by Gholamhassan Najafi, Barat Ghobadian and Talal F. Yusaf, in Renewable and Sustainable Energy Reviews, Volume 15, Issue 8, October 2011, Pages 3870–3876, http://www.dx.doi.org/10.1016/j.rser.2011.07.010

NASA (1989). "The Fuel Cell in Space: Yesterday, Today and Tomorrow", by Marvin Warshay and Paul R. Prokopius, paper given at the Grove Anniversary (1839–1989) Fuel Cell Symposium, sponsored by the Royal Institution, London, 18–21 September, 1989.

National Grid (2009). "The potential for Renewable Gas in the UK", A paper by National Grid, January 2009. Accessed online.

National Grid (2014a). "Short Term Operating Reserve", by National Grid, 2014. Accessed online.

National Grid (2014b). "UK Future Energy Scenarios", by National Grid, 2014. Accessed online.

National Grid (2014c). "Provisional Auction Result: T-4 Capacity Market Auction 2014", by National Grid, 18 December 2014. Accessed online.

National Grid (n.d.). "BioSNG demonstration project technology". Accessed online.

NEA (2011). "Technical and Economic Aspects of Load Following With Nuclear Power Plants", Nuclear Energy Agency (NEA/AEN), Organisation for Economic Co-operation and Development (OECD), 2011.

Neff and Coleman (2014). "EIA outlook: Reversal in U.S. oil import dependency", by Shirley Neff and Margaret Coleman, in Energy Strategy Reviews, Volume 5, December 2014, Pages 6–13, http://www.dx.doi.org/10.1016/j.esr.2014.10.007

Nemeth (2012). "Should You Convert Your Car to Natural Gas?", by Michael Nemeth, in Popular Mechanics, 10 February 2012. Accessed online.

NERA (2009a). "Renewable Heat Technologies for Carbon Abatement: Characteristics and Potential", by NERA Economic Consulting, Entec and Element Energy, A final report to the Committee on Climate Change, July 2009, Section 4, Section 4.2, Section 4.5.

NERA (2009b). "The UK Supply Curve for Renewable Heat", by NERA Economic Consulting and AEA, Study for HMG DECC, Document: URN 09D/689, July 2009.

NERA (2010). "Design of the Renewable Heat Incentive", by NERA Economic Consulting, Study for HMG DECC, Document: URN 10D/544, February 2010.

NERA (2011). "Achieving deployment of renewable heat", by NERA Economic Consulting and Element Energy, Final report for The Committee on Climate Change, April 2011.

NET Power (2014). "Carbon Capture & Utilisation from Power Generation", by Rodney J. Allam, 8 Rivers Capital, presentation to NEPIC International Bioresources Conference, 19 June 2014.

NETL (2007). "Carbon Dioxide Capture from Existing Coal-Fired Power Plants", published by NETL, November 2007. Accessed online.

NETL (2014). "Current Syngas Capacity By Region", Gasification Database, National Energy Technology Laboratory (NETL). Accessed online.

NETL (n.d.). "Lurgi Dry-Ash Gasifier", by National Energy Technology Laboratory (NETL). Accessed online.

Nezhad et al. (2011). "Gas In Place Determination in Abnormally Pressured Masjed-Soleyman-Jurassic Gas Reservoir", by Abolfazl Bagheri Nezhad, Reza Mohebbifar and Sepehr Aghahoseini, International Journal of Research and Reviews in Applied Sciences (IJRRAS), ISSN: 2076-734X, Volume 6, Issue 4, March 2011.

NFCRC (2002). "Hybrid Fuel Cell/Gas Turbine Systems: Hybrid – 220kW SOFC/Micro Turbine Generator System", by National Fuel Cell Research Center, Summer 2002. Accessed online.

NFCRC (2004). "Fuel Cell/Gas Turbine Hybrid Systems", by Professor Scott Samuelson, National Fuel Cell Research Center, published by ASME International Gas Turbine Institute, 2004. Accessed online.

Nguyen (2011). "System analysis of chemical and carbonate looping processes in IGCC power plants for CO2 separation", by Tuong-Van Nguyen, MSc Thesis, Sustainable Energy Systems, Department of Energy and Environment Division of Energy Technology, Chalmbers University of Technology, Göteborg, Sweden 2011.

NGVA (2014). "European map of CNG and LNG filling stations", by Natural and Biogas Vehicle Association (NGVA). Accessed online.

NGV Network (2014). "Presentations", by NGV Network. Accessed online.

Ni et al. (2006). "An overview of hydrogen production from biomass", by Meng Ni, Dennis Y. C. Leung, Michael K. H. Leung and K. Sumathy, in Fuel Processing Technology, Volume 87, Issue 5, May 2006, Pages 461–472, http://www.dx.doi.org/10.1016/j.fuproc.2005.11.003

Nijboer (2010). "The Contribution of Natural Gas Vehicles to Sustainable Transport", by Michiel Nijboer, published by International Energy Agency (IEA), Working Paper, 2010.

Nisbet et al. (2014). "Methane on the Rise – Again", by Euan G. Nisbet, Edward J. Dlugokencky and Philippe Bousquet, in Science, Volume 343, Issue 6170, Pages 493–495, 31 January 2014, DOI: http://www.dx.doi.org/10.1126/science.1247828

NL (2011). "Energy Report 2011", published by the Government of The Netherlands, Ministry of Economic Affairs, Agriculture and Innovation, The Hague, June 2011, Page 18, "Strengthening the gas hub". Accessed online.

NL EA (2014). "SDE+ 2014", by Netherlands Enterprise Agency, "Instructions on how to apply for a subsidy for the production of renewable energy", Febryar 2014. Accessed online.

NNFCC (2014). "Case Studies", by National Non-Food Crops Centre (NNFCC), The Official Information Portal on Anaerobic Digestion. Accessed online.

NREAP UK (2010). "National Renewable Energy Action Plan for the United Kingdom: Article 4 of the Renewable Energy Directive 2009/28/EC", published by UK Government, under the obligation of Article 4 of the Renewable Energy Directive 2009/28/EC, 1 July 2010.

NREL (2004). "PV FAQs: What is the energy payback for PV?", National Renwewable Energy Laboratory (NREL), Document: DOE/GO-102004-1847, January 2004. Accessed online.

NREL (2013). "Wind LCA Harmonization", National Renwewable Energy Laboratory (NREL), Document: NREL/FS-6A20-57131, June 2013. Accessed online.

Ofgem (2012). "Gas Security of Supply Report", by HMG, Office of Gas and Electricity Markets (Ofgem), the UK Government's Gas and Electricity Market Regulator, November 2012, "Risks and resilience appendix".

Ofgem (2013). "Electricity Capacity Assessment Report 2013", Office of Gas and Electricity Markets (Ofgem), UK Government, 27 June 2013.

Ofgem (2014). "Electricity Capacity Assessment Report 2014", Office of Gas and Electricity Markets (Ofgem), UK Government, 30 June 2014.

OGJ (1996). "NEWS Iran seeking help in regaining prerevolution oil and gas flow", by Oil and Gas Journal, Volume 94, Issue 8, 19 February 1996. Accessed online.

OGJ (2007). "Heat-stable salts reduce H2S leaking from amine absorbers", by Oil and Gas Journal, Volume 105, Issue 22, 11 June 2007. Accessed online.

OGJ (2008). "Contaminants key to refinery offgas treatment unit design", by Sheng-Yi Chuang, Rajeev Nanda and John Rizopoulus, in Oil and Gas Journal, Volume 106, Issue 35, 15 September 2008. Accessed online.

OilGas (n.d.) "North Pars Gas Field", published by Oil, Gas and Petrochemical Industries of Iran. Accessed online.

Omidvar (2011). "Prospect of Iran Natural Gas Export Projects", by Hedayat Omidvar, "Pars LNG Project", presented at Jordan International Energy Conference (JIEC), Amman, Jordan, 20–22 September 2011.

OPEC (2009). "Annual Statistical Bulletin: 2008", by Organization of the Petroleum Exporting Countries (OPEC), 2009, ISSN: 0475-0608.

OPEC (2012). "World Oil Outlook", published by the Organization of the Petroleum Exporting Countries (OPEC), 2012.

OPEC (2013). "World Oil Outlook", published by the Organization of the Petroleum Exporting Countries (OPEC), 2013.

OPEC (2014). "World Oil Outlook", published by the Organization of the Petroleum Exporting Countries (OPEC), 2014.

Orhan et al. (2014). "Gasification of Oil Refinery Waste for Power and Hydrogen Production", by O. Yuksel Orhan, G. Ís and E. Alper, K. McApline, S. Daly,

M. Sycz and A. Elkamel, Proceedings of the 2014 International Conference on Industrial Engineering and Operations Management, Bali, Indonesia, 7–9 January 2014.

Ortenzi et al. (2008). "Experimental tests of blends of hydrogen and natural gas in light-duty vehicles", by Fernando Ortenzi, Maria Chiesa, Riccardo Scarcelli and Giovanni Pede, in International Journal of Hydrogen Energy, Volume 33, Issue 12, June 2008, Pages 3225–3229, http://www.dx.doi.org/10.1016/j.ijhydene.2008.01.050

O'Sullivan and Paltsev (2012). "Shale Gas Production: Potential versus Actual GHG Emissions", by Francis O'Sullivan and Sergey Paltsev, published by Massachusetts Institute of Technology (MIT), Joint Program on the Science and Policy of Global Change, Report Number 234, November 2012.

Palmer (2013). "Household Solar Photovoltaics: Supplier of Marginal Abatement, or Primary Source of Low-Emission Power?", by Graham Palmer, in Sustainability, Volume 5, Issue 4, Pages 1406–1442, http://www.dx.doi.org/10.3390/su5041406

Patel and Salo (2007). "Carbona Biomass Gasification Technology", by Jim Patel and Kari Salo, Carbona Corporation, presentation to TAPPI 2007 International Conference on Renewable Energy, Atlanta, GA, 10–11 May 2007, Slides 14 and 17. Accessed online.

Patterson (2012). "Exploitation of Unconventional Fossil Fuels: Enhanced Greenhouse Gas Emissions", by Judith Patterson, in "Greenhouse Gases – Emission, Measurement and Management", edited by Dr Guoxiang Liu, published by InTech, 14 March 2012, ISBN: 978-953-51-0323-3, http://www.dx.doi.org/10.5772/1797

Patzek and Croft (2010). "A global coal production forecast with multi-Hubbert cycle analysis", by Tadeusz W. Patzek and Gregory D. Croft, in Energy, Volume 35, Issue 8, August 2010, Pages 3109–3122, http://www.dx.doi.org/10.1016/j.energy.2010.02.009

Peng et al. (2013). "Review on life cycle assessment of energy payback and greenhouse gas emission of solar photovoltaic systems", by Jinqing Peng, Lin Hu and Hongxing Yang, in Renewable and Sustainable Energy Reviews, Volume 19, March 2013, Pages 255–274, http://www.dx.doi.org/10.1016/j.rser.2012.11.035

Perry and Fuller (2002). "A Historical Perspective of Fuel Cell Technology in the 20th Century", by M. L. Perry and T. F. Fuller, in Journal of The Electrochemical Society, Volume 149, Issue 7, Pages S59–S67, 2002, http://www.dx.doi.org/10.1149/1.1488651

Persson et al. (2006). "Biogas Upgrading to Vehicle Fuel Standards and Grid Injection", by Margareta Persson, Owe Jönsson and Arthur Wellinger, IEA Bioenergy, Task 37, "Energy from Biogas and Landfill Gas", December 2006. Accessed online.

Peter and Lehmann (2008). "Renewable Energy Outlook 2030 – Energy Watch Group Global Renewable Energy Scenarios", by Stefan Peter and Harry Lehmann, Energy Watch Group, presented at the 24th European Photovoltaic Solar Energy Conference, 21–25 September 2009, Hamburg, Germany.

Phillips (2014). "Electricity market reform: the Capacity Market explained", by James Phillips, 4 February 2014. Accessed online.

Pihl et al. (2010). "Highly efficient electricity generation from biomass by integration and hybridization with combined cycle gas turbine (CCGT) plants for natural gas", by Erik Pihl, Stefan Heyne, Henrik Thunman and Filip Johnsson, in Energy, Volume 35, Issue 10, October 2010, Pages 4042–4052, http://www.dx.doi.org/10.1016/j.energy.2010.06.008

Platt et al. (2014). "A New Approach to Electricity Markets: How New, Disruptive Technologies Change Everything", by Reg Platt, Jack Williams, Anne Pardoe and Will Straw, a report of the Institute for Public Policy Research (IPPR).

Pliny (c. 77) (1855). "Naturalis Historia", by Gaius Plinius Secundus, from a translation by John Bostock, "The Natural History. Pliny the Elder", published by Taylor and Francis, Red Lion Court, Fleet Street, 1855. Chapter 110 (106), "Places which are always burning", Plin. Nat. 2.110.

POGC (n.d.). "Phases 6, 7 & 8", published online by Pars Oil and Gas Company. Accessed online.

POST (2004). "The Future of UK Gas Supplies", The Parliamentary Office of Science and Technology, Number 230, October 2004. Accessed online.

Poyry (2014). "UK Shale Gas – where are we now?", by Pöyry. Box 5, "Applications under the Production Phase". Accessed online.

Praxair (2011). "Global Hydrogen Growth Investor Day: Global Growth Outlook", presentation by Steve Angel, 22 June 2011, Slide 19, "Hydrogen Demand Growth". Accessed online.

PressTV (2013a). "Iran launches largest NGL plant in Persian Gulf", published online by PressTV, 14 February 2013. Accessed online.

PressTV (2013b). "Siri Island to turn into Iran's gas export hub: Official", published online by PressTV, 2 September 2013. Accessed online.

Priem (2013). "Hydrogen Production Storage and Fuel Cells", by Thierry Priem, presentation, 3 October 2013. Accessed online.

Priyakorn et al. (2011). "Modeling of Solid Oxide Fuel Cell with Internal Reforming Operation Fueled by Natural Gas", by F. Priyakorn, N. Laosiripojana and S. Assabumrungrat, in Journal of Sustainable Energy & Environment, Volume 2, Issue 4, October–December 2011, Pages 187–194.

PR Newswire (2011). "MarketsandMarkets: Global Hydrogen Generation Market Worth $118 Billion by 2016", 2 December 2011. Accessed online.

PR Newswire (2013). "Global Syngas Market & Derivatives (Methanol, Ammonia, Hydrogen, Oxo Chemicals, N-Butanol, DME) Market 2013–2018", 26 December 2013. Accessed online.

Proll and Hofbauer (2009). "Chemical Looping for CO2 ready power and syngas generation", by Tobias Pröll and Hermann Hofbauer, presentation to Topsøe Catalysis Forum 2009, "Catalysis in new environmental processes", at Munkerupgaard, 27–28 August 2009.

Proll and Hofbauer (2011). "Chemical Looping Combustion and Reforming", by Tobias Pröll and Hermann Hofbauer, Proceedings of the 9th European Conference on Industrial Furnaces and Boilers (INFUB-9), Estoril, Portugal, 26–29 April 2011, Section "Application of chemical looping technology".

PTT (n.d.). "Natural Gas Information", published by PTT plc. Accessed online.

PURE (2008). "Promoting Unst Renewable Energy (PURE) Project", by PURE Energy Centre, 2008. Accessed online.

PwC (2010). "100% renewable electricity: A roadmap to 2050 for Europe and North Africa", by PricewaterhouseCoopers LLP, 2010. Accessed online.

Rahimpour et al. (2012). "Feasibility of flare gas reformation to practical energy in Farashband gas refinery: No gas flaring", by Mohammad Reaza Rahimpour and Seyyed Mohammad Jokar, in Journal of Hazardous Materials, Volumes 209–210, 2012, Pages 204–217, http://www.dx.doi.org/10.1016/j.jhazmat.2012.01.017

Rana et al. (2007). "A review of recent advances on process technologies for upgrading of heavy oils and residua", by Mohan S. Rana, Vicente Sámano, Jorge Ancheyta and J. A. I. Diaz, in Fuel, Volume 86, 2007, Pages 1216–1231, http://www.dx.doi.org/10.1016/j.fuel.2006.08.004

RASE (2014). "Refuelling the Countryside", by Royal Agricultural Society of England (RASE), April 2014. Accessed online.

Rasmussen (2012). "Technologies relevant for gasification and methanation in Denmark", by Niels Bjarne Rasmussen, Danish Gas Technology Centre, September 2012, Section 2.3.

Rau et al. (2013). "Direct electrolytic dissolution of silicate minerals for air CO2 mitigation and carbon-negative H2 production", by Greg H. Rau, Susan A. Carroll, William L. Bourcier, Michael J. Singleton, Megan M. Smith and Roger D. Aines, in PNAS, Volume 110, Issue 25, 18 June 2013, Pages 10095–10100.

Raugei et al. (2012). "The energy return on energy investment (EROI) of photovoltaics: Methodology and comparisons with fossil fuel life cycles", by Marco Raugei, Pere Fullana-i-Palmer and Vasilis Fthenakis, in Energy Policy, Volume 45, June 2012, Pages 576–582, http://www.dx.doi.org/10.1016/j.enpol.2012.03.008

Ravindra et al. (2006). "Assessment of air quality after the implementation of compressed natural gas (CNG) as fuel in public transport in Delhi, India", by Khaiwal Ravindra, Eric Wauters, Sushil K. Tyagi, Suman Mor and René Van Grieken, in Environmental Monitoring and Assessment, Volume 115, Issues 1–3, Pages 405–417, 13 April 2006, http://www.dx.doi.org/10.1007/s10661-006-7051-5

Ravishankar (2013). "Why Shell won ADNOC's Bab Sour Gas field contract", by Jyotsna Ravishankar, published by Arabian Industry, ITP Publishing Ltd, 9 September 2013. Accessed online.

Ray (2014). "IGCC with Carbon Capture: A Comparison of Technologies and Efficiencies for Underground Storage of Hydrogen- and Methane-Rich Gases for Flexible Power Generation", by Dr Ruby Ray, presentation to IChemE Gasification Conference, Rotterdam, The Netherlands, 12 March 2014.

REA (2009). "REA response to the draft "UK Supply Curve for Renewable Heat" report by NERA Economic Consulting", by Renewable Energy Association, August 2009.

Reed (2013). "Shell Makes Big Bet on Boom in Natural Gas", by Stanley Reed, in New York Times, 1 May 2013. Accessed online.

Reed (2014). "Shell Earnings Buoyed by Strength in Liquefied Natural Gas", by Stanley Reed, in New York Times, 30 October 2014. Accessed online.

Rehman et al. (2013). "Hydrogen Production from Water Vapor Plasmolysis Using DBD-Corona Hybrid Reactor", by Fahad Rehman, Wameath S. Abdul Majeed and William B. Zimmerman, in Energy Fuels, Volume 27, Issue 5, Pages 2748–2761, http://www.dx.doi.org/10.1021/ef301981f

Rettger et al. (2004). "Gasification in the Canadian Oil Sands: The Long Lake Integrated Upgrading Project", by Phil Rettger, Jim Arnold, Arjan Jonckers and Doug Cathro, paper presented to Gasification Technologies – 2004, 3–6 October 2004.

Rettger et al. (2006). "The Long Lake Integrated Upgrading Project: Status Report and Discussion of Soot Processing", by Phil Rettger, Jim Arnold, Bruce Brandenburg and Chad Felch, paper presented to Gasification Technologies – 2006, 1–4 October 2006.

Reuter (2013). "Power to Gas: Microbial Methanation, a Flexible and Highly Efficient Method", by Dr Monika Reuter, of MicrobEnergy GmbH, part of Viessmann Group, presentation to the Hannover Messe, 2013, 8 April 2013.

Rhodes (2009). "Heating Value of Wood Gas and Coal (Town) Gas", by Professor Chris Rhodes, 12 February 2009. Accessed online.

Roberts et al. (2013). "Perinatal Air Pollutant Exposures and Autism Spectrum Disorder in the Children of Nurses' Health Study II Participants", by Andrea L. Roberts, Kristen Lyall, Jaime E. Hart, Francine Laden, Allan C. Just, Jennifer F. Bobb, Karestan C. Koenen, Alberto Ascherio and Marc G. Weisskopf, in Environmental Health Perspectives, Volume 121, Issue 8, August 2013, http://www.dx.doi.org/10.1289/ehp.1206187

Rochelle et al. (2002). "Research Results for CO2 Capture from Flue Gas by Aqueous Absorption/Stripping", by Gary T. Rochelle, George S. Goff, J. Tim Cullinane and Stefano Freguia, Department of Chemical Engineering, The University of Texas at Austin, presented at the Laurance Reid Gas Conditioning Conference, 25–27 February 2002.

Rodriguez et al. (2014). "Transmission needs across a fully renewable European power system", by Rolando A. Rodríguez, Sarah Becker, Gorm B. Andresen, Dominik Heide and Martin Greiner, in Renewable Energy, Volume 63, Pages 467–476, 2014, http://www.dx.doi.org/10.1016/j.renene.2013.10.005

Roeb et al. (2013). "Solar Thermal Water Splitting", by Martin Roeb, Nathalie Monnerie, Anis Houaijia, Dennis Thomey and Christian Sattler, in "Renewable Hydrogen Technologies", edited by Luis M. Gandía; Gurutze Arzamendi and Pedro M. Diéguez, ISBN: 978-0080444291.

Rogers (2013). "Shale and Wall Street: Was the Decline in Natural Gas Prices Orchestrated?", by Deborah Rogers (Deborah Lawrence, Deborah Lawrence Rogers), published by Energy Policy Forum, February 2013. Accessed online.

Romm (2013). "The Hype About Hydrogen: Fact and Fiction in the Race to Save the Climate", by Joseph Romm, published by Island Press, 2004, ISBN 13: 9781559637046, Page 71.

Rostrup-Nielsen and Hansen (2011). "Steam Reforming for Fuel Cells", by Jens R. Rostrup-Nielsen and John Bøgild Hansen, Chapter 4 in "Fuel Cells: Technologies for Fuel Processing", edited by Dushyant Shekhawat, James J. Spivey and David A. Berry, published by Elsevier Science, 18 March 2011, ISBN: 978-0444535634, Section 4.1.

Rozmarynowska and Oldakowski (2012). "Implications of new regulation regarding sulphur content in ship's fuel on maritime transport sector within Baltic Sea Region", by Monika Rozmarynowska and Bogdan Oldakowski, July 2012.

Rutledge (2011). "Estimating Long-Term World Coal Production With logit and probit Transforms", by Professor David B. Rutledge, International Journal of Coal Geology, Volume 85, Issue 1, Pages 23–33, 1 January 2011, http://www.dx.doi.org/10.1016/j.coal.2010.10.012

Rutledge (2013). "Projections for Ultimate Coal Production from Production Histories Through 2012", by Professor David B. Rutledge, paper presented at 2013 Geological Society of America (GSA) Annual Meeting, Denver, Colorado.

Sabatier (1912). "The Method of Direct Hydrogenation by Catalysis", Nobel Lecture of Paul Sabatier, 11 December 1912, for the award of the Nobel Prize in Chemistry 1912. Accessed online.

Sabatier (1926). "How I Have Been Led to the Direct Hydrogenation Method by Metallic Catalysts", by Paul Sabatier, Industrial and Engineering Chemistry, Volume 18, Issue 10, Pages 1005–1008, October 1926, http://www.dx.doi.org/10.1021/ie50202a005

Sabatier and Senderens (1902). "Hydrogénation directe des oxydes du carbone eu présence de divers métaux divisés", by Paul Sabatier and J.-B. Senderens, in Comptes Rendus Hebdomadaires des Séances de l'Académie des Sciences, Janvier à Juin 1902, Séance du 24 March 1902, Pages 689–691, Note de Messieurs Paul Sabatier et J.-B. Senderens.

Santos-Alamillos et al. (2014). "A methodology for evaluating the spatial variability of wind energy resources: Application to assess the potential contribution of wind energy to baseload power", by F. J. Santos-Alamillos, D. Pozo-Vázquez, J. A. Ruiz-Arias, V. Lara-Fanego and J. Tovar-Pescador, in Renewable Energy, Volume 69, Pages 147–156, 2014, http://www.dx.doi.org/10.1016/j.renene.2014.03.006

SBC (2013). "Hydrogen-Based Energy Storage Solutions", by Schlumberger Business Consulting (SBC) Energy Institute, presentation to IEA Workshop on Hydrogen Technology Roll-Out in Europe, 10 July 2013. Accessed online.

SBC (2014). "Hydrogen-Based Energy Conversion", by Schlumberger Business Consulting (SBC) Energy Institute, presentation, February 2014. Accessed online.

Schaaf et al. (2014). "Methanation of CO2 – storage of renewable energy in a gas distribution system", by Tanja Schaaf, Jochen Grünig, Markus Roman Schuster, Tobias Rothenfluh and Andreas Orth, in Energy, Sustainability and Society, Review, published by Springer, 24 December 2014, http://www.dx.doi.org/10.1186/s13705-014-0029-1

Scheepers (2013). "Green Gas in the Dutch Gas Supply", by Martin J. J. Scheepers, presentation to EDGaR-DVGW Joint Conference, Arnhem, 7 February 2013, Document: ECN-L--13-002. Accessed online.

Schindler et al. (2006). "Where will the Energy for Hydrogen Production come from? Status and alternatives", by J. Schindler, R. Wurster, M. Zerta, V. Blandow and W. Zittel, for European Hydrogen Association (EHA), 2006.

Schivelbusch (1988). "Disenchanted Night: Industrialization of Light in the Nineteenth Century", by Wolfgang Schivelbusch, published by University of California Press, 1988 (also in an edition published 23 June 1998), ISBN: 978-0520203549.

Schlesinger et al. (1965). "Conversion of Fossil Fuels to Liquid Fuels", by M. D. Schlesinger, G. U. Dinneen and S. Katell, paper presented to the Symposium on Fuel and Energy Economics, American Chemical Society (ACS), Fuel Division, Spring 1965, Detroit, 9 (2), Pages 120–126.

Schulz et al. (2006). "Methane from Wood: Assessment of Wood-Based Synthetic Natural Gas Technologies with the Swiss MARKAL Model", by Thorsten F. Schulz, Leonardo Barreto, Socrates Kypreos and Samuel Stucki, of Paul Scherrer Institut(e) (PSI), Research Paper: 2006/02, April 2006.

SDC (2010). "SDC (2010). "Energy from Waste Potential in Scotland", by Charlie Blair, Phil Matthews, Valerie Caldwell and Maf Smith, of Sustainable Development Commission (SDC) Scotland, January 2010, Section 3.4, "Sensitivity", "Composition of MSW".

SE4ALL (2014). "SE4ALL: Finance Working Group", report prepared by World Bank, Bank of America Merrill Lynch and BNDES (Brazilian Development Bank), for Sustainable Energy for All (SE4ALL), a project of the United Nations (UN), Advisory Board Meeting Draft, 1 June 2014.

Seiffert et al. (2009). "Bio-SNG – Demonstration of the production and utilization of synthetic natural gas (SNG) from solid biofuels", by Michael Seiffert, Stefan Rönsch, Ralf Schmersahl, Martin Zeymer, Stefan Majer, Christian Pätz, Martin Kaltschmitt, Reinhard Rauch, Barbara Rehling, Harald Tremmel, Daniel Ulrich, Serge Biollaz and Tilmann Schildhauer, European Commission, Sixth Framework Programme, TREN/05/FP6EN/S07.56632/019895, Final Project Report D 1.4.

Sep and Wolff (2004). "Residue Upgrading with the Shell Gasification Process for LOTOS Refinery Gdansk", by Karol Sep and Joachim Wolff, presentation to Gasification Technologies – 2004, 2–4 October 2004, Washington, D.C. Accessed online.

Shahsavand (2008). "Investigating various cooling scenarios to increase the capacity of an Iranian natural gas treating unit", by A. Shahsavand, Chemical Engineering Department, Faculty of Engineering Ferdowsi University of Mashhad.

Shell (2008). "Shell energy scenarios to 2050", Royal Dutch Shell, 2008. Accessed online.

Shell (2011). "Signals and Signposts: Shell Energy Scenarios to 2050", Royal Dutch Shell, 2011. Accessed online.

Shell (2013). "New Lens Scenarios: A Shift in Perspective for a World in Transition", Royal Dutch Shell, March 2013. Accessed online.

Shell (2014). Letter, signed by Dr J. J. Traynor, Executive Vice President, Investor Relations, Royal Dutch Shell plc, 16 May 2014. Accessed online.

SHFCA (2014). "World Hydrogen demand for refining to rise 3.5% annually to 2018", by Scottish Hydrogen Fuel Cell Association (SHFCA). Accessed online.

Shoko et al. (2006). "Hydrogen from coal: Production and utilisation technologies", by E. Shoko, B. McLellan, A. L. Dicks and J. C. Diniz da Costa, in International Journal of Coal Geology, Volume 65, 2006, Pages 213–222, http://www.dx.doi.org/10.1016/j.coal.2005.05.004

Siemens (2002). "Energy Fuel Cells: Gentle Revolution", by Siemens Global, Spring 2002. Accessed online.

Siemens (2011). "Biogas Plants", "Pictures of the Future", by Siemens Global, Magazine Spring 2011, "The New Age of Electricity". Accessed online.

Silverman et al. (2012). "The Diesel Exhaust in Miners Study: A Nested Case – Control Study of Lung Cancer and Diesel Exhaust", by Debra T. Silverman, Claudine M. Samanic, Jay H. Lubin, Aaron E. Blair, Patricia A. Stewart, Roel Vermeulen, Joseph B. Coble, Nathaniel Rothman, Patricia L. Schleiff, William D. Travis, Regina G. Ziegler, Sholom Wacholder and Michael D. Attfield, in Journal of the National Cancer Institute, Volume 104, Issue 11, 5 March 2012, Pages 855–868, http://www.dx.doi.org/10.1093/jnci/djs034

Singhal and Kendall (2003). "High-Temperature Solid Oxide Fuel Cells: Fundamentals, Design and Applications", by Subhash C. Singhal and Kevin Kendall, published by Elsevier Science, 8 December 2003, ISBN: 978-1856173872.

SKM Enviros (2011). "Analysis of Characteristics and Growth Assumptions Regarding AD Biogas Combustion for Heat, Electricity and Transport and Biomethane Production and Injection into the Grid", by SKM Enviros, a report commission by HMG DECC, Reference Number: 09/06/2010, 3 May 2011.

Skorobogatov et al. (2000). "Conventional and Unconventional Natural Gas Resources in Russia and the Strategy of their Development in the First Half of the 21st Century", by V. A. Skorobogatov, V. A. Istomin and V. S. Yakushev, 1 July 2000.

Soderbergh et al. (2010). "European energy security: An analysis of future Russian natural gas production and exports", by Bengt Söderbergh, Kristofer Jakobsson and Kjell Aleklett, in Energy Policy, Volume 38, Issue 12, December 2010, Pages 7827–7843, http://www.dx.doi.org/10.1016/j.enpol.2010.08.042

Soetaert and Vandamme (2009). "Biofuels", edited by Wim Soetaert and Erik Vandamme, published by Wiley-Blackwell, 6 February 2009, ISBN 13: 978-0470026748.

Sorrell et al. (2010). "Global oil depletion: A review of the evidence", by Steve Sorrell, Jamie Speirs, Roger Bentley, Adam Brandt and Richard Miller, in Energy Policy, Volume 38, Issue 9, Pages 5290–5295, September 2010, http://www.dx.doi.org/10.1016/j.enpol.2010.04.046

Sorrell et al. (2012). "Shaping the global oil peak: A review of the evidence on field sizes, reserve growth, decline rates and depletion rates", by Steve Sorrell, Jamie Speirs, Roger Bentley, Richard Miller and Erica Thompson, in Energy, Volume 37, Issue 1, January 2012, Pages 709–724, http://www.dx.doi.org/10.1016/j.energy.2011.10.010

SourceWatch (2012). "Integrated Gasification Combined Cycle (IGCC)", by SourceWatch, 19 June 2012. Accessed online.

Sovacool (2014). "Cornucopia or curse? Reviewing the costs and benefits of shale gas hydraulic fracturing (fracking)", by Benjamin K. Sovacool, in Renewable and Sustainable Energy Reviews, Volume 37, September 2014, Pages 249–264, http://www.dx.doi.org/10.1016/j.rser.2014.04.068

Spath and Dayton (2003). "Preliminary Screening – Technical and Economic Assessment of Synthesis Gas to Fuels and Chemicals with Emphasis on the Potential for Biomass-Derived Syngas", by P. L. Spath and D. C. Dayton, National Renewable Energy Laboratory (NREL), Report Number: NREL/TP-510-34929, December 2003.

Specht et al. (2009). "AER-Gas II: Biomass Fluidised Bed Gasification with in-situ Hot Gas Cleaning", CORDIS project coordinator, Dr Michael Specht, Publishable Final Activity Report.

Spulber (2010). "Geogenic Emissions of Methane in Transylvania and their Environmental Implications", by Liana Spulber, PhD Thesis, Babes-Bolyai University, Faculty of Environmental Science, 2010. Chapter I, "General Geological Setting".

Sreedharan (2012). "CFD Analysis of Coal and Heavy Oil Gasification for Syngas Production", by Vikram Sreedharan, PhD Thesis, Aalborg University, Denmark, 2012. Accessed online.

Statoil (2013). "Energy Perspectives: Long-term macro and market outlook", Statoil, June 2013. Accessed online.

Steinberg (2004). "The Highly Efficient Integrated Plasma Fuel Cell (IPFC) Energy Cycle for Conversion of Fossil and Biomass Fuels to Electric Power Generation and Hydrogen and Liquid Transportation Fuel Production with Reduced or Zero CO2 Emission", by Meyer Steinberg, in American Chemical Society, Division of Fuel Chemistry, Preprints 2004, Volume 49, Issue 2, Pages 737–738.

Stephenson and MacKay (2014). "Scenarios for Assessing the Greenhouse Gas Impacts and Energy Input Requirements of Using North American Woody Biomass for Electricity Generation in the UK", by Dr Anna L. Stephenson and Professor David J. C. MacKay, published by HMG DECC, July 2014.

Stevens (2012). "The 'Shale Gas Revolution': Developments and Changes", by Paul Stevens, Report to Chatham House: EERG BP 201204, August 2012. Accessed online.

Strabo (7 BC) (c.18 AD) (1903). "Geographika", by Strabon, Book XVI, Chapter I, Section 4 (16.1.4), in the translation of H. C. Hamilton, edited by W. Falconer, "The Geography of Strabo: Literally translated, with notes, in three volumes", George Bell & Sons, London, 1903.

Strezov and Evans (2014). "Biomass Processing Technologies", by Vladimir Strezov and Tim J. Evans, published by CRC Press, 22 July 2014, ISBN: 978-1466566163.

Sutikino and Turini (2012). "Gasifying coke to produce hydrogen in refineries", by Tek Sutikino and Kevin Turini, in Digital Refining, PTQ magazine, PTQ Q3, 2012. Accessed online.

SXCoal (2014). "China halves 2020 shale gas output target to 30 bcm on tough challenges", published by China Coal Resource, 8 August 2014. Accessed online.

Symes and Cronin (2013). "Decoupling hydrogen and oxygen evolution during electrolytic water splitting using an electron-coupled-proton buffer", by Mark D. Symes and Leroy Cronin, in Nature Chemistry, Volume 5, May 2013, Pages 403–409, http://www.dx.doi.org/10.1038/NCHEM.1621

Tarr (2009). "Manufactured Gas, History of", by Joel A. Tarr, in "Concise Encyclopedia of the History of Energy", edited by Cutler J. Cleveland, Academic Press, 1 October 2009, ISBN: 978-0123751171. "M", Pages 153–162.

Themelis and Mussche (2014). "2014 Energy and Economic Value of Municipal Solid Waste (MSW), including Non-Recycled Plastics (NRP), Currently Landfilled in the Fifty States", by Nickolas J. Themelis and Charles Mussche, Columbia University, Earth Engineering Centre, 9 July 2014.

ThermoFisher (n.d.). "Blast Furnace Top Gas Application Summary", by ThermoFisher Scientific. Accessed online.

Thermopedia (2011). "Coke-Oven Gas", by J. A. Lacey, Thermopedia, http://www.dx.doi.org/10.1615/AtoZ.c.coke-oven_gas

Thomas (2010a). "Water Gas Plants", by Dr Russell Thomas, Table 3. Accessed online.

Thomas (2010b). "Producer Gas Plants", by Dr Russell Thomas, Table 1. Accessed online.

Thwaite (1889). "Gaseous Fuel: including Water Gas: Its Production and Application: A Lecture delivered at The Association Hall, Peter Street, Manchester On March 29, 1889", by B. H. Thwaite, to the Manchester and Salford Noxious Vapours Abatement Association, published by Whittaker and Company, 2 White Hart Street, Paternoster Square, 1889, Page 14.

TINA Bioenergy (2012). "Bioenergy", by Low Carbon Innovation Coordination Group (LCICG), Technology Innovation Needs Assessment (TINA), Summary Report, September 2012.

TREC-STEP (2011). "Nuon Magnum", by European Union Cluster for Clean Coal Technologies (CCT) and Carbon Capture and Storage (CCS), Fact Sheet,

Tiruchirappalli Regional Engineering College (TREC), Science and Technology Entrepreneurs Park (STEP), Technologies for the Indian Thermal Power Sector.
Trieb and Muller-Steinhagen (2009). "The DESERTEC Concept: Sustainable Electricity and Water for Europe, Middle East and North Africa", by Dr. Franz Trieb and Professor Dr. Hans Müller-Steinhagen, Chapter 3 in "Clean Power from Deserts: The DESERTEC Concept for Energy, Water and Climate Security", a Club of Rome inititative, WhiteBook, 4th edition, published by Trans-Mediterranean Renewable Energy Cooperation TREC, published by Protext Verlag, Bonn, February 2009, ISBN: 978-3-929118-67-4.
Truck & Bus Builder (2011). "Dual Fuel – a growing characteristic of the British truck market", in Truck & Bus Builder briefing, May 2011, "Future Fuels & Power", "Fuel Profile", Pages 17–19. Accessed online.
TU Wien (n.d.). "Actual projects", by Institut für Verfahrenstechnik, Umwelttechnik und Technische Biowissenschaften (VT), at TU Wien, "Gasification and Gas Cleaning", "Projects". Accessed online.
UCS (2013). "Gas Ceiling: Assessing the Climate Risks of an Overreliance on Natural Gas for Electricity", Union of Concerned Scientists (UCS), released 18 September 2013. Accessed online.
UKGOV (2013a). "Autumn Statement 2013", Government of the United Kingdom, Document: Cm 8747, published by Her Majesty's Treasury, December 2013.
UKGOV (2013b). "Running hours during winter 2012/13 for plants opted-out of the Large Combustion Plant Directive (LCPD)", Government of the United Kingdom, "Special Feature – Large Combustion Plant Directive".
UKPIA (2012). "The impact of IMO MARPOL sulphur regulations on UK refining", by United Kingdom Petroleum Industry Association, 2012. Accessed online.
UNFCCC (n.d.). "GHG emission profiles", United Nations Framework Convention on Climate Change (UNFCCC). Accessed online.
UNFCCC (1997). "Kyoto Protocol to the United Nations Framework Convention on Climate Change", 1997. Accessed online.
USGS (2008). "Circum-Arctic Resource Appraisal: Estimates of Undiscovered Oil and Gas North of the Arctic Circle", by USGS, USGS Factsheet 2008–3049. Accessed online.
USGS (2013). "National Oil and Gas Assessment 2013 Assessment Updates", 2013. Accessed online.
van der Meijden (2012). "MILENA gasification bed materials", by Christiaan van der Meijden. Accessed online.
van der Ploeg et al. (2007). "Shell Coal Gasification Process for Power", by Rik van der Ploeg, Gerard Grootveld, Matthieu Fleys, Lisanne Deprez, Rien van Haperen and Robert de Kler, presentation to 2nd International Freiburg Conference on IGCC & XtL", held 8–12 May 2007, by the Institute of Energy Process Engineering and Chemical Engineering (IEC) at TU Bergakademie, Freibug, and The German Centre for Gasification Technologies.
Vandeweijer et al. (2011). "Monitoring the CO2 injection site: K12-B", by Vincent Vandeweijer, Bert van der Meer, Cor Hofstee, Frans Mulders, Daan D'Hoore and Hilbrand Graven, in Energy Procedia, Volume 4, 2011, Pages 5471–5478, http://www.dx.doi.org/10.1016/j.egypro.2011.02.532
Van Eijs et al. (2006). "Correlation between hydrocarbon reservoir properties and induced seismicity in the Netherlands", by R. M. H. E. Van Eijs, F. M. M. Mulders, M. Nepveu, C. J. Kenter and B. C. Scheffers, in Engineering Geology,

Volume 84, Issues 3–4, 16 May 2006, Pages 99–111, http://www.dx.doi.org/10.1016/j.enggeo.2006.01.002

Vasagar (2014a). "Eon to spin off its fossil fuel assets as big losses loom", by Jeevan Vasagar, in Financial Times, 30 November 2014. Accessed online.

Vasagar (2014b). "RWE rejects following Eon plan to split business", by Jeevan Vasagar, in Financial Times, 1 December 2014. Accessed online.

Vasagar (2014c). "Eon split driven by technology", by Jeevan Vasagar, in Financial Times, 2 December 2014. Accessed online.

Vassilopoulos (2013). "Capacity Mechanisms in EU Power Markets", by Philippe Vassilopoulos, presentation by IHS CERA, Paris, 16 April 2013. Accessed online.

Venkatesh et al. (2011). "Uncertainty in Life Cycle Greenhouse Gas Emissions from United States Natural Gas End-Uses and its Effects on Policy", by Aranya Venkatesh, Paulina Jaramillo, W. Michael Griffin and H. Scott Matthews, in Environmental Science and Technology, Volume 45, Issue 19, Pages 8182–8189, 16 August 2011, http://www.dx.doi.org/10.1021/es200930h

Verma et al. (2004). "A Re-Look on The Subsurface of Himalayan Foothills In View of New Acquisition And Reprocessing of Seismic Data", by N. K. Verma, G. S. Rana, Vinod Ranjan and Jitendra Kumar, 5th Conference & Exposition on Petroleum Geophysics, Hyderabad, India, 2004, Pages 810–817.

Verostko and Forsythe (1974). "A Study of the Sebatier-Methanation Reaction", by C. E. Verostko and R. K. Forsythe. Accessed online.

Vidal (2014). "Shell hits back at 'carbon bubble' claims", by John Vidal, in The Guardian newspaper, 20 May 2014. Accessed online.

Viscopedia (n.d.). "Bunker Oil – Marine Fuel Oil", by Viscopedia. Accessed online.

Wang and Krupnick (2013). "A Retrospective Review of Shale Gas Development in the United States", by Zhongmin Wang and Alan Krupnick, Resources for the Future (RFF), Document: RFF DP 13-12, April 2013.

Wang and Massoudi (2013). "Slag Behavior in Gasifiers. Part I: Influence of Coal Properties and Gasification Conditions", by Ping Wang and Mehrdad Massoudi, in Energies, Volume 6, Issue 2, 2013, Pages 784–806, http://www.dx.doi.org/10.3390/en6020784

Wang et al. (2014). "The intensification technologies to water electrolysis for hydrogen production – A review", by Mingyong Wang, Zhi Wang, Xuzhong Gong and Zhancheng Guo, in Renewable and Sustainable Energy Reviews, Volume 29, 2014, Pages 573–588, http://www.dx.doi.org/10.1016/j.rser.2013.08.090

Waples (2012). "The Natural Gas Industry in Appalachia: A History from the First Discovery to the Maturity of the Industry", by David A. Waples, McFarland & Co. Inc., 31 March 2005, ISBN-10: 0786420774, ISBN: 978-0786420773. "The Natural Gas Industry in Appalachia: A History from the First Discovery to the Tapping of the Marcellus Shale", David A. Waples, McFarland & Co. Inc., 2nd edition, 15 May 2012, ISBN: 978-0786470006, Pages 8–9.

Watkins (2011). "Heating Services in Buildings", by David E. Watkins, published by John Wiley & Sons, ISBN: 978-0-470-65603-7, 2011. Accessed online.

Watt Committee (1982). "The European Energy Scene", by The Watt Committee on Energy, Report Number 11, 1 April 1982, ISSN: 0141-9676 published by CRC Press, ISBN: 978-0946392124, Sections: "Government and the energy sector", "The changing energy scene", Pages 2–5.

Wawrzinek and Keller (2007). "Industrial Hydrogen Production & Technology", by Dr Klemens Wawrzinek and Claude Keller, of Linde Engineering, presentation

to Karlsruhe FuncHy-Workshop, 21 November 2007, Slide 4, "Installed capacity worldwide".
WBCSD (2012). "The Energy Mix: Low-Carbon Pathways to 2050", World Business Council for Sustainable Development (WBCSD), 23 May 2012. Accessed online.
WEC (2007a). "Deciding the Future: Energy Policy Scenarios to 2050", World Energy Council, 2007, Executive Summary, ISBN: 09-46-12129-X. Accessed online.
WEC (2007b). "Energy Scenario Development Analysis: WEC Policy to 2050", World Energy Council, 2007. Accessed online.
WEC (2013). "World Energy Perspective: Cost of Energy Technologies", by World Energy Council (WEC), 2013, ISBN: 978-0-94612-130-4. Accessed online.
Weijermars (2014). "US shale gas production outlook based on well roll-out rate scenarios", by Ruud Weijermars, in Applied Energy, Volume 124, 1 July 2014, Pages 283–297, http://www.dx.doi.org/10.1016/j.apenergy.2014.02.058
Welfle et al. (2014). "Securing a bioenergy future without imports", by Andrew Welfle, Paul Gilbert and Patricia Thornley, in Energy Policy, Volume 68, May 2014, Pages 1–14, http://www.dx.doi.org/10.1016/j.enpol.2013.11.079
WHO (2013). "Global Status Report on Road Safety 2013", by World Health Organization (WHO), 2013, ISBN: 978-92-4-156456-4. Accessed online.
Wiggert et al. (2005). "Monsoon-driven biogeochemical processes in the Arabian Sea", by J. D. Wiggert, R. R. Hood, K. Banse and J. C. Kindle, in Progress in Oceanography, Volume 65, Issues 2–4, May–June 2005, Pages 176–213, http://www.dx.doi.org/10.1016/j.pocean.2005.03.008
Wigley (2011). "Coal to gas: the influence of methane leakage", by Tom M. L. Wigley, in Climatic Change, Letters, October 2011, Volume 108, Issue 3, Pages 601–608, 26 August 2011, http://www.dx.doi.org/10.1007/s10584-011-0217-3
Willauer et al. (2013). Extraction of Carbon Dioxide and Hydrogen from Seawater by an Electrochemical Acidification Cell Part IV: Electrode Compartments of Cell Modified and Tested in Scaled-Up Mobile Unit", by Heather D. Willauer, Dennis R. Hardy, Frederick W. Williams and Felice DiMascio, Naval Research Laboratory, Washington, D.C., Document: NRL/MR/6300--13-9463, 3 September 2013.
Willauer et al. (2014). "Method for the Continuous Recovery of Carbon Dioxide from Acidified Seawater", by Heather D. Willauer, John Barile, Felice DiMascio, Dennis R. Hardy and Frederick Williams, in Patent US 8,663,365, "Background of the Invention", paragraph 2.
Williams (1981). "A History of the British Gas Industry", by Trevor Illtyd Williams, published by Oxford University Press, 1 July 1981, ISBN: 978-0198581574. Published book.
Wokaun and Wilhelm (2011). "Transition to Hydrogen: Pathways Towards Clean Transportation", by Alexander Wokaun and Erik Wilhelm, published by Cambridge University Press, October 2011, ISBN: 9780521192880, Appendix A, Page 224.
World Bank (2013). "Toward a Sustainable Energy Future for All: Directions for the World Bank Group's Energy Sector", World Bank, Washington, D.C., 9 July 2013.
WRAP (2010). "European Legislation", WRAP, 2010. Accessed online.
Wurster (2001a). "Hydrogen Activities in Germany and the European Union", by Reinhold Wurster, presentation, June 2001, Slide 23, "Industrial hydrogen experience across Europe".
Wurster (2001b). "Hydrogen Activities in the European Union and Germany", by Reinhold Wurster, presentation to H2NET Symposium, 13 June 2001, Slide 23, "Industrial hydrogen experience across Europe".

WWEA (2014). "World Wind Resource Assessment Report", by World Wind Energy Association, Technical Committee, lead author Dr Jami Hossain.
WWF (2011). "The Energy Report: 100% Renewable Energy by 2050", World Wildlife Fund (WWF), January 2011. Accessed online.
Yang et al. (2014). "China's fuel gas sector: History, current status, and future prospects", by Chi-Jen Yang, Yipei Zhou and Robert B. Jackson, in Utilities Policy, Volume 28, 2014, Pages 12–21, http://www.dx.doi.org/10.1016/j.jup.2013.11.002
Yee (2011). "UK signs £2bn Qatari gas deal", by April Yee, in The National, United Arab Emirates, 25 April 2011. Accessed online.
Yergin (1991) (1992) (2009). "The Prize: The Epic Quest for Oil, Money and Power", by Daniel Yergin, published in paperback by Free Press, A Division of Simon & Schuster Inc., 19 January 2009, ISBN: 978-1847376466, Pages 187–188. Published book.
Youngukkasem et al. (2013). "Rapid Biogas Production by Compact Multi-Layer Membrane Bioreactor: Efficiency of Synthetic Polymeric Membranes", by Supansa Youngsukkasem, Hamidreza Barghi, Sudip K. Rakshit and Mohammad J. Taherzadeh, in Energies 2013, Volume 6, Issue 12, Pages 6211–6224, ISSN: 1996-1073, http://www.dx.doi.org/10.3390/en6126211
Zadakar et al. (2011). "Risk Analysis of Flare Flame-out Condition in a Gas Process Facility", by O. Zadakar, R. Abbassi, F. Khan, K. Karimpour, M. Golshani and A. Vatani, in Oil & Gas Science and Technology, Revue IFP Energies Nouvelles, Volume 66, Issue 3, 2011, Pages 521–530, http://www.dx.doi.org/10.2516/ogst/2010027
Zhan et al. (2006). "High-rate electrochemical partial oxidation of methane in solid oxide fuel cells", by Zhongliang Zhan, Yuanbo Lin, Manoj Pillai, Ilwon Kim and Scott A. Barnett, in Journal of Power Sources, Volume 161, Issue 1, 20 October 2006, Pages 460–465, http://www.dx.doi.org/10.1016/j.jpowsour.2006.04.139
Zhang et al. (2009). "Health Effects of Real-World Exposure to Diesel Exhaust in Persons with Asthma", by Junfeng (Jim) Zhang, James E. McCreanor, Paul Cullinan, Kian Fan Chung, Pamela Ohman-Strickland, In-Kyu Han, Lars Järup and Mark J. Nieuwenhuijsen, in Health Effects Institute, Research Report 138, 27 February 2009.
Zhenggang Nie et al. (2013). "Full chain analysis and comparison of gas-fired power plants with CO2 capture and storage with clean coal alternatives", by Zhenggang Nie, Anna Korre and Sevket Durucan, in Energy Procedia, Volume 37, 2013, Special Issue, "GHGT-11", edited by Tim Dixon and Kenji Yamaji, Pages 2840–2847, Figure 5, http://www.dx.doi.org/10.1016/j.egypro.2013.06.169
Ziegler (1920). "Popular Oil Geology", by Victor Ziegler, Consulting Geologist, formerly Professor of Geology and Mineralogy, Colorado School of Mines, Second Edition, published by John Wiley & Sons Inc., Chapman & Hall, Pages 5–6.
Zwart et al. (2004). ""Green Gas" as SNG (Synthetic Natural Gas) a Renewable Fuel with Conventional Quality", by M. Mozaffarian, R. W. R. Zwart, H. Boerrigter, E. P. Deurwaarder and S. R. A. Kersten, ECN, August 2004, Report: ECN-RX--04-085.
Zwart et al. (2006). "Production of Synthetic Natural Gas (SNG) from Biomass Development and operation of an integrated bio-SNG system", by R. W. R. Zwart, H. Boerrigter, E. P. Deurwaarder, C. M. van der Meijden and S. V. B. van Paasen, by ECN, Report: ECN-E--06-018, Non-confidential version.

Index

2P (proven-plus-probable reserves), 28, 34

AAD (Advanced Anaerobic Digestion), 41, 103
acid gas, sour gas, 3, 10, 57–58, 119–123, 129–130, 209
acidification
  electrochemical acidification, 123
  ocean acidification, 123
ACT (Advanced Conversion Technologies), 196–198, 200
activated carbon, 105
activation energy, 114, 118, 157–158
active zone, 136
AD (Anaerobic Digestion), 41, 55–58, 62–64, 71, 78, 85–86, 102, 117, 149–150, 153
advanced biofuels, 23, 192
  advanced biogas, 1
  see also AAD (Advanced Anaerobic Digestion), ACT (Advanced Conversion Technologies)
AER (Absorption-Enhanced Reforming), 151–152
AFC (Alkaline Fuel Cell), 154
algaculture, 122
  algae, microalgae, 23, 41, 65, 122–123, 150
  Harmful Algal Bloom (HAB), 122
  macroalgae, 122
  seagrass, 122
  seaweed, 122
Allam Cycle, 152, 155
  Rodney Allam, 152
ammonia, 65, 82, 101, 144, 186–187, 204
  bioammonia, 69, 211
APR (Aqueous Phase Reforming), 122, 163
Arctic, 58
  Arctic gas, 9, 14, 34, 126
  Arctic oil, 14, 34
ash, 135–136, 162
  carbon ash, 142
  coal ash, 147
asset, 5, 28, 46, 171
  asset class, 46, 181
  see also grids, infrastructure
ASU (Air Separation Unit), 111, 142, 162

backup (back up), 6, 15, 47–48, 53, 65, 165, 172–173, 206, 211
  see also standby
balancing services
  see load balancing
baseload, 6, 68–69, 165, 173, 178–180
battery, 64
  gas battery, 88
  chemical battery, 51
  see also BEV (Battery Electric Vehicle)
BECCS (Bioenergy Carbon Capture and Storage), 95, 107
BEV (Battery Electric Vehicle), 6, 67
BGL (British Gas/Lurgi), 83
bi-fuel, 65
BIGCC (Biomass Integrated Gasification Combined Cycle), 140
bio-oil, 156, 162, 192
  see also biofuel, biotar, pyrolysis
bioammonia, 69, 211
biocarbon, 99
  see also biochar, biocoal, biocoke, char, coke, petcoke
biochar, 41, 97–99, 136, 152, 164, 209
  see also biocarbon, biocoal, biocoke, char, coke, petcoke
biocoal, 117, 201, 211
  see also biocarbon, biochar, biocoke, char, coke, petcoke
biocoke, 117, 201, 211
  see also biocarbon, biochar, biocoal, char, coke, petcoke

biodiesel, 23, 40, 64–65
see also diesel
bioenergy, 23, 54, 64, 70, 94–98, 196–198
see also biomass
bioethanol, 23, 64–65
see also biofuel, ethanol
biofuel, 4, 19, 23, 40, 63–65, 122, 192
see also advanced biofuels
biogas, xvi–xvii, 1–2, 9, 16, 19, 41, 49, 52–58, 62–64, 73, 78, 85, 94, 102, 117, 121, 150–153, 171–176, 181, 184, 194, 197–198, 200–201
see also biohydrogen, biomethane, bioSNG, Renewable Gas
biohydrogen, 41, 85, 162, 181
see also biogas, biomethane, Renewable Gas
biomass, xvi, 1–2, 9–11, 16, 19, 23, 40–41, 49–58, 64–65, 69–71, 82–87, 98–99, 101–104, 107, 111, 117–118, 121–125, 132, 135–140, 143–152, 155–164, 171, 177, 184–186, 189, 196–200, 203–204, 208–211
see also bioenergy
biomethane, 9, 49, 64, 85, 94, 150–153, 161, 172, 181–185
see also biogas, biohydrogen, bioSNG, Renewable Gas
biophotolysis, 163
bioplastics, 41, 149
see also plastics
biopower, 196
see also green power, renewable electricity
biorefinery, 24, 50, 65, 70, 131
biochemical refinery, 24
bioSNG, 85, 194, 197–201
see also biogas, biomethane, biosyngas
biosyngas, 62–65, 69, 73, 85, 155
see also biohydrogen, bioSNG
biotar, 152
bitumen, 127
bituminous sands, 8, 35, 193
black carbon, 91
Blue Water Gas, 78
bond
energy bond, 30, 194
energy efficiency bond, 30
Green Gas Certificate, 183
Renewable Gas bond, 194
bottom of the barrel, 190
bottoms, 143
tops, 144
see also fuel oil, heavy oil, residua, ROG (Refinery Off Gas)
BtG (Biomass-to-Gas, Bioenergy-to-Gas), 149–150
BtH (Biomass-to-Hydrogen, Bioenergy-to-Hydrogen), 161
BtL (Biomass-to-Liquids, Bioenergy-to-Liquids), 145
by-product hydrogen (byproduct hydrogen), 187–191

Calcium Looping (CaL), 151–152
Capacity Mechanism, 178–180, 209, 212
capacity payments, 15
capital, 7, 12, 23–31, 104, 170–171, 194, 210–211
capital expenditure (capex), 30, 45, 70, 104, 128–130, 168
carbon
see carbon market, carbon price, carbon tax
see also biocarbon, biochar, biocoal, biocoke, carbon dioxide, carbon monoxide, char, coke, petcoke
carbon-neutral, 59, 64, 98, 124, 161–163, 204
carbon-rich gas, 4, 41, 49–52, 73, 149
carbon bubble, 31, 194
carbon burden, 96–98, 103
carbon capture, carbon dioxide capture, xvi–xvii, 6, 41, 53–58, 62, 96, 101–103, 107, 110–113, 122–125, 131, 139–142, 145–147, 151–155, 159, 163, 180
biogas capture, 71
gas capture, 72
Natural Gas capture, 87
see also carbon recycling, CCS (Carbon Capture and Storage), DAC (Direct Air Capture)

carbon dioxide, (CO_2), carbon dioxide emissions, xvi–xvii, 2–6, 9–11, 15–19, 26, 29–30, 39–43, 46–47, 50, 54–58, 61–62, 67–69, 72, 78, 82–88, 91–103, 106–126, 129–135, 139–171, 185–187, 190–192, 200–205, 209–210
see also carbon dioxide leakage, carbon leakage, CCS (Carbon Capture and Storage), CCU (Carbon Capture and Utilisation)
carbon dioxide storage
temporary, 50, 108
carbon intensity, carbon intensive, carbon-intensive, xvi, 15, 43, 192–193
carbon leakage, carbon dioxide leakage, 155, 192
see also carbon slip
carbon market, carbon trading, emissions trading, 6, 10, 30–31, 40–41, 169, 192
see also carbon price, carbon tax
carbon price, emissions pricing, 6, 40–42
see also carbon market, carbon tax
carbon recycling, 20, 40–41, 50–57, 68–69, 72–75, 86–87, 96–100, 103–104, 113, 118, 121–125, 131, 140–142, 145–146, 151–159, 164, 180–181, 184, 200, 205, 210
see also carbon capture, carbon dioxide capture, CCS (Carbon Capture and Storage), DAC (Direct Air Capture)
carbon rejection, 129, 187, 205
see also carbon leakage, carbon slip
carbon slip, 158–159, 192
see also carbon leakage
carbon tax, carbon taxation, 6, 42
see also carbon market, carbon price
carbonic acid, 116
carbonisation, 100
of coal, 79–82, 88, 100, 147
see also pyrolysis
Carburetted Water Gas (CWG), 79
catalysis, catalytic, catalyst, 1, 41, 51, 71, 78, 88, 100, 103–104, 109–110, 114, 131, 134, 137–139, 144, 147–148, 156–158, 162–165, 191, 208–209
see also FCC (Fluid Catalytic Cracking)
CAWE (Carbon-Assisted Water Electrolysis), 105
CCGT (Combined Cycle Gas Turbine), 15, 74, 113, 179–181, 195–196
see also NGCC (Natural Gas Combined Cycle)
CCHP (Combined Cooling Heating Power), 62
see also CHP (Combined Heat and Power)
CCR (Carbon Capture and Reutilisation)
see CCU (Carbon Capture and Utilisation)
CCS (Carbon Capture and Storage, Carbon Capture and Sequestration), xvi, 6, 10, 22, 45, 53–54, 83, 93–95, 101–103, 113, 118, 125, 130, 139, 145, 169
CCU (Carbon Capture and Utilisation), 10, 93–94, 96, 101–103, 107, 125, 130, 139, 145
CCUS (Carbon Capture Utilisation and Storage, Carbon Capture Utilisation and Sequestration)
see CCU (Carbon Capture and Utilisation), CCS (Carbon Capture and Storage)
centralised energy, 6–7, 48, 52–53, 56, 61–62, 68–69, 75, 89, 93, 96, 104, 130, 142, 151, 159, 162, 176, 180–183, 210
see also decentralised energy, DG (Distributed Generation)
CfD (Contracts for Difference), 179–180, 200
CH_4
see methane
char, 100, 136, 148, 162
see also biocarbon, biochar, biocoal, biocoke, carbon, coke, petcoke
Chemical-Looping Reforming (CLR), 156–158
China, Chinese, 2, 12, 18–19, 27, 36, 53, 77, 127, 150, 171, 177, 193, 200

CHP (Combined Heat and Power), 40, 62, 72, 75, 144, 150, 172, 175–176, 182, 185, 196
micro-CHP, microCHP, 61–62
see also CCHP (Combined Cooling Heat and Power), DC (District Cooling), DG (Distributed Generation), DH (District Heating)
circulating bed, 136, 151
CLC (Chemical Looping Combustion), 40, 110, 113, 142, 151, 156–158
Clean Coal, 140
see also IGCC (Integrated Gasification Combined Cycle)
CLG (Chemical Looping Gasification), 163
climate change, xv–xvi, 1–7, 13–17, 20–26, 30–33, 38, 47, 53, 60–62, 69, 83–85, 91–92, 98, 123, 129–132, 150, 167–170, 174, 177, 186, 193, 204–205, 210
CLR (Chemical-Looping Reforming), 156–158
CNG (Compressed Natural Gas), 64–67, 173–175, 193, 198–199
CO_2
see carbon dioxide
CO_2 Acceptor Process, 110, 163
coal, xvi, 2, 6–7, 12, 15–18, 22–31, 35, 39–40, 47–52, 57–61, 69, 77–88, 92–93, 100–103, 113, 135, 140, 145–148, 152–155, 158, 161–162, 165, 171–173, 179–181, 186, 191–194, 201, 206, 209–211
fuel switching (from coal to gas), 2, 12–14, 22, 47, 60–61, 85, 92, 146, 161, 165, 171–172, 209
fuel switching (from coal to biomass), 40, 69, 147–148, 171, 186
fuel switching (from coal to renewable electricity), 181
see also Clean Coal, IGCC (Integrated Gasification Combined Cycle)
coal seam gas (CSG)
see coalbed methane
coalbed methane (CBM), 9, 14, 37, 126–127, 169, 184
cogeneration, 62, 72
see CHP (Combined Heat and Power), trigeneration
coke, 78–81, 100, 190, 201, 205
coking, coker, delayed coker, delayed coking, 84, 187, 190–191, 205
see also carbonisation, of coal, destructive distillation, petcoke, pyrolysis
commodity prices
see prices, of commodities
compressed air, 72
condensate, gas condensate, 8–9, 35–38, 72
CRG (Compressed Renewable Gas), 64
crude oil, crude petroleum oil, petroleum oil, 3–4, 8, 19, 25, 28–29, 32–34, 43, 51, 60, 63, 69, 82–84, 88–89, 106, 127–130, 143–146, 177, 183, 190–193, 204–205
CtG (Coal to Gas, Coal-to-Gas), 146
CtL (Coal to Liquids, Coal-to-Liquids), 145

DAC (Direct Air Capture), 122–124
see also carbon capture, carbon dioxide capture, CCS (Carbon Capture and Storage)
dark fermentation, 104, 163
Dash for Gas, Dash to Gas, 9, 85, 182
DC (District Cooling), 62, 175
DCFB (Dual Circulating Fluidised Bed), 151
DCFC (Direct Carbon Fuel Cell), 99, 104, 108, 147–149
deactivated catalyst, 158
see also catalysis
decarbonise, decarbonisation, decarbonised, xvi–xvii, 2, 9–10, 19, 25, 31, 41, 47–48, 53, 59–62, 65–67, 71, 75, 92, 121, 164, 168, 171–172, 176–177, 210–211
decentralised energy, decentralisation, 24, 39–40, 69, 75, 85, 162, 172, 180–185
see also DG (Distributed Generation, Decentralised Generation)
deepwater, deepwater gas, 9
depletion, fossil fuel depletion, decline, 3, 7–8, 13–14, 16–17, 26, 31–36, 69

Natural Gas depletion, shale gas depletion, 8, 38, 59, 81, 126, 181
oil depletion, crude oil depletion, shale oil depletion, 8, 37, 60, 82, 127
see also peak Natural Gas demand, peak Natural Gas supply, peak oil demand, peak oil supply
destructive distillation, 79–80, 88, 100, 147
see also carbonisation, pyrolysis
desulphurisation, 138, 187, 191
low sulphur fuels, 143, 187, 190
see also HDS (Hydro Desulphurisation)
DG (Decentralised Generation, Distributed Generation), 182
see also decentralised energy
DH (District Heating), 40, 62, 150, 175
see also CHP (Combined Heat and Power)
diesel, diesel fuel, 23, 29, 60, 63–67, 75, 84, 146
see also biodiesel
digestate, 41
see also AD (Anaerobic Digestion), AAD (Advanced Anaerobic Digestion), biochar, dark fermentation
discovery, fossil fuel discovery, gas discovery, oil discovery, 29, 34, 80–81, 125–126
upstream, 30
see also E&D (Exploration and Discovery)
disinvest, disinvestment, divest, divestment, 16, 194
see also investment, reinvestment
distillate, distillation, fraction, fractionation, 60, 80, 84, 87, 115, 143–145, 187, 190, 204
divest, divestment
see disinvest, disinvestment
DME (Di-Methyl Ether), 65
DSM (Demand Side Management), 48
see also energy demand
DSR (Demand Side Response), 48
see also energy demand
dual fuel, 65, 174, 190–191
E&D (Exploration and Discovery), F&D (Finding and Development), 26, 81
ECA (Emissions Control Area), SECA (Sulphur Emissions Control Area), 187
EfW (Energy from Waste), 73, 148–149
see also WtE (Waste to Energy)
EGR (Enhanced Gas Recovery), 169
see also EOR (Enhanced Oil Recovery), tertiary recovery
electrocatalysis, 71
electrochemical
see fuel cell
electrolysis, electrolyser, xvii, 49, 58, 71–73, 104–105, 108, 114, 117, 133–135, 143, 154, 158, 161, 164, 176, 207–208
see also fuel cell
emissions pricing
see carbon pricing
emissions trading
see carbon trading
EMR (Electricity Market Reform), 178
energy
energy conservation, 24, 53, 168–170, 176, 182–185, 212
energy demand, 6–9, 25, 29–30, 139, 167–168, 172, 181–182, 184
energy efficiency, energy efficient, 5, 15–17, 21, 25, 30–31, 40, 45, 63, 75, 136, 182
energy investment, see investment
energy security, supply security, grid security, 2, 5–9, 14–15, 27, 46, 68–70, 82, 88, 106, 167, 170, 173, 181–185, 210–211
energy crops, 23, 150
EOR (Enhanced Oil Recovery), 83–84, 119
see also tertiary recovery
EPOx (Electrochemical Partial Oxidation), 101
see also POx, POX
EPS (Emissions Performance Standard), 171–172
EROEI (Energy Return on Energy Invested), 21, 26, 34, 39, 46, 57

EROEI – *continued*
EROI (Energy Return on Investment), 21, 46
ERR (Economically Recoverable Resources), 34–36, 126
see also 2P (proven-plus-probable reserves), TRR (Technically Recoverable Resources)
ethanol, oxygenates, 112
see also bioethanol
excess electricity, excess generation, excess power, surplus, spare, xvii, 46, 49, 70–74, 104, 113, 133–134, 160–161, 173, 200, 207–208
see also exports, renewable electricity, solar power, wind power
exports, of energy, of electricity, 12, 30, 35, 61, 74, 83, 119–124, 133, 160–161, 182, 200, 207
see also imports

FBH (Fluidised Bed Hydrogenation), 84
FCC (Fluid Catalytic Cracking), 78, 84, 191
FCEV (Fuel Cell Electric Vehicle), HFCV (Hydrogen Fuel Cell Vehicle), HFC vehicle, 65–66
filling station
see HFS (Hydrogen Filling Station)
financeable, financeability, 25–26, 46–47, 129
First of a Kind (FOAK), 19, 183, 200
FiT (Feed-in Tariff), CfD FiT (Contracts for Difference Feed-in Tariff), 179, 199–200
flare, flaring, venting, 3, 11–12, 48, 57, 59, 77, 89, 107, 120–121, 159, 209
see also fugitive emissions
flash pyrolysis
see pyrolysis
fluidised bed, circulating bed, 84, 136, 151, 163
fluidising medium
see gasification agent
fossil fuels
conventional, 3, 7–10, 13–14, 19, 29, 32, 34–38, 54, 57, 125–126, 128–130
reserves, reserves growth, xvi–xvii, 3, 7, 18, 22, 25, 28–29, 32–38, 58, 83, 88, 119, 122, 125–129
resources, 3, 5, 7–8, 10–14, 18–26, 29–38, 47, 51, 59–60, 87, 106, 126–130, 144, 147, 169, 178, 192–194, 203
unconventional/non-conventional, 3, 7–9, 13–14, 18–19, 22–24, 29, 32, 35–38, 45–47, 51, 54, 59, 125–130, 170
see also growth
FQD (Fuel Quality Directive), 192
fraction, fractionation
see distillate, distillation
fuel cell
see AFC, DCFC, PEMFC, PEM FC, PEFC, SOEC, SOFC, Superwind
fuel oil
see HFO (Heavy Fuel Oil), MDO (Marine Diesel Oil), MGO (Marine Gas Oil), RGO (Residual Gas Oil)
fuel switching, from coal to biomass
see coal
fuel switching, from coal to renewable electricity
see coal
fuel switching, from coal to Natural Gas
see coal
fugitive emissions, 2, 28, 39, 48, 58, 72, 91–93, 130, 169, 185
see also flaring, venting

gain
gas processing gain, 17
refinery gain, 205
gas engines, 68, 71–72, 75, 79, 85, 139, 150, 172
see also gas turbines
gas storage, xvi–xvii, 1, 6, 9–11, 40, 48, 52, 56–58, 65–75, 79, 85–86, 105–111, 114–118, 125, 134–137, 141, 145, 155–159, 161, 165, 172–176, 180–181, 193, 209–211
gas turbines, aeroderivatives, 15–17, 41, 52, 68–69, 71–74, 112–113, 138–139, 153, 156, 179, 195–196
see also gas engines, SOFC-GT

gasification, gasifier, gasifying, 19, 40–41, 49–50, 54–65, 69, 73, 79–89, 100–103, 106–112, 115–118, 121–122, 135–155, 158, 162–165, 185, 188, 191, 196–200, 203, 209
biomass gasification, 41, 65, 69, 86, 136, 197, 199–200, 209
coal gasification, 49, 135
complete gasification, total gasification, 80–82, 100–101, 147
flash gasify, 149
incomplete gasification, 100
indirect gasification, 152–153, 197
see also ACT (Advanced Conversion Technologies), circulating bed, fluidised bed, POx, POX (Partial Oxidation), pyrolysis
gasification agent, gasifying medium, oxidising medium, oxidant, fluidising medium, 87, 100–102, 108, 111, 131, 148, 152, 155, 162
gasoline, 29, 60, 84, 146, 174–175, 208
see also naphtha, petrol
GCH (Gas Central Heating), 80
Germany, German, 5, 17–19, 74, 82–83, 105, 150, 160–161, 165, 173, 181, 189, 206–207, 210
global warming, 1, 12, 15, 25, 50, 91, 121, 124, 129, 148, 174, 177, 190
green electricity, green energy, green power, 28, 134, 146, 206
see also renewable electricity
green gas, xvi–xvii, 16, 134, 210
see also Renewable Hydrogen, Renewable Methane
green power, green electricity, 134, 146, 206
see also renewable electricity
greenhouse gas, xv, 2–3, 15, 21, 38–39, 43, 85, 91–94, 97–98, 121, 129–132, 146–147, 186, 192, 203
grid capacity, grid reinforcement, 172
grid injection, 57, 62, 68, 116, 145, 154, 199–200
grid investments, 27
grid security, 211
grids, xvi, 2–6, 9, 12, 15–28, 47–48, 51–58, 61–62, 65–69, 73–75, 79–85, 98, 102–106, 111, 116, 119, 123, 134, 137–141, 145–146, 153–154, 160–161, 165, 171–186, 193, 199–201, 206–211
growth
economic growth, 17, 21, 27, 42, 45
energy sector growth, 5–6, 28, 42–43, 46, 146, 149, 161, 164, 168, 180
demand growth, 177, 193
supply growth, 5–9, 13–14, 36–37, 43, 61, 124, 160
reserves growth, 5, 32–34, 128
see also fossil fuels, reserves, reserves growth
GtL (Gas to Liquid, GTL), 65, 102, 121, 145

HCNG (Hydrogen Compressed Natural Gas), 65
HDS (Hydro Desulphurisation), 187
see also desulphurisation
heavy oil, 8, 14, 32–35, 52, 84, 127, 143–145, 169, 174, 190, 193, 205, 208
HENG (Hydrogen-Enriched Natural Gas), 65
HFCV (Hydrogen Fuel Cell Vehicle), HFC
see Fuel Cell Electric Vehicle
HFO (Heavy Fuel Oil), 187, 205
HFS (Hydrogen Filling Station)
see HRS (Hydrogen Refuelling Station)
HGV (Heavy Goods Vehicle), commercial vehicles, heavy duty vehicle, light duty vehicle, lorry, tanker truck, truck, 65–67, 174–176, 183
HIA (Hydrogen Implementing Agreement), 70, 186, 191–192
horizontal drilling, hydraulic fracturing, 2, 35, 69, 126
HRS (Hydrogen Refuelling Station), HFS (Hydrogen Filling Station), filling station, 64–67, 141, 173–176, 186, 191
hybrid fuel cell system, 99, 138, 153–154, 196
hybrid vehicles, hybrid drive vehicles, 63

hydraulic fracturing
see horizontal drilling
hydrocracking, 143, 190
see also hydroprocessing
hydrodesulphurisation
see HDS (Hydro Desulphurisation)
hydrogasification, 84–86, 111, 148, 191, 197
see also steam hydrogasification
hydrogen (H_2), xvii, 1–2, 11, 29, 41, 49–58, 64–72, 75, 78, 82–87, 97–119, 122–124, 129, 133–137, 140–151, 154–165, 173–176, 180, 186–193, 200, 203–211
captive hydrogen, 186, 207
see also biohydrogen, HFCV, Hydrogen Economy, Natural Hydrogen, Renewable Hydrogen, Transitional Hydrogen
Hydrogen Economy, 4, 19, 49, 53–55, 66, 70, 86, 99
hydrogen-rich gas, 52–53, 65, 99–100, 141, 149, 165
hydrogen evolution (HER, hydrogen evolution reaction), 105, 163, 208
hydrogen sulphide (H_2S), 10, 17, 57, 116, 119–121, 126
hydroprocessing, 204
see also hydrotreating, hydrogasification
hydrotreating, 143, 191
see also hydrogasification, hydroprocessing
Hythane, 49, 67, 112

ICE (Internal Combustion Engine), 66–67, 173
IED (Industrial Emissions Directive), 171
IGCC (Integrated Gasification Combined Cycle), 84, 87, 140, 196
BIGCC (Biomass Integrated Gasification Combined Cycle), 140
see also Clean Coal
imports, of energy, of electricity, 12, 16, 30, 35, 38, 46, 58–60, 118, 128, 181–182, 185, 200
see also exports
independent power, off-grid, 52, 68, 173
India, 19, 53, 64, 150, 171, 177
indigenous energy, 14, 35–36, 127, 170, 182–184, 200, 207, 210
infrastructure, 5–7, 12–13, 15–17, 21–22, 27–28, 32, 46, 52–53, 59–60, 64–66, 75, 80, 83, 93, 103, 121, 129, 147, 168, 171, 178–180, 193, 209, 211
insulation, building energy efficiency, home energy efficiency, 1, 15, 45, 61–62, 67, 172, 184–186
interconnector, 12, 181
internally reforming fuel cells
see SOFC, IR-SOFC
investment, reinvestment, investment funds, investors, 2–8, 12–19, 21–46, 59, 81, 128, 161, 164, 168–174, 177, 181–183, 186, 193–194, 205, 209–211
see also divest, disinvest
IOC (International Oil Company), 29–30
Iran, 34, 118–123, 126, 177

jet fuel, 29

kerogen, 8, 127
kerosene, kerosine, 29

landfill, 48, 73, 78, 149, 153
LCOE (Levelised Cost of Energy, Levelised Cost of Electricity), 46, 194–195, 201
LCPD (Large Combustion Plant Directive), 171
light sourcing, 192–193
liquefied Renewable Methane, 193
see also Renewable Gas
LMG (Liquefied Mixed Gas), 67
see also Renewable Gas
LNG (Liquefied Natural Gas), 12–14, 80, 87, 119, 178, 181, 193, 200
see also Natural Gas
load balancing, load-balancing, grid balancing, balancing services, 41, 47, 153, 173, 183, 209

see also DSM (Demand Side Management), DSR (Demand Side Response), SSM (Supply Side Management), STOR (Short-Term Operating Reserve)
locked in, lock in, 13–15, 23
low carbon gas, low carbon gas fuels, 1–4, 9–10, 16, 19, 24, 54–56, 59–61, 119–122, 146–150, 168–170, 176, 181, 194, 210
low carbon energy, xvii, 6, 26, 33, 42, 73, 168, 179
low carbon fuels, Low Carbon Fuel Standard, 16, 192
LPG (Liquefied Petroleum Gas), 63–65, 68, 178
see also Natural Gas, propane

manufactured gas, 1–4, 11, 16–19, 45, 50, 62, 74, 77–89, 98, 118, 173, 176, 185, 200–201, 204
MARPOL (Marine Pollution Convention), 174, 187
MCFC (Molten Carbonate Fuel Cell), 138, 150, 157
MDO (Marine Diesel Oil), 190
membrane, membrane reactor, 88, 104
methane-rich gas, 53–55, 64–67, 81, 86, 131, 139, 155
methanation, methanate, methanisation, methanator, 52–58, 64–71, 75, 82, 96–98, 105–111, 114–115, 117–118, 121, 125, 130–132, 137–139, 144–145, 150–151, 155–159, 165, 209
see also Sabatier reaction
Methane Economy, 55
methane hydrates, 9, 14, 126–127
methanol, 65
MGO (Marine Gas Oil), 187, 190
MIA (Methane Implementing Agreement), 193
microwave, 118, 209
MSW (Municipal Solid Waste), 41

nameplate capacity, 134, 160
see also rated capacity
naphtha, 78, 80, 84, 87
see also gasoline, petrol
Natural Carbon Dioxide, 39
Natural Gas, 1–4, 6–19, 22, 25, 28–29, 32–35, 38–39, 46–61, 64–89, 92–93, 101–103, 106, 112, 116–132, 140–146, 152–156, 159–161, 165–166, 169–172, 175–187, 190–200, 203–212
Natural Hydrogen, 11, 78
Natural Methane, 11
NET Power, 152, 155–157
NG (Natural Gas)
see Natural Gas
NGCC (Natural Gas Combined Cycle)
see CCGT (Combined Cycle Gas Turbine)
NGLs (Natural Gas Liquids), 8, 60, 112, 121, 204
see also total liquids
NGV (Natural Gas Vehicle), 6, 17, 64–67, 119, 131
nitrogen (N_2), 17, 57, 78, 82, 85–86, 101, 120, 141, 150–151, 171, 204
nitrous oxide (N_2O), 91
NOC (National Oil Company), 28–30
NOx (Nitrogen dioxide (NO_2) and Nitric oxide (NO)), 171
NREAP (National Renewable Energy Action Plan), 73
nuclear decommissioning, 7, 24, 27, 39, 68, 206
nuclear power, nuclear energy, atomic electricity, atomic energy, xv–xvi, 6–7, 17–18, 22–23, 26–27, 39, 45, 48, 53, 60–61, 68–69, 169–170, 177–180, 200, 206, 211
Nuclear Renaissance, 7, 26, 27, 69

ocean acidification
see acidification
ocean fuels, 124
OCGT (Open Cycle Gas Turbine), 179
offgas, off-gas, off gas, 75, 78, 84, 108, 121, 144, 153, 201
see also ROG (Refinery Off Gas)
oil shock, 83
oil window, 128
operational expenditure, opex, 128–130
see also capital expenditure, capex

oxyfuel, oxy-fuel, 113, 140–142, 151–153
oxygen (O_2), 86, 100–102, 108–113, 117, 127, 131–135, 138–142, 151, 155–157, 162–164
oxygenates, 112

P2G (Power to Gas, power-to-gas, PtG), xvii, 70, 73–74, 133, 161, 200, 208, 211
parasitic load, parasitic energy demand, 139–141, 159, 191
particles, particulates, particulate emissions, 11, 21, 63, 91, 105, 129, 135, 144, 147, 171, 193
Peak Fossil Fuels, 21, 31
peak geological oil, 35
Peak Natural Gas, Peak Sweet Natural Gas, Peak High Value Natural Gas, 7–9, 38–39, 59, 74, 88
  peak Natural Gas supply, 35
Peak Oil, Peak Crude, Peak Crude Oil, Peak Sweet Crude, Peak Sweet Crude Oil, 7, 32, 60, 63, 88, 177
  peak oil demand, peak demand, 7, 177, 183
  peak oil supply, 29, 34–35, 60, 63
  peak crude oil production, peak oil production, 51
peaker plant, peaking, peak demand, 48, 154, 164, 210
PEC (photoelectrocatalysis), 51, 104
PEMFC (Proton Exchange Membrane Fuel Cell), PEFC, PEM FC, 104–105, 135, 164, 176, 207
permeability, of hydrogen, 86
petcoke, petroleum coke, 84, 190–193, 205
petrol, 63, 75, 174–175
  see also gasoline, naphtha
petroleum, petroleum oil, crude petroleum oil, crude oil, 7–8, 25, 29, 38, 51, 60, 69, 78, 83, 87–89, 106, 127–128, 177, 204
petroleum production, petroleum oil production, petroleum supply, 3, 32, 48, 63, 82, 183, 204
petroleum refinery, petrorefinery, petrochemical, 4, 11, 19, 24, 56, 78, 80–82, 142–146, 187, 190–192, 204–205
photocatalysis, 71
photoelectrocatalysis, 104
photoelectrochemical, 51
photofermentation, 163
photovoltaic, PV, xvii, 46, 52
  see also solar power
pipeline quality, pipeline quality gas, gas quality, gas quality standards, 49, 59, 88, 106, 111, 116, 151, 173
plasma, plasmolysis, 104, 118, 149, 201
plastics, 41, 131, 149, 164
  see also bioplastics
platinum, 71
post-combustion, 112, 140–142, 147, 159, 169
  see also oxyfuel, pre-combustion
power-to-gas, power to gas
  see P2G
POX, POx (Partial Oxidation), 100–101
  see also EPOx
pre-combustion, 112, 140–141
  see also oxyfuel, post-combustion
prices, of commodities, of fossil fuel commodities, 15, 21, 26, 29–31, 37, 43, 80, 87, 177, 200, 206, 211–212
  see also carbon price
prices, of energy, 28, 31, 40–43, 61, 184
prices, of fossil fuel products, 29, 43
prices, of power, 31, 104
prices, of Renewable Gas, 134, 200
product gas, 96, 100–103, 106, 111, 118, 136–139, 209
  reformate gas, 102
  see syngas
product slate, yield, 143, 187, 190–191
propane, 68, 81, 178
PSA (Pressure Swing Adsorption), 145
PtG (Power to Gas, Power-to-Gas)
  see P2G (Power to Gas)
pyrolysis, 100, 136, 149, 162
  fast pyrolysis, flash pyrolysis, 162
  see also bio-oil, gasification

rated capacity, 160
  see also nameplate capacity

reactive, reactivity, 86, 100, 136
unreactive, 135
recharge, recharging, charging, 64
home compressor, 67
home recharging, home charging, home car charging, 172, 175
refinery
see biorefinery, petroleum refinery
refinery capacity, refinery processing capacity, 29, 32
refinery gain
see gain
reformate gas
see product gas
reforming
see SMR (Steam Methane Reforming, Steam Reforming of Methane, SMR, SRM)
Renewable Carbon, 50, 95–98, 148, 106, 119, 122, 125, 132, 210–211
renewable electricity, green electricity, green power, xvii, 2, 5–7, 10, 15–17, 22, 27, 31, 39, 45–48, 51–52, 58, 68, 71–75, 85, 96, 113, 117, 133–134, 143, 154, 160–161, 165, 168–177, 180–181, 200, 207–208, 211–212
constrained, 70, 74, 160
curtailed, 70, 74
shed, of load, 70
zero cost electricity, 6, 200
see also load balancing, solar power, wind power
Renewable Gas Obligation, 183–184
Renewable Hydrogen, 10, 19, 41, 49–52, 55–57, 64, 68, 70–75, 87, 96–99, 104–105, 114, 117–125, 131–139, 142–145, 148–150, 155–164, 175–176, 186, 192, 200–211
Renewable Methane, 4, 19, 49, 52, 55–58, 64–70, 75, 101–103, 119, 122–123, 132–134, 142, 161, 175–176, 193–194, 200–201, 205–207, 211
reserves
see fossil fuels
reserves growth
see fossil fuels
residua, residuum, resid, resids, residues, 84, 143–144, 170, 187, 190–193, 203–205
resources
see fossil fuels
retrofit, 28, 40, 141
RFO (Residual Fuel Oil), 190
RHI (Renewable Heat Incentive), 199–200
RMG (Renewable Mixed Gas), 67
ROG (Refinery Off Gas), 84, 143–144, 153, 204–205
light refinery gas, 143–144
see also offgas
Russia, 13, 18, 33–34, 98, 177

Sabatier reaction, 82, 87, 103, 114, 138, 145, 156–158
see also methanation
SAGD (Steam Assisted Gravity Drainage), 127
Saudi Arabia, 18, 34
SC (Supercritical), SCW (Supercritical Water), SCWF (Supercritical Water Gasification), 111, 122, 163
SE-SHR (Sorption-Enhanced Hydrogasification), 148
see also steam hydrogasification
seagrass
see algaculture
seawater, 118, 123–124, 135
seaweed
see algaculture
SECA (Sulphur Emissions Control Area)
see ECA (Emissions Control Area)
SEWGS (Sorption-Enhanced Water Gas Shift), 152
SG (Steam Gasification), 79, 87, 103, 162, 191
see also SMR/SRM (Steam Methane Reforming)
SGHP (Shell Gasification Hydrogen Process), SGP (Shell Gasification Process), 144
shale gas
see fossil fuels, unconventional/non-conventional

shale oil
see fossil fuels, unconventional/non-conventional
shift, Water Gas Shift, Water Gas Shift Reaction, WGS, WGSR, 56, 82, 103, 106, 110, 140, 145, 152, 157
SMR/SRM (Steam Methane Reforming, Steam Reforming of Methane, SMR, SRM), methane reforming, steam reforming, 82, 102, 143–144, 156–158, 162, 187, 209
see also gasification
SMR (Small Modular Reactor), 7, 169, 211
SNG (Synthetic Natural Gas, Synthesised Natural Gas, Substitute Natural Gas, Supplemental Natural Gas, Simulated Natural Gas), 80–86, 102–103, 111, 165
see also bioSNG
SOEC (Solid Oxide Electrolysis Cell), 58, 105, 154–155, 176
SOFC (Solid Oxide Fuel Cell), 101, 105, 113, 138, 141, 150, 153–157, 175–176, 179
IR-SOFC (Internally Reforming SOFC), 101, 175
see also SOFC-GT (SOFC Gas Turbine)
SOFC-GT (SOFC Gas Turbine), 138, 141, 151–154
see also gas turbines
solar farm, solar park, solar roof, 22, 45, 71, 143
solar power, xvii, 5, 15, 22, 27, 39, 42, 45–49, 62, 74, 85, 104, 114, 118, 122–123, 131–133, 160–161, 166, 170, 177–182
see also renewable electricity, wind power
sour gas, sour/acid gas
see acid gas
space cooling, coolth, 62, 150
see also DC (District Cooling)
space heating, 6–7, 10–12, 53, 59–62, 98, 119, 123, 131, 175, 210
see also CHP (Combined Heat and Power), DH (District Heating)
SR (Steam Reforming)
see SMR/SRM (Steam Methane Reforming)
SSM (Supply Side Management), 48
see also energy demand
standby, stand-by, mothball, 15, 48, 165, 178
see also backup
steam gasification, 79, 87, 103, 162, 191
see also SG (Steam Gasification), SMR/SRM (Steam Methane Reforming), steam hydrogasification
steam hydrogasification, 86, 111, 148
see also hydrogasification
STOR (Short-Term Operating Reserve), Balancing Services, 48, 173
stranded assets, 29, 33
strike prices, 180
subsidies, subsidy, 15, 28, 31, 37, 43–46, 59, 165–166, 170, 174, 178–180, 183–184, 194, 198–200, 206, 209, 212
sulphur (S), 10, 57, 121, 138, 143, 177, 187, 190
sulphur dioxide (SO_2), 171
Superwind, 114
see SOFC, wind power
sweet gas, sweetening, sweetened, dry gas, 121, 145
see also fossil fuels
sweet crude, sweet crude oil
see fossil fuels
Switzerland, 195–197, 200
syncrude, 35, 143
synfuels, synthetic fuels, 51, 65, 81–83, 115, 124, 145
syngas (synthesis gas), xvii, 56, 67, 71, 80–87, 99, 101–111, 115–117, 121, 136–137, 140–145, 152–158, 165, 188, 211
see also biosyngas, synthetic gas, synthetic methane
Synthane, 67
synthetic gas, 81–83, 121
see also biosyngas, syngas, synthetic methane
synthetic methane, 81, 194, 201
see also biosyngas, syngas, synthetic gas

tar, tars, tar formation, 86–87, 136, 163, 209
 biotar
 tar moderant, 209
tar sands, oil sands, bituminous sands, 8, 14, 32, 35, 51, 69, 127, 143, 169, 190, 193, 208
tertiary recovery, 84
 see also EGR (Enhanced Gas Recovery), EOR (Enhanced Oil Recovery)
THAI (Toe-to-Heel Air Injection), 127
thermocatalysis, 71
thermochemical, 41, 51, 56, 80, 86, 97–106, 118, 135–137, 140–141, 144, 150–153, 161–164, 200, 203–204
 see also gasification
thermodynamics, thermodynamically, 84, 129, 141, 158–159
thermolysis, 162
 see also hydrogen evolution
torrefaction, 111
total liquids, 8
 see also NGLs (Natural Gas Liquids)
Town Gas, 49, 55, 77–81, 87, 100
 Coal Gas, 77–80, 100
 see also carbonisation, of coal, destructive distillation
Transitional Gas, xvii, 2, 50, 99, 132, 193, 203
 see also hydrogen
Transitional Hydrogen, 206
transport, transportation, mobility, 6, 11, 23, 29, 40–41, 48, 51, 54, 59, 62–69, 72, 75, 84, 96–98, 102–104, 107, 115, 119, 164, 173–174, 178, 183, 186–187, 191–193, 206, 210
trigeneration, 62, 150, 176
 see CHP (Combined Heat and Power)
trilemma, 167, 175, 185
TRR (Technically Recoverable Resources), 34–36, 127

UAE (United Arab Emirates), 18, 121
UCG (Underground Coal Gasification), 14, 127, 169
UKCS (United Kingdom Continental Shelf), 184–185
unburnable carbon, 31, 130, 193
unconventional, unconventional fossil fuels, non-conventional fossil fuels, unconventional oil, unconventional gas
 see fossil fuels
uranium, 23–27, 69, 206

variability, variable, intermittency, xvi–xvii, 5–6, 15, 22, 41, 47–51, 55, 75, 85, 104, 111, 118, 128, 134, 154, 160, 165, 172–173, 178–180
Venezuela, 18, 32, 127

water splitting, 58, 68, 133–135, 163–164
 electrolysis, xvii, 49, 71–73, 104–105, 108, 114, 117, 133–135, 143, 154, 158, 161, 164, 176, 207–208
 see also Renewable Hydrogen
Water Gas Shift
 see shift
Water Gas Shift Reaction
 see shift
WGS (Water Gas Shift)
 see shift
WGSR (Water Gas Shift Reaction)
 see shift
WindGas, 70
 see also Power-to-Gas, wind power
wind farm, windfarm, 41
wind power, 5, 19, 42, 46, 53, 61, 70, 74, 85, 104, 133, 160–161, 170, 173
 see also renewable electricity, solar power
wood, 49, 69, 83, 91, 118, 137, 148, 162
 wood pellet, 186
WtE (Waste to Energy), 73, 148–149
WtH (Waste to Hydrogen), 163

XtL (X to Liquid, XTL), 145

yield
 see product slate
Young Carbon, 53, 123

zero carbon, v, 2, 15, 24–25, 186, 193
zero waste, 143

Printed and bound in Great Britain by
CPI Group (UK) Ltd, Croydon, CR0 4YY

SBR